PLUG-IN HYBRID VEHICLE TECHNOLOGY

PLUG-IN HYBRID VEHICLE TECHNOLOGY

Design and Build a Plug-In Electric Hybrid Vehicle for a Carbon-Constrained World.

William H. Kemp

William H. Kemp

Copyright © 2007 by William H. Kemp

All rights reserved. No part of this book may be transmitted in any form by any means, electronic or mechanical, including photocopying and recording, or by any information storage or retrieval system without written permission from the author, except for brief passages quoted in a review.

Library and Archives Canada Cataloguing in Publication

Kemp, William H., 1960-

    The zero-carbon car : building the car the auto industry can't get right / William H. Kemp.

Includes index.

ISBN 978-1505839869
    1. Alternative fuel vehicles. I. Title.

Library of congress Control Number: 2015900364

CreateSpace Independent Publishing Platform, North Charleston, SC

TL216.5.K44 2007   629.22'9   C2007-906422-1

DISCLAIMER:

The author, contributors, and publishers assume no liability for personal injury, property damage, consequential damage or loss, from using the information in this book, however caused.

The views expressed in this book are those of the author personally and do not necessarily reflect the views of contributors or others who have provided information or material used herein.

Acknowledgments

The writing of this book and development of the demonstration Zero-Carbon Car required the technical knowledge and expertise of many individuals and I would like to take a moment to thank those who provided input, feedback and support in this project:

George Argiris for writing the PLC engine control software of the ZCC and design of the voltage-controller system; ASTM International for assistance with biodiesel quality standards; Gary Baker of Alternative Designs Inc. for his 3-D modeling and machining skills; Dr. Marc Dube, P.Eng., University of Ottawa Chemical Engineering Department; Fischer Panda Generators; General Motors Inc.; Jeff Goodman, Iogen Corporation for his assistance with cellulosic ethanol data and procurement; Mark Harris for his work on the biodiesel production system; Honda Motor Company; Steve Howell, MARC-IV Consulting and Chair of the ASTM D 6751 Biodiesel Quality Committee; Jeff Knapp for his work in developing and producing zero-carbon biodiesel; Richard Lane of RV Consultants Ltd. for his work in fabricating the electric drive system of the ZCC and his superb mechanical skills during the hybridization phase of the project; Mercedes-Benz Canada; Dr. Martin Mittelbach, University of Graz, Austria, for his technical support and kind words of enthusiasm; Christina Moretto, Canadawide Scientific, for her assistance in researching and locating equipment and lab supplies required to make the biodiesel facility a reality; Olympia Battery Company; Ken O'Rielly for writing the ZCC touch screen software and assisting with the hybrid system design; Christine Paquette, Past Executive Director, Biodiesel Association of Canada, for assistance, research, review, and access to her amazing Rolodex; Tesla Motors Inc.; Dr. Andre Tremblay, P.Eng., Professor and Chair, Chemical Engineering Department, University of Ottawa; U.S. Department of Transportation, Bureau of Transportation Statistics; Jamie Wilson for his work building the biodiesel production facility.

If I have missed anyone, please accept my apologies and understand that your assistance is truly appreciated!

A huge thank you must also be given to Joan McKibbin, my editor. Joan worked tirelessly under very demanding deadlines, as usual, and managed not to lose her cool. Thanks Joan!

I am also indebted to those who have had to work directly with me in the preparation of the text. A big thanks goes out to Cam Mather for his excellent line drawings, graphics, and layout skills, not to mention having to work with my demanding timelines; Michelle Mather who provided the mounds of research material referenced in the text; and to Lorraine Kemp for her great help behind the camera. (All images without photo credits are by her.)

I can't thank my wife Lorraine enough for providing most of the photographs in the book, putting up with me during the building and fabrication of the Zero-Carbon Car and not to mention assisting with the miles of electrical wiring. Thanks Lor!

I would be remiss without giving special thanks to Micheline Lane , who made sure the coffee was always hot and cookies fresh during the construction of the Zero-Carbon Car and to Michelle Mather for her excellent work as research assistant. And a final thanks to Jason Nichols, Principal of Kraken Graphics for the cover design.

Table of Contents

Acknowledgements

Preface

Introduction   1

Chapter 1 – Personal Transportation in the Third Millennium   11
  The Myth of Infinite Freedom   11
  The Reality Check – 37 kilometers per Day   16
  Long Distance Travel   20
  The Energy Demand of Transportation   23
  The Limits of Roads and Infrastructure   30
  The Obvious Lessons   35

Chapter 2 – A Brief Review of Energy   41
  Geopolitics and Oil   44
    Canada to the Rescue?   53
    Mexico Perhaps?   55
    The Persian Gulf Region   56
    China   58
    The World   61
  Peak Oil a.k.a. Production Limits   64
    Defining Peak Oil   66
    Understanding the Geophysics of Peak Oil   67
    Tinkering with the Data   71
  Summary   74

Chapter 3 – Energy and the Environment   77
  The Chemistry of Smog and Green House Gases   80
  The Issues of Climate Change   85
    The I.P.C.C. Report   86
  Decarbonizing our Energy Supply   91
  A Carbon-Constraining System   96
  The Paradigm of Reality Convergence   100

Chapter 4 – Transportation Systems in a Carbon Constrained World — 105
- There is no Substitute for Oil — 110
- When Oil-based Transportation comes to an End — 111
- Moving Ourselves — 113
  - Why not walk or bicycle? — 116
  - Inner-City Travel — 120
    - Improved Public Transit — 120
    - Ultra-Low Fuel Consumption Vehicles — 125
    - Neighborhood Vehicles — 132
  - Inter-City Personal Travel and Beyond — 138
    - Flying and Marine Transport — 146
  - Logistics and the Movement of Goods — 148
    - A Sustainable Freight System — 155
    - Local Transport — 157
    - Intercity Transport — 157
    - Marine Transportation — 161
- Summary — 161

Chapter 5 – The Personal Transportation Appliance — 163
- Summarizing the Case for the Zero-Carbon Car — 163
- Personal Transportation Technologies and Psychology — 167
  - Long Life — 172
  - Energy Efficiency (Fuel Economy) — 174
  - Vehicle Mass — 176
  - Virtual Upgrading — 179
- The Myth of Zero-Emission Vehicles — 183
- Summary — 184

Chapter 6 – A Closer Look at Advanced Automotive Technologies — 185
- Hybrid Automobiles — 187
  - Today's Hybrid Power Systems — 187
  - The Nuts and Bolts of Hybrid Technology — 189
    - "Weak" vs. "Strong" Hybrids — 192
    - Hybrid Performance — 194
    - Large Hybrid Vehicles — 195
  - Hybrid Car Summary — 196

A Battery Electric Vehicle (BEV) Primer	199
The Home-Built BEV	201
A Few Details of Fred Green's Car	204
Comparing BEVs to Internal-Combustion-Powered Vehicles	207
BEV Range	208
Recharging Time	211
Operating Cost	211
A Collage of Battery Electric Vehicles	212
The ZENN Neighborhood Vehicle	212
General Motors EV-1 – Advanced before its Time	213
Myer Motors NMG – Disney on Wheels?	214
Dancing with the Tango Commuter Car	215
Tesla Roadster – A Rocket on Wheels	216
BEV Summary	218
The Plug-In Hybrid Electric Vehicle (PHEV)	219
Overview of PHEV Technology and History	219
PHEV Summary	228
The Propaganda of Hydrogen-Powered Vehicles	229
Obstacles on the Road	231
A Few Facts about Hydrogen Technology	234
Hydrogen as a Fuel	234
Hydrogen Storage	238
Hydrogen Distribution	243
Fuel Cell Vehicles	245
Summary	247
Chapter 7 – Introducing the Zero-Carbon Car	**251**
7.1 Getting the Project Started	**251**
The Case for Converting vs. "Scratch Build"	252
Series vs. Parallel Hybridization	253
Overview of the Design	255
The Donor Vehicle	263
Out with the Old	264
7.2 Battery Technologies: an Overview	**267**
The Story of Electrons	267
Conductors and Insulators	269

Batteries, Cells and Voltage	271
Alternating Current	273
Power, Energy and Efficiency	275
How Batteries Work	277
Depth of Discharge	281
Operating Temperature	282
Battery Sizing	284
Hydrogen Gas Production	288
Safe Installation of Batteries	288
Battery Installation in the Zero-Carbon Car	289
The Fine Art of Battery Cable Manufacturing	297
Battery Charging	302
Battery Technology Selection	305
Summary	307
7.3 D.C. Motor and Controller	**309**
Electric Motor Overview	309
Controller Overview	312
DC Electronic Controllers	314
AC Electronic Controllers	315
Regenerative Braking	316
7.4 The Liquid Fuel Power Plant	**317**
The Fisher Panda Generating Unit	317
Fitting the Generating Unit into the Zero-Carbon Car	321
Photographic Collage of the Installation	321
7.5 Engine Management	**333**
The Programmable Logic Controller	333
PLC Structure	335
PLC Input and Output Configuration	337
Program Flow	338
PLC Tutorial	340
User Touch Screen	344
Summary	345
7.6 Integrating the Functions	**347**
Touch Screen Overview	347
PLC and Metering Overview	349
PLC Inputs and Outputs	352
Battery Low Voltage Detection	352
Battery Charged Detection	355

Battery Charging Status and AC Mains Schematic	357
Battery Box Vent Fan(s)	359
Battery Charger Cooling Fan(s)	359
Battery Heating Blankets	359
Vehicle Charger Status	359
Automatic Battery Chargers	359
AC Mains Ground	359
PHEV Operation	360
User Display and Operation	365
Clock Icon	366
Fan Icon	366
Garage Door Icon	366
Home Icon	368
Generator Icon	368
Heater Icon	368
Battery Charger Icon	368
iPod MP3 Player Icon	368
Icy Car Icon	368
Defroster Vent Icon	369
Empty Gas Tank Icon	369
Wrench Icon	369
7.7 The Test Drive	**371**
Chassis Dynamometer Testing	374
Reviewing the Results	378
Summary	382
7.8 Adding Some Bling to the Zero-Carbon Car	**383**
Chapter 8 – Zero-Carbon Electricity	**389**
The North American Electrical Power System	392
Baseload Generation	394
Intermediate Generation	395
Peak Generation	395
Demand Time Shifting	400
Zero-Carbon Coal and Fossil Fuels	401
Moving the Charging Plug around Town	406
The Value of Zero-Carbon Electricity	410
Summary	412

Chapter 9 – Zero-Carbon Liquid Fuels — 415
- An Introduction to Biofuels — 416
  - Ethanol from Food – a Non-Starter — 416
  - Cellulosic Ethanol – the better choice — 421
  - The Downside of Ethanol — 421
  - Biodiesel as a Source of Green Fuel — 421
  - Biodiesel in the Transportation Sector — 426
  - Biodiesel Composition — 429
  - The Pros of Biodiesel — 434
    - Blending — 434
      - Biodiesel Concentration — 435
    - Biodegradability and Nontoxicity — 437
    - High Cetane Value — 437
    - High Lubricity — 437
    - Low Emissions — 437
    - Renewability — 438
    - Low Sulfur — 439
  - The Cons of Biodiesel — 440
    - Oxidation and Bacterial Stability — 440
    - Nitrogen Oxide Emissions — 441
    - Cold Flow Issues — 442
    - OEM Warranty Issues — 444
  - The Diesel Engine — 445
    - Engine Technology Overview — 448
    - Fuel Injection Systems — 452
      - Basic Fuel Injection — 453
      - Common Rail Direct Injection — 454
    - Engine and Vehicle Efficiency — 455
  - The Biodiesel Production Process — 457
  - Small-Scale Biodiesel Production System — 461
    - The WVO Receiver/Dryer — 468
    - The Biodiesel Reaction Tank — 471
    - Sodium Methoxide System — 471
    - Biodiesel Washing System — 475
    - Biodiesel Drying and Final Filtration — 479
- Fuel Dispensing and Storage — 481
  - The Fuel Dispensing Unit — 482

Cold Weather Issues	484
Blending Biodiesel with Petrodiesel	486
Summary	487

Chapter 10 – The Unveiling — 489

Conclusion — 495

Glossary — 499

Appendices — 503
- A. Resource Guide — 503
- B. Engine Management Software Listing — 509
- C. Zero-Carbon Car Specifications — 519
- D. LED Battery Voltage Monitor Schematics — 520
- E. Battery Voltage Monitor PCB Layout — 524

Endnotes — 525

Index — 536

About the Author — 543

Environmental Stewardship — 545

Preface

Plug-In Hybrid Vehicle Technology was originally conceived as a book that would describe the technologies involved in the design and implementation of plug-in hybrid electric cars and how their enhanced fuel economy could help reduce atmospheric pollution and greenhouse gas emissions. Upon further consideration, I realized that the plug-in hybrid car (or any alternate fuel vehicle, for that matter) is a bit like the automotive catalytic converter that was developed several decades ago and heralded as the solution to air pollution caused by the burning of fossil fuels in personal automobiles. It was a great technology when implemented, since it improved existing pollution control systems, but with the pas-sage of time (and increased automobile usage) the benefits have been completely negated.

Although plug-in hybrid technology is slowly coming to fruition and may provide a doubling or tripling of vehicular fuel economy, it is a stopgap measure on the road of personal transportation history. If all of the world's vehicles could suddenly be converted to plug-in hybrid (or other) technology, the total energy consumed to power this fleet would improve somewhat, but growing demand for personal transportation would negate this benefit within a few short decades, as in the case of the catalytic converter of a generation ago. Worse still, there is no guarantee that the primary energy source of the electricity, hydrogen, or other fuel sources used to power these vehicles is from renewable or low- or zero-carbon sources. It was imperative that my project employ only zero-carbon energy sources, measured from cradle to grave.

How much personal freedom and mobility should a person have, especially if it continues to degrade the environment and reduce our natural capital in the long term? My friend (and fellow eco-conspirator) Jeff Knapp likes to say, "If you spoil your nest, sooner or later you will be kicked out." Considering that we don't have any more nests, this is a harrowing prospect.

We may like to think that commuting from the suburbs, living in monster-sized homes, or owning more vehicles than there are drivers to operate them are entitlements that will continue forever. They will not. With human population and income growing exponentially there is simply not enough energy or natural resources to continue building a

Unfortunately, humans are notoriously bad at planning for the future, even when well-being is at risk. History is rife with examples: we smoke even though lung disease and cancer are likely to afflict us; we become obese knowing full well that diabetes and other health complications will ensue; we continue to purchase SUVs and other low-efficiency vehicles even though we know they increase greenhouse gas emissions and contribute to climate change.

If the feedback mechanism of cause and effect were better defined, faster, more visible, humans would be more careful with their choices. We do not require logic or experience to know that picking up a hot pot causes us to immediately drop it and avoid being burned.

Interestingly, we are not always a failure at planning for the future. People invest in retirement savings plans, socking away current income for future times, or purchase fire insurance for a home, where the likelihood of catastrophic destruction is very low. This dichotomy is not easily explained but is an essential issue in the debate about what to do about so many of society's problems, including personal transportation.

My exploration of these issues leads me to believe that personal transportation vehicles, whether plug-in hybrid or fueled with hydrogen or some hitherto unknown clean, exotic fuel, will come to an end in the foreseeable future. The automobile directly and indirectly causes more damage to the ecosystem than most people would care to consider. Air pollution and climate change issues hold the limelight today, but infrastructure materials and their associated costs as well as military "support" to keep the oil flowing are keeping badly needed dollars out of the more democratic mass transit system. If this continues to happen, large segments of the population will be prevented from ever becoming mobile, and the ecosystem will also suffer.

Soon after the attacks on New York City on September 11, 2001, Vice President Dick Cheney is quoted as saying that "the American way of life is non-negotiable." This may be his feeling, but I suspect it is very far from reality. I estimate that the personal automobile is destined for the scrap heap within a very few decades. However, society will maintain its collective denial of the converging issues that are stacking up against limitless personal mobility and will fight with more road construction and suburban development at enormous cost to taxpayers.

Until the day of awakening occurs, if we as a society are going to drive, let us convert to the most efficient technologies that are available today. Consider **The Zero-Carbon Car**.

Introduction

The demand for energy in the United States and Canada is insatiable, and it appears no one has learned from the mistakes of the '70s. Look around: sport utility vehicles and minivans abound, with the result that the total number of vehicles plying the world's roads now exceeds 750 million units[1]. While this may be a dizzying number, total volumes are expected to double by 2030 and triple by 2050, exceeding 2 billion vehicles[2].

Figure 1. North Americans will continue to cram ever more cars into bursting urban cores, while the inflationary effects on house prices and the lack of regulation against sprawl will cause people to scurry to distant suburbs and beyond to seek financial solace.

North Americans will continue to cram ever more cars into the bursting urban core, while the inflationary effects on house prices and the lack of regulation against housing sprawl will cause people to seek financial solace in distant, lower-cost suburbs and beyond. Most of this commuting will, to no one's surprise, be done in a personal automobile. As people are well aware, public transportation systems in North America are spotty at best, nonexistent at worst, and underutilized the majority of the time. Americans use public transit for less than 5% of their transportation trips[3], (a trip being defined as travel from one point to another), while personal vehicles accounted for 87% of trips in 2001[4]. With an average of 1.9 personal vehicles per household and 1.8 drivers[5], Americans have more vehicles than people to drive them!

In the meantime, China, Russia, India, and Brazil are all avidly watching western television programs, envying our entitlements, and salivating for their chance to have a slice of the purported "good life." What better way for members of the developing world to emulate the West than to line up and purchase shiny new automobiles as a way of displaying their newfound wealth and personal freedom? No reasonable

Figure 2. Americans use public transit for less than 5% of their transportation trips, while personal vehicles accounted for 87% of trips in 2001. With an average of 1.9 personal vehicles per household and 1.8 drivers, Americans have more vehicles than people to drive them!

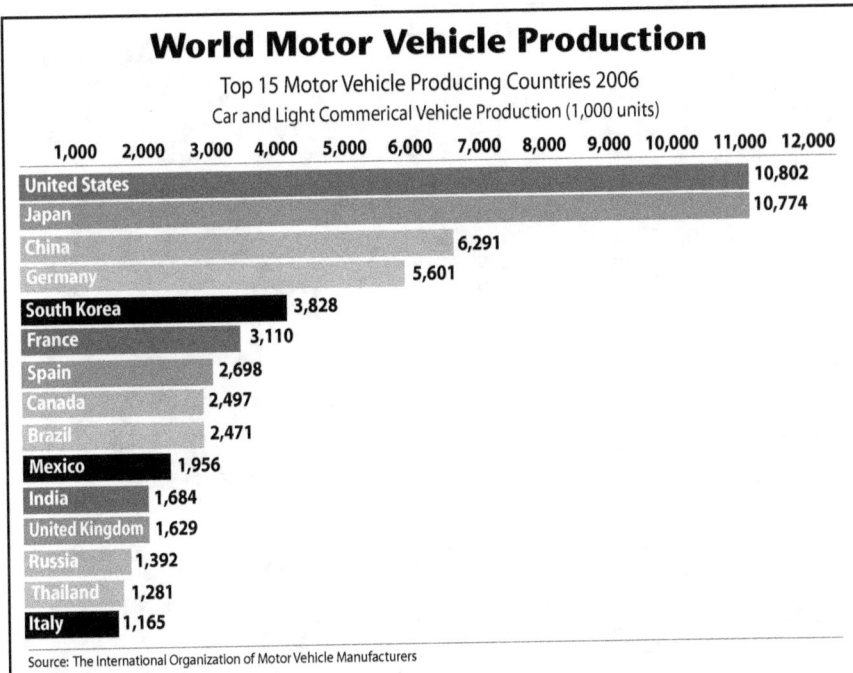

Figure 3. Many people consider China to be the land of the bicycle. While that may appear to be the case to the casual observer, the fact is that China is the world's third largest automobile manufacturer, producing 7.2 million vehicles in 2006, and growing quickly. If current trends continue, the number of vehicles plying the world's roads will expand to 2 billion units by 2050.

person would deprive anyone of the opportunity to share in the financial spoils of a burgeoning, globalized economy. China produced 7.2 million units in 2006[6] and has overtaken Germany in the number-three auto-manufacturing spot, right after the United States and Japan. The world's total vehicular production output is expected to reach 76 million units per year by 2014[7].

Considering that Americans own 785 cars for every 1,000 people and that the Chinese own 28 cars per 1,000[8], not to forget similar demand potential in India and other developing nations, the opportunity for automotive manufacturing growth seems almost limitless. While these potential vehicular production numbers sound very impressive, they are neither sustainable nor possible in the long term due to the interaction of economic, raw material, environmental, and societal pressures.

Figure 4. Society is reaching a point where the costs of providing roads, fuel, parking, safety, and other related direct support items are reaching their upper limits.

Although most people love their automobiles, (even if a considerable amount of time using them is spent in mindless, unenjoyable commuting and traffic), society is reaching a point where the costs of providing roads, fuel, parking, safety, and other related direct support items are reaching their upper limits. Military support to ensure the free flow of oil from the Middle East, climate change and global warming, accident victim liability, and air pollution are indirect costs that are rarely considered when discussing personal automotive transportation.

Economists and governments take a single-minded approach to ever-expanding economic growth, assuming that "externalities" such as natural resource supply and climate change are simple accounting items to be fiddled with at the bottom of a bookkeeper's ledger, if they are considered at all. There is an excellent example of the lack of full cost accounting in a recent article in *The Economist*, where the writer states that "coal-fired boilers generate the cheapest electricity," as if this statement were correct and factual. Every report that I have read indicates that thousands of deaths and long-term respiratory and other health effects are attributed to coal mining, which takes a huge personal toll and burdens society with massive support costs. Environmental degradation occurs at every step in the life cycle of coal, including ecosystem destruction, water source contamination, and the production of carbon dioxide, mercury, sulfur dioxide, and oxides of nitrogen emissions during combustion. Rail and shipping fuel used in the transportation of coal from mine to power plant also contributes to greenhouse gas emissions, hastening climate change.

Figure 5. Economists and governments take a single-minded approach to ever-expanding economic growth, assuming that "externalities" such as natural resource supply and climate change are simple accounting items to be fiddled with at the bottom of a bookkeeper's ledger, if they are considered at all.

Granted, governments are starting to discuss the issues of climate change, fossil fuel depletion, and peak oil, even if the discussion leads to large-scale group denial of the problem or generates halfhearted "solutions" such as the Kyoto Protocol that don't even come close to dealing with the reality of the situation. And society fails to consider how these issues relate to the personal automobile. Consider for a moment that:

- 2.9 million Americans were injured in highway accidents in 2003[9];
- the World Health Organization estimates that the cost of road accidents to society exceeds $500 billion globally each year;
- 42,643 Americans[10] and approximately 250,000 people worldwide were killed in highway accidents in 2003. Road traffic accidents are in the top three causes of death worldwide amongst under-twenty-five-year-olds[11];
- exhaust from cars contains carbon monoxide, nitrogen oxide, ammonia, sulfur dioxide, and particulate matter that cause cancer, asthma, and other respiratory illnesses that cost billions of dollars in treatment;
- automobiles are a major contributor of carbon dioxide emissions and are a leading contributor to global warming;
- road construction increases urban sprawl, wildlife habitat destruction, and loss of community and also results in road congestion;

Figure 6. The reality is that in the coming decades society is going to have to reduce and ultimately eliminate its love affair with the personal automobile and retool its transportation systems if we are to enjoy a livable planet, pass our heritage on to future generations, and have any ability to commute and travel at all.

- only 8% of Americans do not own a personal vehicle[12] and, not surprisingly, households with low mean income own older, more polluting, and less safe vehicles.
- roads and related infrastructure are taking an ever-greater share of municipal property taxes, with some jurisdictions reporting that over 1/3 of their budgets are used for these purposes;
- road and related infrastructure costs have taken money away from mass transit and better city planning (for walking, bicycling, and green space).

These are a few of the personal transportation issues that have been plaguing society for some time but are now starting to bubble up from the fringes and catch the attention of the mainstream population. The reality is that in the coming decades society is going to have to reduce and ultimately eliminate its love affair with the personal automobile and retool its transportation systems if we are to enjoy a livable planet, pass our heritage on to future generations, and have any ability to commute and travel at all.

This transition will not happen quickly, nor will everyone agree with my outlook; that of course is the nature of open debate. What I wish to accomplish is to look at the facts and analyze peer-reviewed research on the converging issues that will eventually force society to drop the personal automobile from its collective lexicon and deal with the new reality of improved transportation, logistics, and support systems—a reality that will require a paradigm shift by all governments of the World.

The second theme of the book is to recognize that cars will not disappear overnight and that society does not have to wait for the "myth of hydrogen" in order to develop low- or even zero-carbon personal transportation vehicles, which can be produced using the technology and infrastructure systems that exist today. To prove this point, **The Zero-Carbon Car** will review a number of technologies that are currently in production as well as several that are just on the verge of commercialization.

The third and final theme of the book is to design and build a plug-in hybrid electric vehicle that is fueled completely with zero-carbon energy, using off-the-shelf technologies that are available today. I have assembled a group of engineering enthusiasts who have pooled their skills to convert an existing gasoline-powered automobile into a series hybrid electric car that uses zero-carbon electricity and liquid fuel for its motive power. The design plans and theory for the vehicle and fuel sources are included in the book and at the accompanying website: www.thezerocarboncar.com.

Many hobbyists and automotive enthusiasts convert standard gasoline-powered vehicles to battery-electric vehicles (BEVs) either at home or with the assistance of others through an electric vehicle club. While most of these conversions work well enough for local, short-range driving, the vehicles are of limited use when it comes to longer trips or when the ability to recharge the battery bank is compromised.

With this in mind, **The Zero-Carbon Car** provides design plans and instructions (including an accompanying website complete with hybrid engine management software and support data) for constructing total electric and zero-carbon, plug-in hybrid automobiles and for upgrading existing BEVs to plug-in hybrid technology.

As the software and design plans are completely open source and royalty free, I am encouraging all potential builders of plug-in hybrid vehicles to use the designs provided and improve upon them as you

Figure 7. The third and final theme of the book is to design and build a plug-in hybrid electric vehicle that is fueled completely with zero-carbon energy, using off-the-shelf technologies that are available today.

see fit. Any new designs will be posted on the www.thezerocarboncar.com website with links to anyone who has developed an equivalent or upgraded technology.

Given the limited budgets and unlimited enthusiasm of amateur vehicle developers, I feel this work will go a long way in showing governments and industry, and possibly shaming them into accepting, that the only impediment to the implementation of large-scale low- or zero-carbon personal vehicle technology is psychological not technological.

1
Personal Transportation in the Third Millennium

The Myth of Infinite Freedom

People love their cars. With over 750 million vehicles[1] currently in use and that number rising quickly, there seems to be no end in sight for automotive manufacturers. A survey conducted by ACNielsen[2] indicates that three of the world's four most populous countries have high levels of consumer aspiration to purchase an automobile, which indicates that global sales volumes will rise dramatically as the economies of these countries continue to grow and fuel greater personal wealth.

Figure 1-1. There is tremendous demand for cars in China and other developing nations. As the economies of these countries grow, more people will be able to purchase vehicles, with increased automobile manufacturing and greater pressure on fuel supplies and road infrastructure.

In North America and other developed areas of the world, the desire to purchase vehicles is lower than in the developing world owing to existing high levels of vehicle ownership. However, despite existing vehicular ownership levels, people in the United States intend to purchase more SUVs in the future, with the result that SUV sales will soon overtake sedan sales[3].

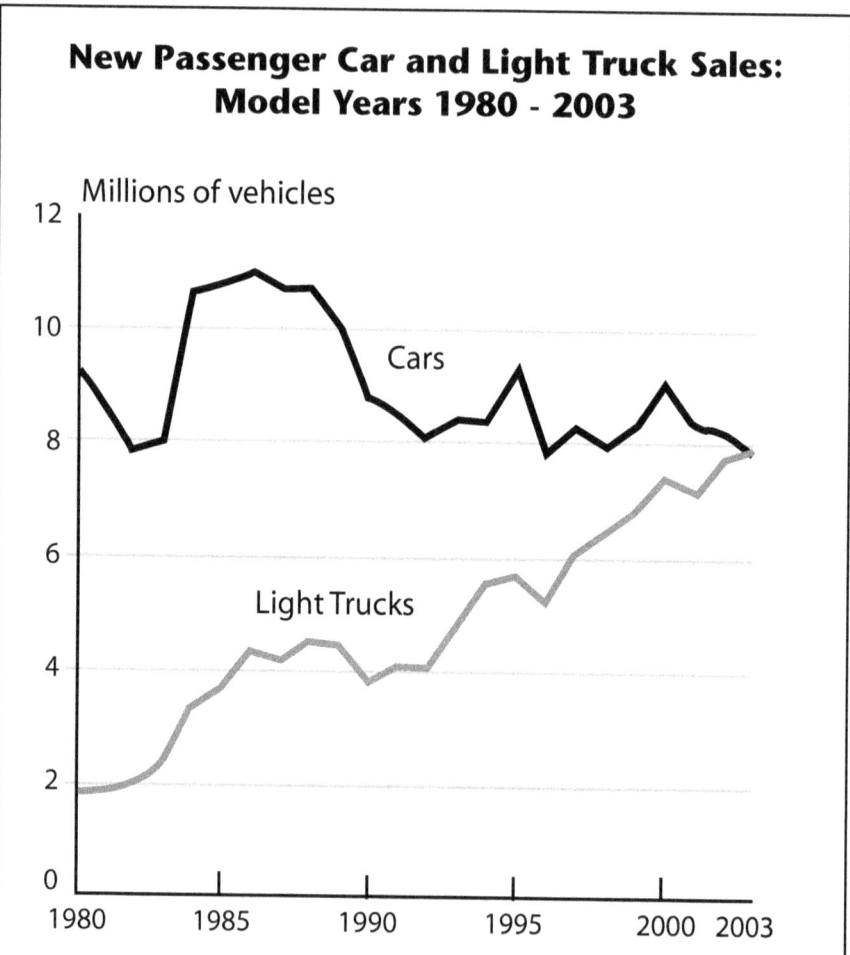

Figure 1-2. Despite existing vehicular ownership levels, people in the United States intend to purchase more SUVs in the future, with the result that SUV sales will soon overtake sedan sales. Source: ACNeilsen, Trends & Insights.

What is driving the increase in automotive sales and the desire for bigger, "better" vehicles? If you believe the automotive ads on television, car ownership is like an aphrodisiac. We are bombarded with images suggesting that we can zoom along in the wide open outdoors in pursuit of happiness while enjoying the company of back-slapping friends and beautiful women.

Durhl Caussey is described as "an award-winning syndicated columnist and member of the Texas Automobile Writers Association";[4] he recently wrote that the 2007 Hummer H2 SUV "provides you the ultimate freedom—you can go just about anywhere, and do anything. You will connect to all those things that make you feel good about yourself." He continues by telling the reader that Hummer drivers "must yearn to journey through cold streams, along valley cliffs, toward mountaintops, while taking on troublesome roads and overgrowth that shrouds untouched and pristine places."

Never mind that if this beast of a vehicle were ever to stray from the day-to-day reality of driving on traffic-clogged freeways it would destroy the very ecosystems Mr. Caussey suggests you visit. Let's be thankful

Figure 1-3. Automotive writers like to promote the myth of infinite freedom, suggesting that drivers of SUVs and Hummers will enjoy journeying through cold streams and mountaintops, when in reality most vehicles never stray much farther than the nearest Wal-Mart parking lot.

that most people who purchase a Hummer or any "off-road" SUV stray no farther than the nearest Wal-Mart parking lot.

But people do get caught up in the hype of infinite freedom. If articles such as this were limited to a select readership, they might be dismissed as a by-product of excess testosterone. Unfortunately, they are not limited to a select few. Advertising in electronic and print media constantly promotes the newfound freedom you will experience when you purchase a shiny $20,000 vehicle (notwithstanding the endless hours of toil required for financing, fuel, and maintenance charges that equate to anything but "freedom") or the status you will achieve and the envy you will arouse when you drive home in a new BMW. (Approximately 60% of all luxury vehicles in the United States are leased[5], with even more purchase-financed because the purchaser isn't actually affluent enough to buy directly.)

The reality is that there are no cold streams to cross, few back roads to zoom along, and no increased libido from pushing a pedal with your right foot. This is all an illusion, all part of the American *dream*. The reality is much more staid. The average person uses a car to commute to work, shop, buy groceries, and run errands on the weekend (at perhaps

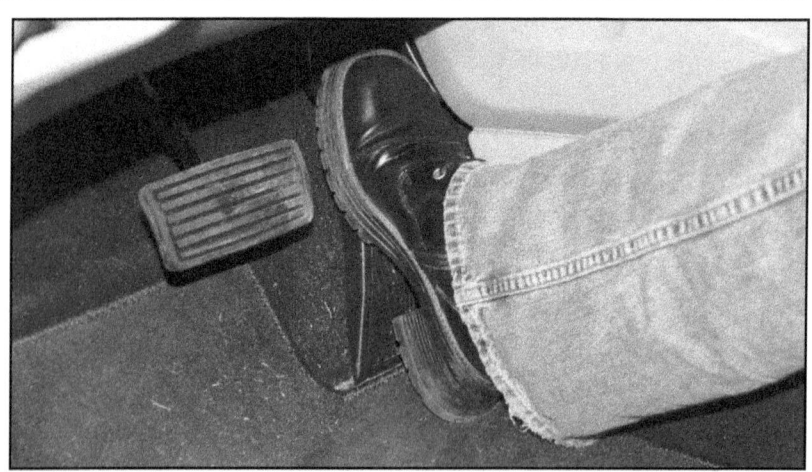

Figure 1-4. The reality is that there are no cold streams to cross, few back roads to zoom along, and no increased libido from pushing a pedal with your right foot. Realize that this is all an illusion and that the reality is much more staid.

half a dozen big-box locations that cannot be reached on foot because of poor urban planning) or run around to meetings and kids' programs during the week. Be honest with yourself and you will realize that your love affair with the car has more to do with tedium than freedom.

If the car isn't the magic carpet proffered by the Madison Avenue ad agencies, then what exactly is it and what do we really do with our vehicles? Humans have developed an interesting affair with material goods, an affair that we use to create our identities and signal to others who we are or, more correctly, who we want them to believe we are. People consider cars a good indicator of our character and wealth: the fellow driving the ten-year-old, slightly rusty Volkswagen must be rather poor, while we assume the man in a suit at the wheel of a current-model Benz to be a successful businessman. The reality might be that the former has a million dollars in the bank (because he isn't worried about what people think and avoids wasting his money on a rapidly depreciating asset), while the latter is uncomfortably in debt trying to stay ahead of the Joneses and appearing to be something he is not.

It is not that the car is the only material item we tend to use as a billboard for our persona; houses, clothes, jewelry, and most other material items are also signals of our personality and the status we want to broadcast. This trait is learned at an early age (promoted by the relentless force of media and peer pressure), ensuring that even children and tweens have the "right" clothes, friends, and model of digital music player.

For adults, the automobile is perhaps something different. For most, it is the second most expensive asset (liability, perhaps?) we own, and because of its pervasiveness in society it becomes the most visible and portable status-signaling device. Anyone can purchase a suit, a counterfeit Rolex, or an imitation Gucci handbag, but it isn't possible to "fake" a Lexus or Porsche, unless having the bank or leasing company own it amounts to the same level of illusion.

But it must be more than status and trips to Wal-Mart that forces people to go into debt and make extensive ongoing payments for insurance, fuel, and maintenance. I believe a great part of our love affair with the car has to do with our desire for privacy and a controlled environment in a crowded world, where normally demure people can "flip the finger" or avoid the social contact and "people in your space" issues that occur during the use of public transit.

In any event, the way society has structured its transportation systems over the past century and the fact that governments have massively funded car manufacturers and road infrastructure have ensured our dependence on the personal automobile.

The Reality Check – 37 Kilometers per Day

If we focus on day-to-day living for a moment, not worrying about long-distance or holiday travel, we find that most Americans have a fairly routine pattern that, on a daily basis, sees the average person traveling 40 miles (64 km) per day, with 35 miles (56 km) of this total occurring in a personal vehicle[6]. As more than one person can travel in a car at the same time, the average 35 *person* miles reduces the actual distance traveled by the vehicle to 23 miles (37 km)[7], a distance that could easily

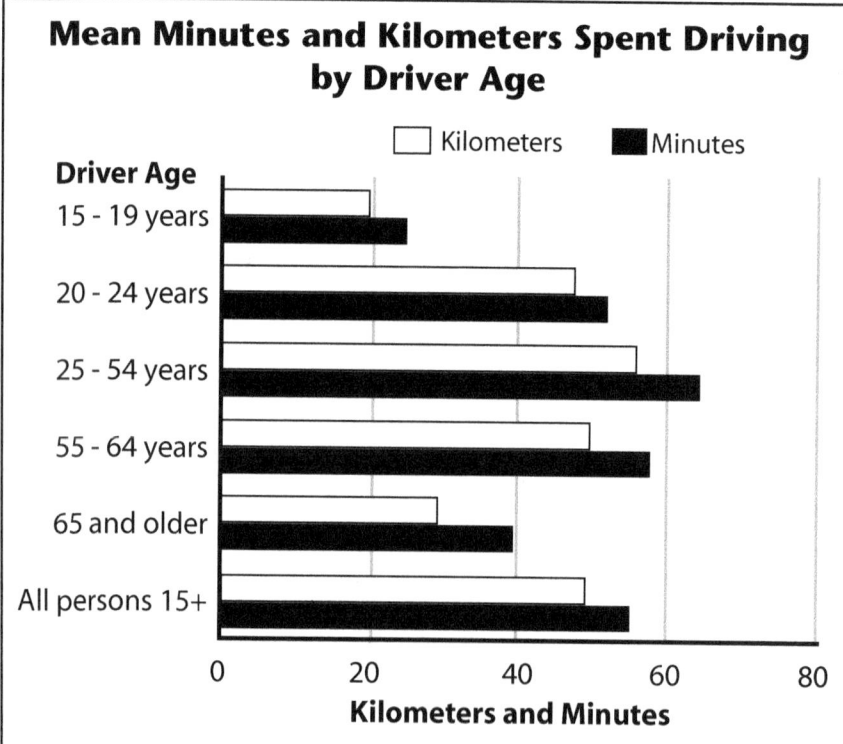

Figure 1-5. The typical American has a fairly predictable travel routine, with working adults taking an average of four trips per day while driving an average of 35 miles (56 km) per day. Source: U.S. Department of Transportation, The 2001 National Household Travel Survey.

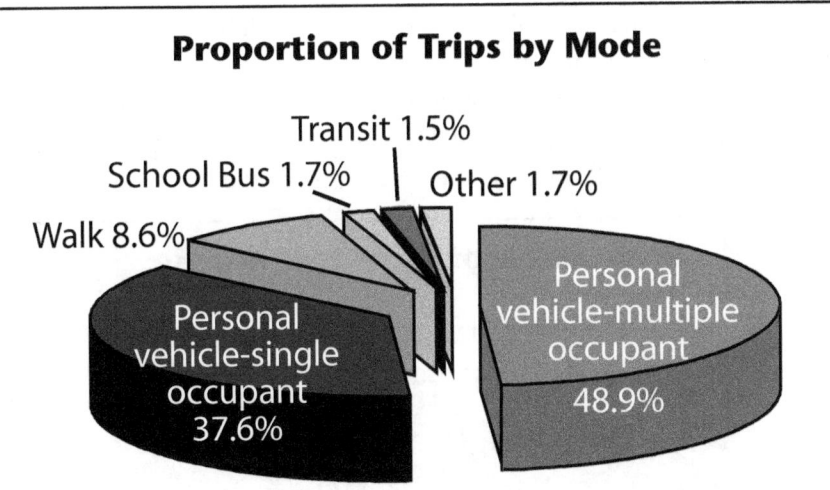

Figure 1-6. 86.5% of all local distance trips of 80 kilometers (50 miles) or less are taken using a personal vehicle. Source: U.S. Department of Transportation, The 2001 National Household Travel Survey.

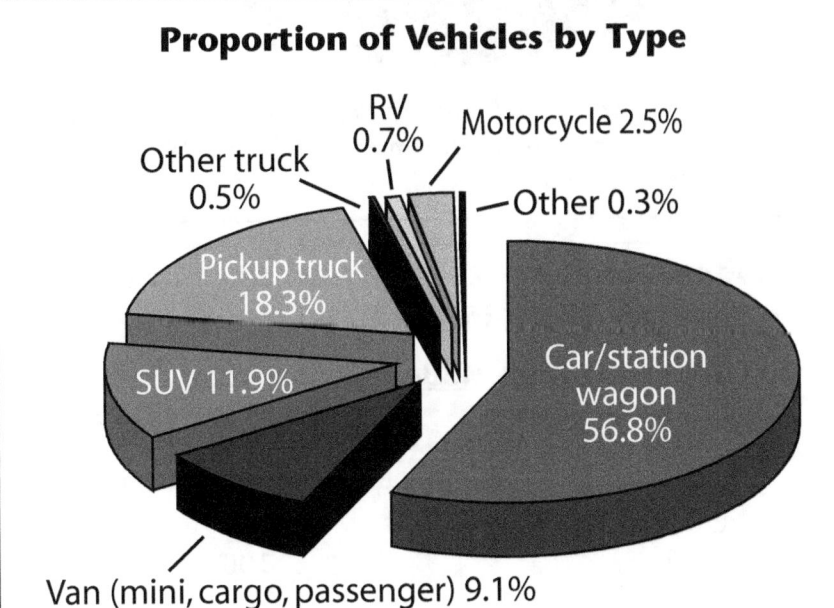

Figure 1-7. Cars, vans, SUVs, and pickup trucks make up the vast majority of vehicular trips. Source: U.S. Department of Transportation, The 2001 National Household Travel Survey.

be accomplished using public transit, were it available as well as socially and aesthetically acceptable.

Although this is a relatively insignificant distance for each trip, the sheer volume of trips taken in 2001 equates to an astonishing 2.3 trillion vehicle miles (3.7 trillion km)[8] over the course of the year. I suspect that

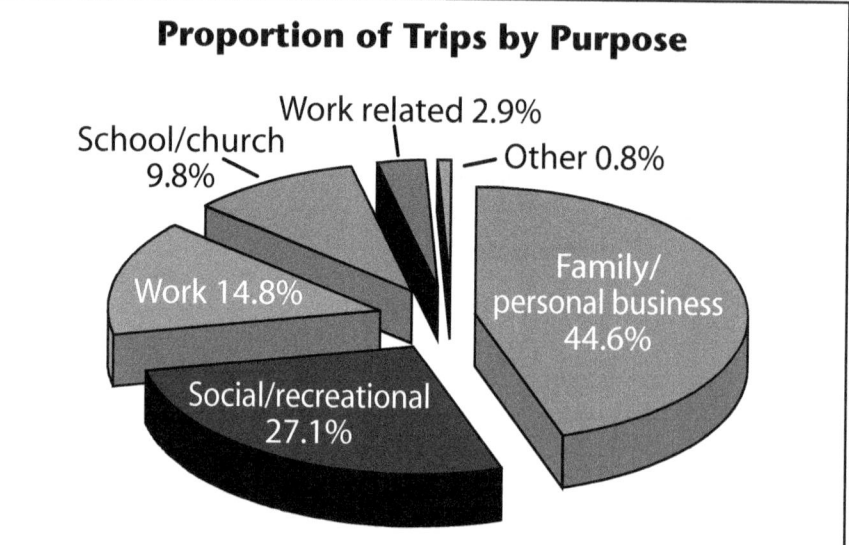

Figure 1-8. *Interestingly, only 18% of all local trips are work related. Source: U.S. Department of Transportation, The 2001 National Household Travel Survey.*

Mode of Transportation Used to Commute to Work in the Past Week

Transportation Mode	Percent
Personal Vehicle	91.2
Transit	4.9
Walk	2.8
Other	1.1
Total	100

Table 1-1. *Although the average American takes four local trips per day and covers a distance of 40 miles (64 km), a distance that could easily be traveled using mass transit, people still overwhelmingly prefer to use their personal vehicles. Source: U.S. Department of Transportation, The 2001 National Household Travel Survey.*

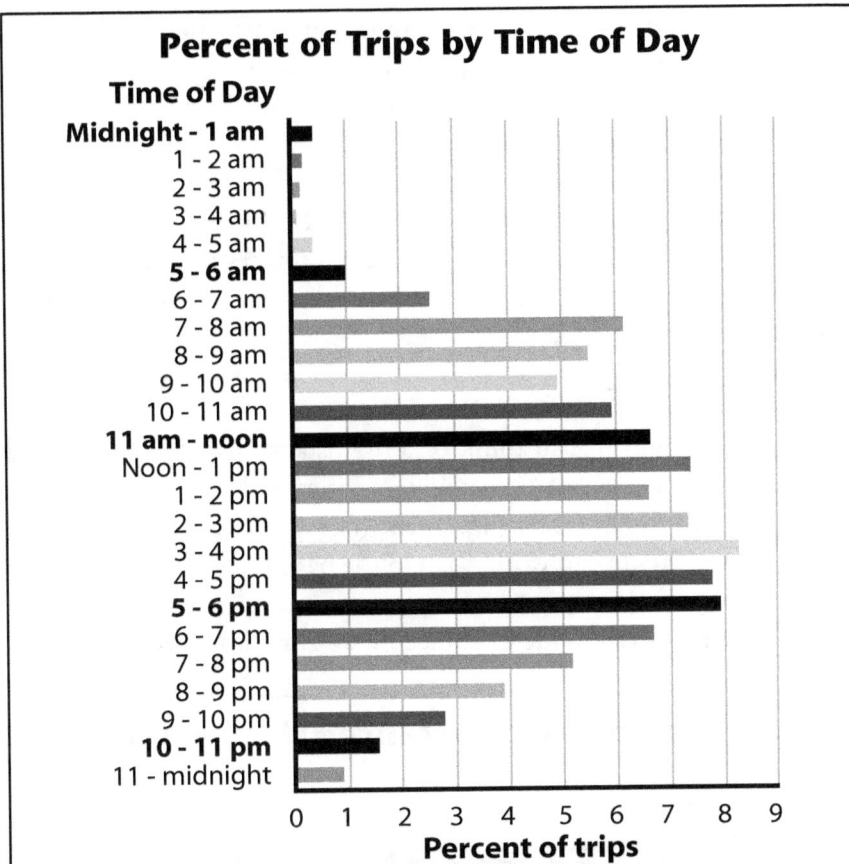

Figure 1-9. Although one would assume that most congestion would occur during the morning and afternoon "rush hour," the escalating number of trips and volume of vehicles have caused congestion to become an all-day affair. Source: U.S Department of Transportation, The 2001 National Household Travel Survey.

most people become "number numb" even trying to imagine a figure of this magnitude, but perhaps that's the point: such distance is unimaginable.

Traffic congestion is a major problem in most cities, caused, naturally enough, by the tremendous number of personal vehicles on the road. Fifty percent of adults in the United States responded in the 2001 National Household Travel Survey that they were "somewhat" to "severely" concerned about highway congestion. Although one would assume that most congestion would occur during the morning and afternoon "rush hour," the escalating number of trips and volume of vehicles have caused congestion to become an all-day affair. My mother commented years ago that as long as she was on the road before 3 pm she would be able to miss the rush-hour traffic. Not any longer. "Rush hour" on that particular highway now extends to approximately 13 hours of the day.

Although I have not reviewed similar data for other parts of the world, one can only surmise that daily trip distances in other countries are less, given the low population density of the United States relative to just about everywhere else.

Long-Distance Travel

Although we might like to think of long-distance travel as being used for vacation time, the rearrangement of cities and suburbs is causing more people to commute *long distances,* (one-way trips that are greater than 50 miles (80 km), to work[9] than in the past. However, the majority (56%) of long-distance trips are for pleasure, with personal vehicles being the transportation mode of choice for all Americans when traveling

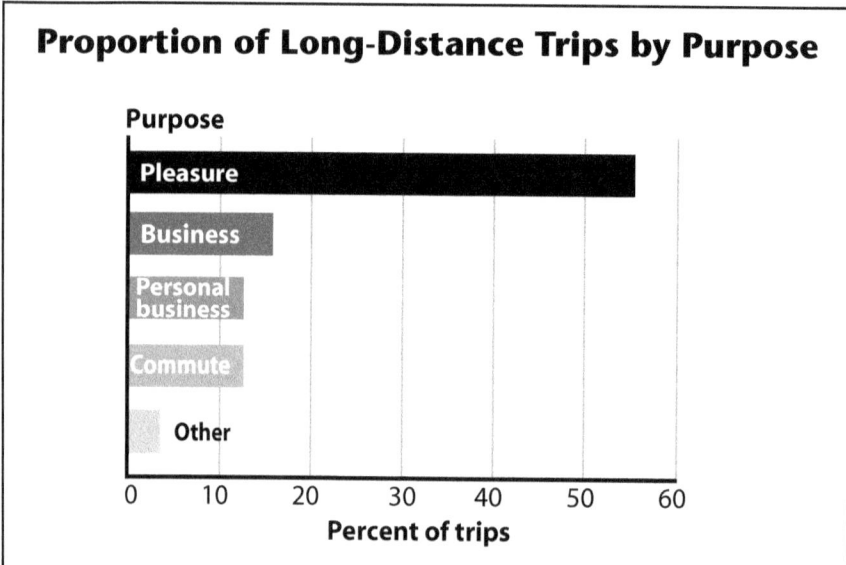

Figure 1-10. The majority of long-distance trips are for pleasure, with personal vehicles being the transportation mode of choice for all Americans, when traveling less than 1,999 miles (3,200 kilometers). Source: U.S. Department of Transportation, The 2001 National Household Travel Survey.

Percent of Long-Distance Trips by Mode and Roundtrip Distance

Miles	100-299	300-499	500-999	1000-1999	2000+	TOTAL
Personal Vehicle	97.2%	94.3%	85.9%	53.9%	22.2%	89.5%
Air	0.2%	1.5%	10.3%	42.4%	74.8%	7.4%
Bus	1.6%	3.4%	3.2%	2.6%	1.4%	2.1%
Train	0.9%	0.7%	0.6%	0.9%	0.8%	0.8%
Other	0.2%	0.1%	0.0%	0.1%	0.8%	0.2%

Table 1-2. The personal automobile dominates all other forms of long distance transportation below 2,000 miles (3,200 kilometers), including air travel, leaving trains, buses and other modes with 3% of the transportation pie. Source: U.S. Department of Transportation, The 2001 National Household Travel Survey.

Figure 1-11. It is not until Americans are traveling more than 2,000 miles (3,219 km) that air travel dominates the transportation scene. Below this distance threshold, the personal automobile dominates.

less than 1,988 miles (3,200 km)[10]. In 2001, an astounding 9 out of 10 long-distance trips were taken in personal vehicles, totaling some 760 billion miles (1.2 trillion km).

I was surprised to discover that air travel comprised only 7% of total long-distance travel, especially as I have personally witnessed the masses of people milling around the nation's airports. One has only to get stuck at LAX or O'Hare during a thunderstorm to wonder if the entire population has been delayed by poor weather.

Long-distance air travel generally means an extended or overnight stay, causing the traveler to incur related costs for accommodation, meals, and car rental. As a result, people with incomes greater than $50,000 travel by air more than twice as often as people who earn half this amount.

Although bus and train travel are the third and fourth most popular means of long-distance travel respectively, they account for less than 3% of all long-distance trips[11].

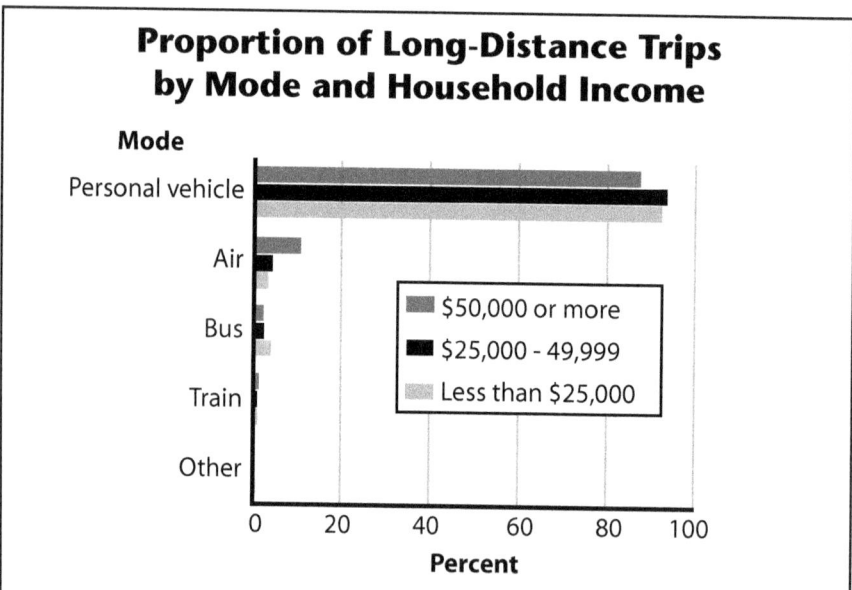

Figure 1-12. Long-distance air travel generally means an extended stay, causing the traveler to incur related expenses for accommodation, meals, and car rental. As a result, people with incomes greater than $50,000 travel by air more than twice as often as people who earn half this amount. Source: U.S. Department of Transportation, The 2001 National Household Travel Survey.

The Energy Demand of Transportation

After a quick turn of a key, we drive our personal vehicles onto the nearest roadway and join a worldwide club that consumes 53 million barrels (2.226 billion gallons/8.427 billion liters) of petroleum products ***per day*** for transportation, while an additional 27 million barrels are used for industry, utilities, and buildings[12].

Figure 1-13. Most of us don't give any thought to the incredible scale and logistics of the oil markets: drilling, extracting, transporting, processing, and retailing the fuel that makes its way into our gas tanks.

Most of us don't give any thought to the incredible scale and logistics of the oil markets: drilling, extracting, transporting, processing, and retailing the fuel that makes its way into our gas tanks. In fact, the closest most people come to even seeing these fuels occurs during their brief transaction at a self-serve gas retailer or when filling their lawn mowers. Nor do most people realize that U.S. petroleum consumption, which stood at 20.8 million barrels per day in 2005, continues to rise unabated even in light of the fact that domestic oil production is falling rapidly and imports make up 60% of the total[13].

Daily Use of Petroleum Worldwide

At present, consumers use 80 million barrels a day (MBD) of petroleum (a barrel contains 42 U.S. gallons). Two thirds of this goes to transportation.

53	**29**	**19**	**5**
MBD for transportation overall	MBD for land transport for people	MBD for land transport for freight	MBD for air transport for people and freight

Table 1-3. The world consumes approximately 80 million barrels of oil per day, while overall transportation consumes 53 million barrels or 66% of the total.

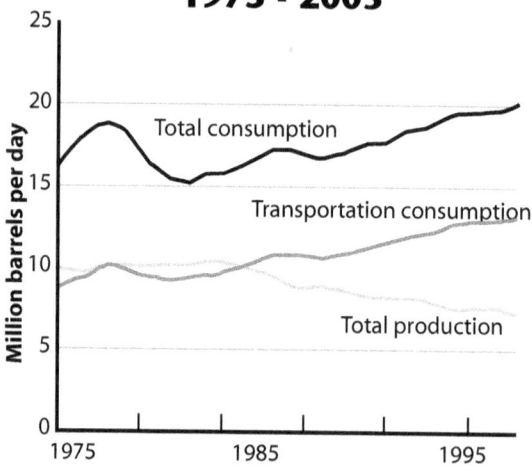

Figure 1-14. Most people do not realize that U.S. petroleum consumption continues to rise unabated even in light of the fact that domestic oil production is falling rapidly. Source: Energy Information Administration, Petroleum Products Consumption (October 2006).

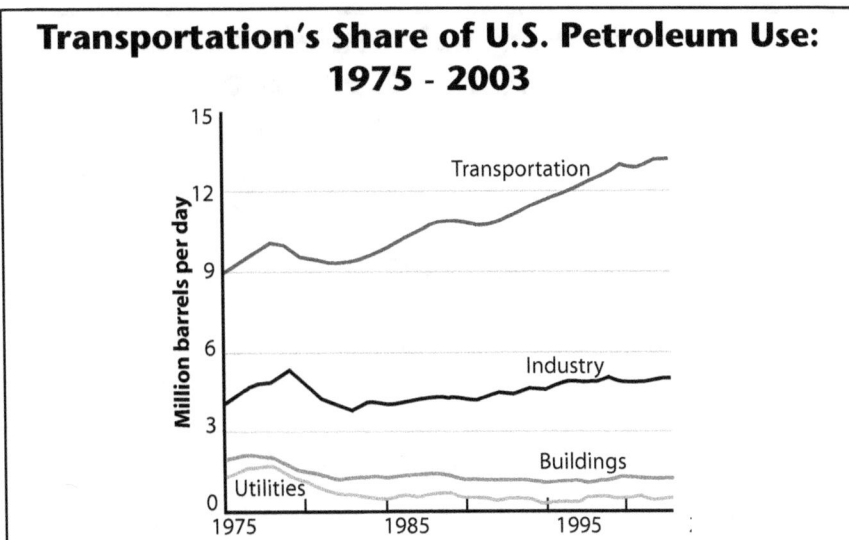

Figure 1-15. U.S. Petroleum consumption, shown by sector. Source: U.S. Department of Transportation, Bureau of Transportation Statistics, Transportation Statistics Annual Report (November 2005).

The vast infrastructure that remains mostly hidden to North Americans comprises vast armies of people, equipment, and capital spread over every region of the world. Exploration and extraction of natural gas, coal, and oil require enormous investments and careful planning to ensure that Mr. and Mrs. Consumer receive their allotment of gasoline per day, along with the other fuels required to maintain their expected lifestyles.

Incredibly, the real cost of supplying these fuels has remained fairly stable compared to overall inflation, especially when compared to fixed costs and other related expenses involved in the purchase and operation of a motor vehicle (Figure 1-16). According to the U.S. Department of Transportation, the cost of driving a vehicle 15,000 miles (24,000 km) per year in 2003 was 53 cents per mile (33 cents per kilometer), an increase of 20% over 1993, when costs were 44 cents per mile (27 cents per kilometer)[14].

The Department of Transportation has converted all cost estimates to "chained dollars in 2000" as a way of eliminating the effects of inflation from the raw data, thus providing an understandable reflection of true cost over time. Figure 1-16 illustrates the fact that gasoline and

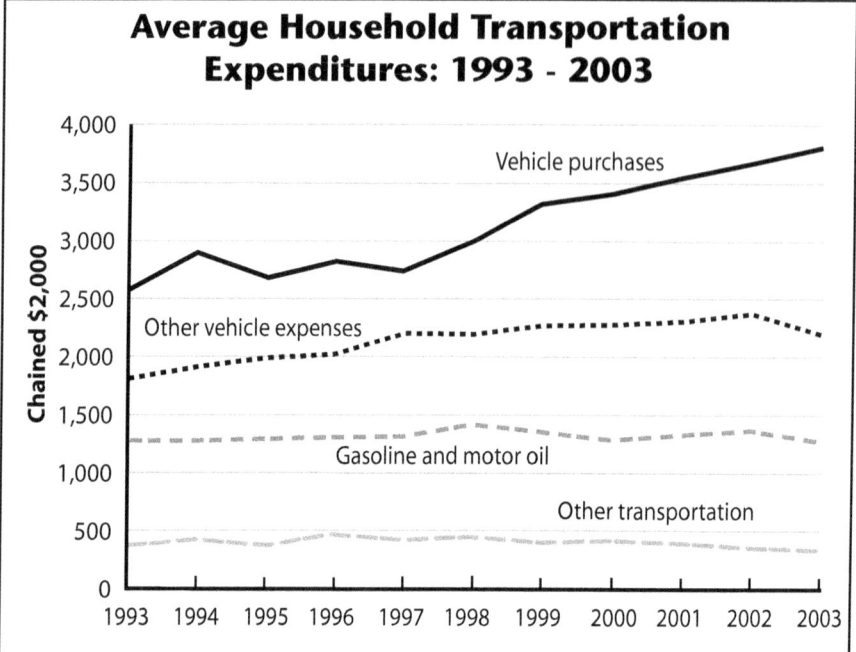

Figure 1-16. U.S. household cost of ownership for a personal automobile driven 15,000 miles (24,000 km) per year. Source: U.S. Department of Transportation, Bureau of Transportation Statistics, Transportation Statistics Annual Report (November 2005).

motor oil form part of the variable operating expenses of a vehicle yet represent only 13%, 6.9 cents per mile (4.3 cents per kilometer), of the total cost of owning a vehicle. Given the incredible importance people have placed on personal transportation, society does not recognize petroleum's overall value, given such historically low prices and its nonrenewable status. We are willing to pay a dollar a liter ($3.79 per gallon) for bottled water, yet howl when the price of gasoline even remotely approaches this level.

The transportation sector of the United States consumed 17% more energy in 2004 than it did a decade earlier. This annual growth rate of 1.2% will continue as long as the economy remains healthy, the population continues to grow, and no supply or price shocks throw a roadblock in the way. Given that the population of the United States was approximately 300 million people in 2007 and is expected to grow to over 403 million people in 2050[15], an increase of 34%, transportation energy consumption, road congestion, greenhouse gas, and exhaust emissions will also rise unless federal and state governments provide the leadership to move towards a more sustainable path.

Governments worldwide are facing similar or worse issues of energy demand growth in the transportation sector but are loath to implement the necessary policy configurations to curb this consumption and create a sustainable path. Rather than trying to restrain automotive growth, implement alternative transportation means, or implement energy-efficiency rules, the world's collective answer appears to be "find more energy." This short-term, reflex answer is what voters want to hear, and politicians (who are anxious to be re-elected) delay the difficult but inevitable energy-constraint policies that must be developed. U.S. President George W. Bush summarized the concept of never-ending growth and energy consumption vividly in his January, 2007 State of the Union Address:

"It's in our vital interest to diversify America's energy supply -- the way forward is through technology. We must continue changing the way America generates electric power, by even greater use of clean coal technology, solar and wind energy, and clean, safe nuclear power. (Applause.) We need to press on with battery research for plug-in and hybrid vehicles, and expand the use of clean diesel vehicles and biodiesel fuel. (Applause.) We must continue investing in new methods of producing ethanol -- (applause) -- using everything from wood chips to grasses, to agricultural wastes.

We made a lot of progress, thanks to good policies here in Washington and the strong response of the market. And now even more dramatic advances are within reach. Tonight, I ask Congress to join me in pursuing a great goal. Let us build on the work we've done and reduce gasoline usage in the United States by 20 percent in the next 10 years. (Applause.) When we do that we will have cut our total imports by the equivalent of three-quarters of all the oil we now import from the Middle East.

To reach this goal, we must increase the supply of alternative fuels, by setting a mandatory fuels standard to require 35 billion gallons of renewable and alternative fuels in 2017 -- and that is nearly five times the current target. (Applause.) At the same time, we need to reform and modernize fuel economy standards for cars the way we did for light trucks -- and conserve up to 8.5 billion more gallons of gasoline by 2017.

Achieving these ambitious goals will dramatically reduce our dependence on foreign oil, but it's not going to eliminate it. And so as we continue to diversify our fuel supply, we must step up domestic oil production in environmentally sensitive ways. (Applause.) And to further protect America against severe disruptions to our oil supply, I ask Congress to double the current capacity of the Strategic Petroleum Reserve. (Applause.)

America is on the verge of technological breakthroughs that will enable us to live our lives less dependent on oil. And these technologies will help us be better stewards of the environment, and they will help us to confront the serious challenge of global climate change. (Applause.)"

While some of these points are worthy of consideration, the vast majority of the ideas are designed as fuel-swapping programs, akin to highway lane changers who always think the other lane is moving faster and is therefore the one to be in. I of course do not believe any American politician who assumes he or she can reduce energy imports, let alone overall transportation energy consumption, by tinkering with the fuel supply mix. Mr. Bush's speech is simply a program to produce more fuel (from various sources which may or may not be sustainable) and maintain the status quo, which reminds me of an alcoholic on a binge, trying to empty every last bottle to feed his addiction. Incidentally, isn't this exactly what Mr. Bush said when he made the comment that "Americans are addicted to oil"?

Diversifying the national energy supply is not an intrinsically bad idea. What is a bad idea is not accepting that increased use of of personal automobiles will lead to continued expansion of roads and related infrastructure (increasing living costs, raw material consumption, environmental degradation, and greenhouse gas emissions along the way) as well as expanded suburbs and exurbs, which inherently require more driving and fuel consumption, creating a vicious feedback cycle that will be difficult to correct without massive societal upheaval. In his book *The Long Emergency*, author James Kunstler predicts that excessive North American energy consumption will lead to supply disruptions and spiraling prices, making the suburbs impossible to reach and transforming them into vacant ghettos no one will want to live in.

There is also more energy consumed in the life cycle of an automobile than the gasoline required to make it move. We have to extract the raw materials necessary to feed the petrochemical industry that makes the electronics, rubber, and plastics (not to forget steel and aluminum) which are used by the subcontractors in the fabrication of auto parts. These factories also consume energy to produce various components, transport end products, and dispose of waste remnants of the production cycle.

The auto manufacturers consume yet more energy fabricating the final vehicle, while at the far end of the automobile's life cycle a recycler or scrap dealer will use a last gulp of energy for its disposal.

Research conducted by Heather L. MacLean and Lester B. Lave of Carnegie Mellon University[16] reviewed the energy consumed over the lifetime of a typical automobile and found that approximately 75% of the vehicle's "energy cycle" was attributed to the fuel used in its opera-

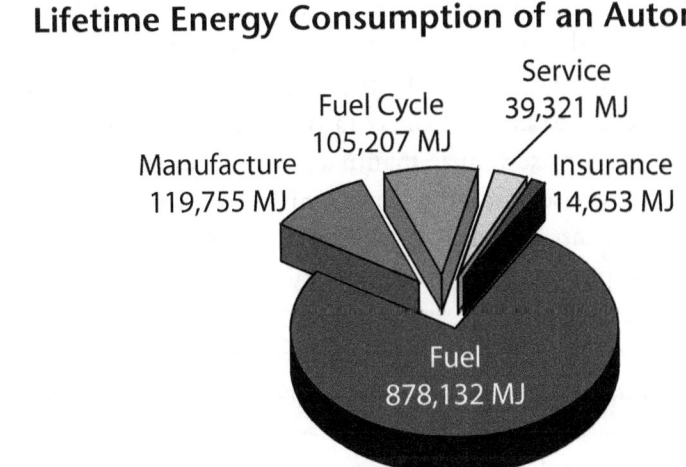

Figure 1-17. The energy consumed over the lifetime of a typical automobile is shown in this pie chart. Approximately 25% of a vehicle's life cycle embedded energy and materials are often derived from or through the use of hydrocarbons; I seriously doubt that engine blocks or transmissions will ever be produced from hydrogen, ethanol, or wind. Source: Heather MacLean & Lester Lave, "A Life-Cycle Model of an Automobile."

tion. The remaining 25% was attributed in descending order to vehicle manufacture, fuel cycle (comprising oil extraction, refining, and delivery), servicing, and lastly, the energy used by the automotive insurance industry for buildings and related infrastructure. I should also point out that emissions of carbon dioxide would be ratio metric to the energy consumption data determined in the study.

The Limits of Roads and Infrastructure

It is well known that members of today's highly technical scientific community must specialize in ever more advanced and focused areas of expertise in order to compete and to advance the state of the art. Perhaps this explains the collective lack of common sense applied by the scientific community to the issue of personal transportation.

Associate Professor Mark Z. Jacobson and his colleagues typify this problem when they suggest in a June 24, 2005 article for the journal *Science* that cleaning America's air and improving transportation problems would be as simple as the government funding an "Apollo Program" for wind-based hydrogen energy that would, in turn, fuel North American cars.

My concern at this point is not with hydrogen (there will be more on that later) but rather with the authors' lack of understanding and inability to grasp the big picture. In the mid to long term, increased (or continued) access to any personal transportation fuel, low carbon or not, is going to increase vehicle manufacturing volumes (thereby increasing green house gases and other toxic emissions[17]) while supporting the development of more roads and fostering a displaced and continuously commuting, practically nomadic society.

Roads have emerged as a serious environmental problem, even though they are initially planned and developed with the opposite in mind, that is to improve the mobility and economic status of the people who use them. In many cases, this result is achieved. Roads provide access to markets, jobs, and the transportation of goods in the short term. However, consider the following negative impacts of roads:
- increased traffic congestion due to increased road development and use;
- movement of goods by road decreases use of rail for long-haul transport, leading to a disproportionate amount of road surface damage;
- consumption of vast amounts of land that was previously used for parks, farming, and natural habitats for wildlife;

- increase in road accident fatalities (There are approximately 18% fewer deaths on American roads *each year* than casualties in the entire Vietnam War.);
- increased number of injuries, often requiring lifelong extensive rehabilitation which results in a huge financial burden to family and society alike;
- demand for a disproportionate share of government funding, especially at the expense of improved mass transit and freight infrastructure systems;
- increased obesity due to the "need" to drive everywhere;
- increased municipal sewer, water, and other "under road" infrastructure costs;
- development of vast areas of land for parking lots.

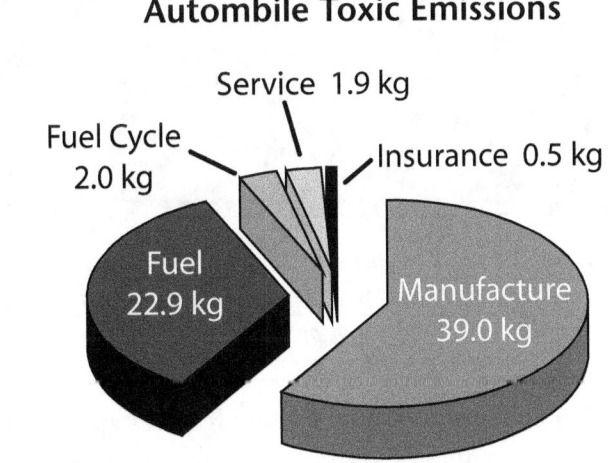

Figure 1-18. The fuel required to power a car over its operating life predominates the vehicle's energy life cycle. However, toxic emissions (shown in this chart) do not correlate with energy consumption. Therefore, reducing the carbon-fuel consumption of a vehicle will not have as sizeable effect on the vehicle's overall environmental footprint as would be desired. Source: Heather MacLean & Lester Lave, "A Life-Cycle Model of an Automobile.".

Roads, bridges, traffic signs, storm sewers, rock crushing, and just about everything else related to the building of infrastructure cannot be fabricated out of hydrogen or electricity. In fact, nothing but good old-fashioned oil and oil byproducts can be used in laying asphalt and concrete. As with vehicles, this infrastructure must be maintained and will require more energy and materials along the way.

Figure 1-19. Roads, bridges, traffic signs, storm sewers, rock crushing, and just about everything else related to the building of infrastructure cannot be fabricated out of hydrogen or electricity.

Each mile (1.6 km) of four-lane highway requires approximately a quarter-million metric tonnes of sand and gravel[18] that may have come from environmentally sensitive areas. Cement kilns require vast amounts of energy, with low-grade fuels including shredded tires increasingly being used. Ironically, as the cost of traditional fossil-fuel energy increases, companies do not necessarily move to renewables or the cleanest energy source. Instead, they gravitate towards the one with lowest first cost. The cost of air pollution and greenhouse gas emissions does not enter into the accounting books, so the evaluation of what really is the lowest-cost energy source is, for the most part, badly skewed in favor of the dirtiest sources.

Roads and related infrastructure are already approaching their physical and financial limits in many parts of the world and the costs for development and maintenance are taking a disproportionate share of the tax dollars. Author and city councillor Clive Doucet, in his book *Urban Meltdown*, has this to say about road construction costs in Ottawa, the capital city of Canada and home to 800,000 people:

"Most people have no idea of how great a burden roads are on their tax dollar. We (Ottawa) are presently rebuilding 7.7 kilometers (4.8 miles) of urban highway in my city, which will add two lanes of road capacity. It will cost taxpayers $67 million to add these two extra lanes, and this is cheap because there is no land to purchase.... To put this $67 million in perspective, the total cost of all the money spent for new community infrastructure in Ottawa is $19.3 million. That's right – two lanes of 4.8 miles is costing more than three times the entire budget for parks, community centers, swimming pools, ice rinks and daycares for 800,000 people. This is typical for every city, large or small, in North America."

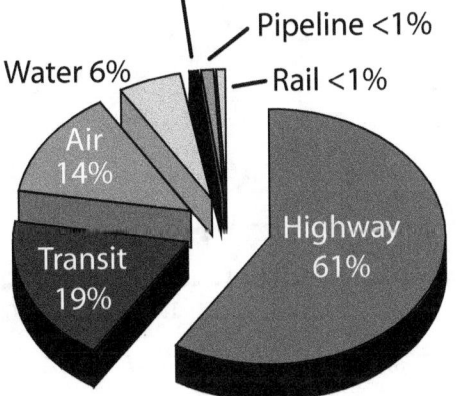

Figure 1-20. Spending on highways received the largest share of U.S. public funds, although offsetting income from fuel taxes and other cash receipts will skew which transportation mode actually receives subsidization. U.S. Department of Transportation, Bureau of Transportation Statistics, Transportation Statistics Annual Report (November 2005).

Society cannot continue to invest in roads leading to low-density suburbs that are simply sleeping quarters for the commuting masses. We cannot build our way out of congestion problems, since more roads lead to more cars, repeating an endless cycle. Likewise, trucks and freight transportation represent 76% of road weight loading, creating a disproportionate demand for maintenance[19], when better, faster, cheaper, and far more energy-efficient means of moving products and people are available. Investment in roads and related infrastructure is an inefficient

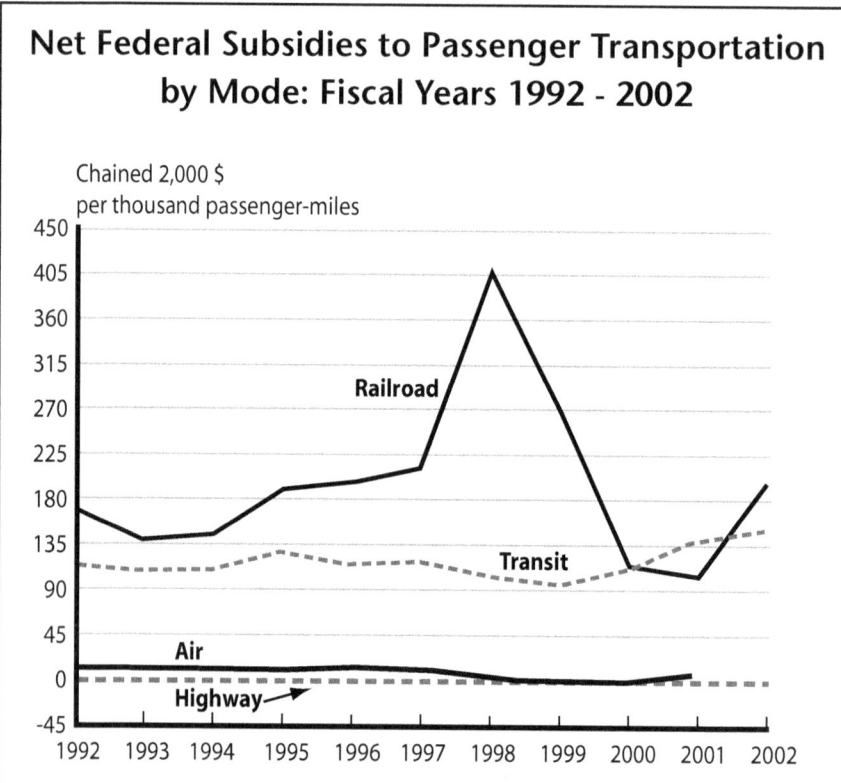

Figure 1-21. The "dollars per passenger mile" system used to evaluate U.S. federal government subsidies to the transportation sector results in highways producing a net profit to the government, while passenger rail systems received the highest subsidy of $405 per passenger mile. This is to be expected, as usage density is distorted in favor of personal vehicles at the expense of other modes of transportation. U.S. Department of Transportation, Bureau of Transportation Statistics, Transportation Statistics Annual Report (November 2005).

use of public monies, skewing transportation investments towards those able to afford the luxury of a personal automobile while eradicating investment in mass transit and a freight logistics system that would benefit all of society.

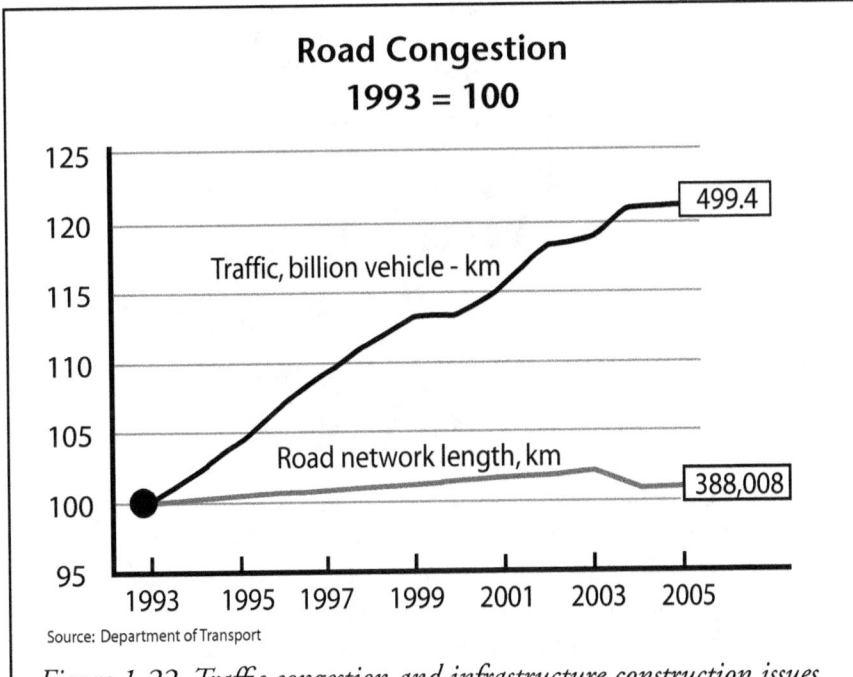

Figure 1-22. *Traffic congestion and infrastructure construction issues are not much different in the U.K. than in the North America. The U.K. Department of Transport figures show investment in roads having stalled over the last decade while vehicular mileage has increased dramatically.*

The Obvious Lessons

With increased population growth and everyone believing they have a right to endless personal mobility, energy consumption, automotive growth, and urban sprawl will never stabilize or decline, which they must do to ensure a clean atmosphere and energy security in both developed and developing countries.

Governments have long subsidized the personal automobile, truck transportation, and road infrastructure over all other means of surface

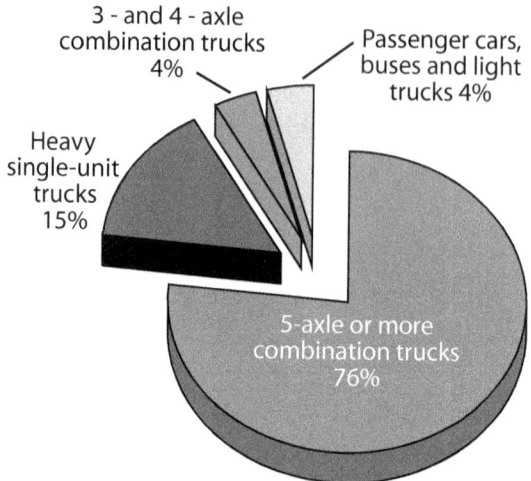

Figure 1-23. Highway vehicle loading is a method of determining the damage caused by large trucks, buses, and automobiles. Only 4% of road damage can be attributed to passenger vehicles, indicating that moving freight off the roadways would reduce infrastructure damage, leading to better road surfaces and reduced maintenance costs. Source: U.S. Department of Transportation, Bureau of Transportation Statistics, Transportation Statistics Annual Report (November 2005).

transport, even though the vast majority of all trips are within range of walking, biking, and public transit, all of which are more environmentally and financially sustainable.

It is also important to understand that switching to *any* alternative fuel will not necessarily make an impact on the total energy consumption used in transportation; it merely moves it to other locations and sources of supply. For example, corn-based ethanol requires massive amounts of fertilizer, insecticides, fuel for planting and harvesting, irrigation, plant transportation, distillation, tanker trucking, and energy for its retail point of sale. If oil and petrochemicals were unavailable, there would be no corn available to produce ethanol in the first place, nor would there be any roads, bridges, or parking lots.

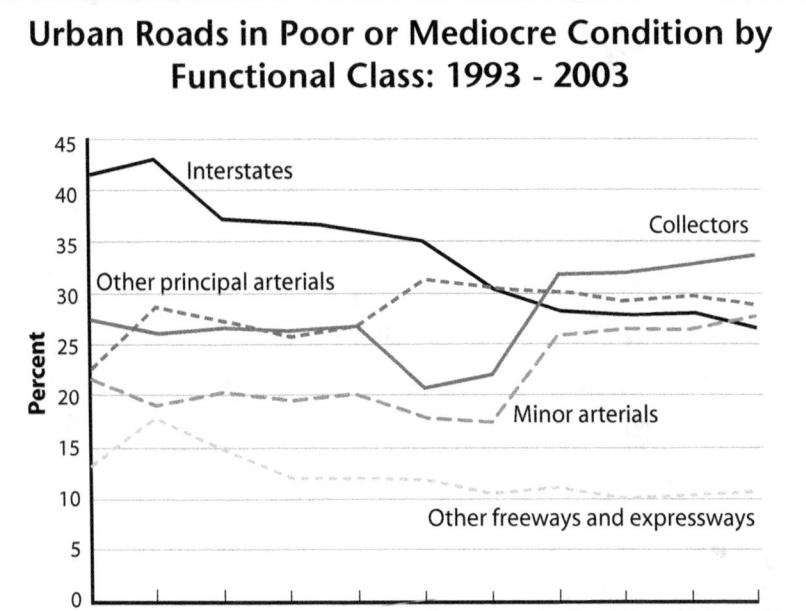

Figure 1-24. Although interstate highways have improved in condition over the last decade, the condition of other road categories has declined. In many areas of the world, roads are being built faster than governments can repair them. Source: U.S. Department of Transportation, Bureau of Transportation Statistics, Transportation Statistics Annual Report (November 2005).

Massive changes in attitude will be required to modify the way the average industrialized person travels. There are numerous issues looming in the background, including energy supplies, costs, and atmospheric carbon limits, which are going to force society to drop its love affair with the personal vehicle. Governments will have to redevelop our transportation systems by employing a number of policies:

- Develop a carbon taxation policy for *personal* energy consumption that immediately charges $500 per tonne of carbon emitted into the atmosphere. Raise this to $1,000 per tonne as society redeploys using more environmentally sound and sustainable technologies.
- Use the proceeds of the carbon tax to improve energy efficiency for socially marginalized people. (Do not help pay rising fuel bills for those people least able to afford them. This is nothing more than a

Figure 1-25. Along with roadways, bridges are also straining (literally and figuratively) for support. The number of bridges in the United States that are structurally deficient has reached alarming levels. U.S. Department of Transportation, Bureau of Transportation Statistics, Transportation Statistics Annual Report (November 2005). Photo courtesy of Tom Ruen (http://en.wikipedia.org/wiki/User:Tomruen)

bonfire for dollars. Instead, provide financial assistance to retrofit homes and subsidize the purchase of the most advanced energy-efficient appliances as well as providing better access to mass transit.)
- Stop subsidizing auto and road industry; ensure that the polluter pays for environmental degradation.
- Ban or tax cars in areas where air quality or congestion is excessive. Consider the London "congestion tax" as a working model.
- Set minimum fuel efficiency standards that do not have loopholes. (A carbon tax program should move the auto industry in this direction automatically.)
- Halt new road infrastructure development and focus on maintaining existing roads. As oil prices escalate as a result of carbon taxation, supply constraints, and geopolitical issues, car commuting will be forced into retreat, reducing the demand for roads.

- Stop long-haul truck transportation and move to better freight and logistic systems.

The coming chapters will explore these issues in detail and will present the case for the end of the personal automobile as well as the steps and technologies required to retool the personal car into a more sustainable appliance in the interim.

2
A Brief Review of Energy Issues

North Americans love their cars—as well as their trucks, minivans, and sport utility vehicles (SUVs). We learned earlier that with increasing personal wealth and access to cheap credit, urban vehicular density has exploded. Simultaneously, cities have expanded, sprawling across the landscape as a result of our upwardly mobile culture and our desire to live in the suburbs.

The results are not pretty. As a direct result of this increased mobility, the population density of North American cities has declined nearly 50% over the last forty years, decimating historic downtown core areas and local family businesses. As the sprawl continues, city infrastructure costs continue to climb unabated, with an ever larger share of budgets swallowed up for road maintenance and expanded sewer and water pipelines. And yet as we move farther and farther away from each other we are deprived of our sense of community, with the result that people don't get to know their neighbors and are reluctant to car pool. Couple this with highly subsidized gasoline prices and the result is traffic congestion, with a high percentage of single-occupant vehicles clogging roadways.

Urban sprawl directly contributes to environmental degradation: dramatically increased fuel consumption for heating and cooling larger suburban homes, toxic runoff from road surfaces, and increased landfill

Figure 2-1. North Americans love their cars, and apparently they also enjoy traffic jams, long commutes, and the summer construction cycle. Civic planners pay only lip service to controlling the urban vehicular density which directly contributes to daily scenes such as this. This scene is played out in most parts of the world.

waste from the higher levels of consumption required to fill our large houses all contribute to creating an energy-efficiency nightmare.

Urban sprawl takes its toll in other ways. As the population density of a city decreases, mass transit costs rise exponentially as a result of the need to service a larger area. For example, a city that is 20 miles (32 km) in diameter has an area of 314 mi^2 (804 km^2). If the city size doubles to a diameter of 40 miles (64 km), the land area increases to a whopping 1257 mi^2 (3217 km^2). This 400% increase in land area necessitates a corresponding increase in the number of buses, rail lines, and other services required for this area, even though population growth and density do not increase at the same rate. Naturally, low-density or rural mass transit systems are more expensive to operate than their high-density urban counterparts. Because of increased commuting time and distance, rural/low-density transit systems are often underutilized in favor of personal vehicles, provoking protest from taxpayers and opposi-

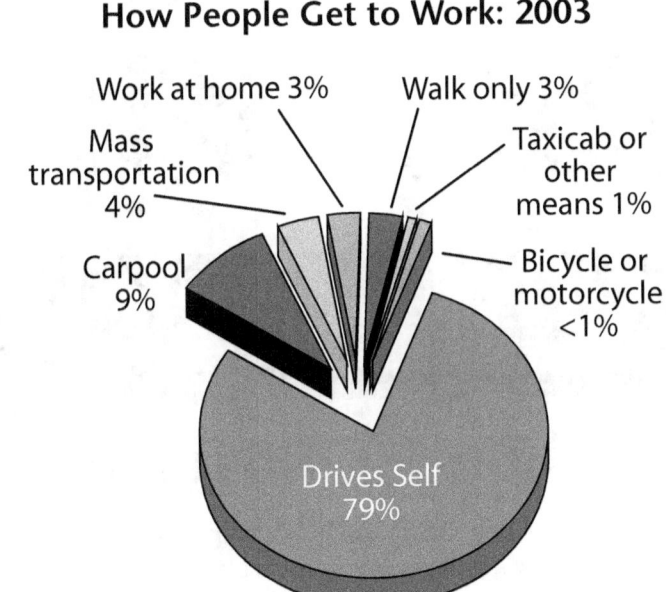

Figure 2-2. In 2003, 79% of people who commuted to work, drove alone, while only 9% car-pooled. The remaining 12% either worked at home or found other transportation means. U.S. Department of Transportation, Bureau of Transportation Statistics, Transportation Statistics Annual Report (November 2005).

tion politicians that these transit systems are merely a money sponge for badly needed social program dollars. As is often the case, the perceived problem is misidentified. The heavy cost of subsidizing all forms of energy and especially personal vehicle fuels along with the linked issues of urban sprawl, roads, and related infrastructure while disregarding the environmental and medical impacts are the real problem.

The human desire to have the unbridled personal freedom that a car supposedly provides is not enjoyed by all people; the car is a rich man's toy. According to a survey by Colliers International, a real estate agency, the typical parking rate in the City of London was U.S. $1,198 **per month** in 2007[1], about the same cost as an apartment in most places of the world.

Politicians create transportation policy in favour of the voting middle classes and above, quietly ignoring the elderly and single-parent and

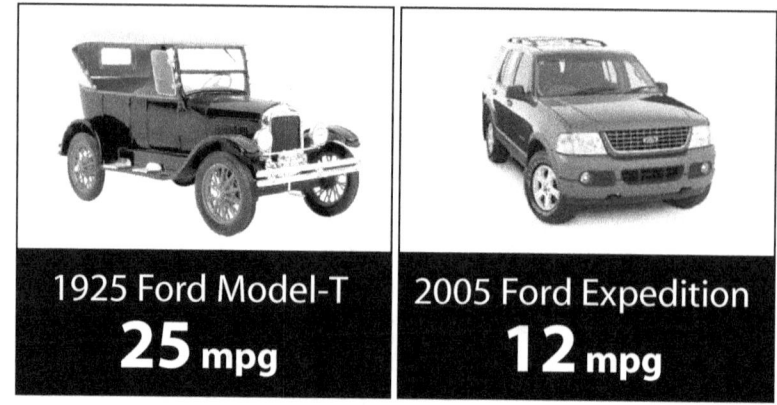

Figure 2-3. Cars, trucks, and SUVs are more about style, ego, and one-upmanship than about transportation. Does anyone really need a 300-horsepower SUV for a city commute? The Ford Motor Company seems to be driving in reverse, building massive, low-efficiency vehicles that generally carry nothing heavier than one commuter.

financially marginalized families as well as those who cannot or are unable to drive for other reasons.

Society, as well as its governments, believes that the only solution to the transportation of people and goods is through more cars and trucks, more streets and more highways, all the while using more of the earth's natural capital and fuel.

These heavily interlinked issues create a need to examine natural resource supply and depletion as well as energy efficiency to help better understand where energy policy should be pointed and how it relates to the personal automobile. Currently, most politicians are pointing straight ahead, with a business-as-usual approach that will only cause hardship and misery for future generations.

Geopolitics and Oil

"The study of the effects of economic geography on the powers of the state" is one definition of the word "geopolitics," a simple term used to describe one of the most complex issues of modern society—a society addicted to and desperate for oil needed for energy and feedstocks for a vast array of petrochemical products worth hundreds of billions of dollars. Every country in the world uses oil and other petrochemicals, yet the dispersion of supply and the consumption of these natural resources require a worldwide trading system. When resource depletion, climate change, and hostilities enter into the mix, the outcome is anything but certain. Predicting future energy supply and policy outcomes may be difficult if not impossible; perhaps with a review of these developments, a level of certainty as to future trends can be determined.

If you were of driving age in 1973 you will no doubt recall the long lineups and short supply of gasoline at your local filling station. Even if gasoline was available, OPEC raised the price of oil from $4.90 a barrel to $8.25 a barrel in that year. None of this had anything to do with diminishing world oil supplies, but came about as a result of Arab indignation over American support of Israel and the misguided energy policies of the Nixon administration. Given the geopolitical issues associated with having to rely on other countries to provide the energy to run the American domestic economy, one would assume that the need

Figure 2-4. It was not lack of supply but rather the misguided energy policies of the Nixon administration coupled with the Arab oil embargo over American support of Israel that led to severe oil supply shortages in the early 1970s. (Source: U.S. National Archives & Record Administration)

for energy self-sufficiency would be obvious, with a corresponding move towards mandatory conservation, efficiency, and regulatory programs. Unfortunately, this assumption is dead wrong.

Energy consumption in the United States and Canada is insatiable. Lulled into a false sense of security by falling oil prices and seemingly endless supplies, sport utility vehicles (SUVs) and minivans abound, with the result that automotive fuel economy has completely stagnated domestically, while Europeans (who are a bit more enlightened in this area owing to a lack of natural resources as well as effective climate change policies) have been able to power their vehicles approximately 30% more efficiently. Need and want have become confusing signals to most of the North American middle class, who have grown up be-

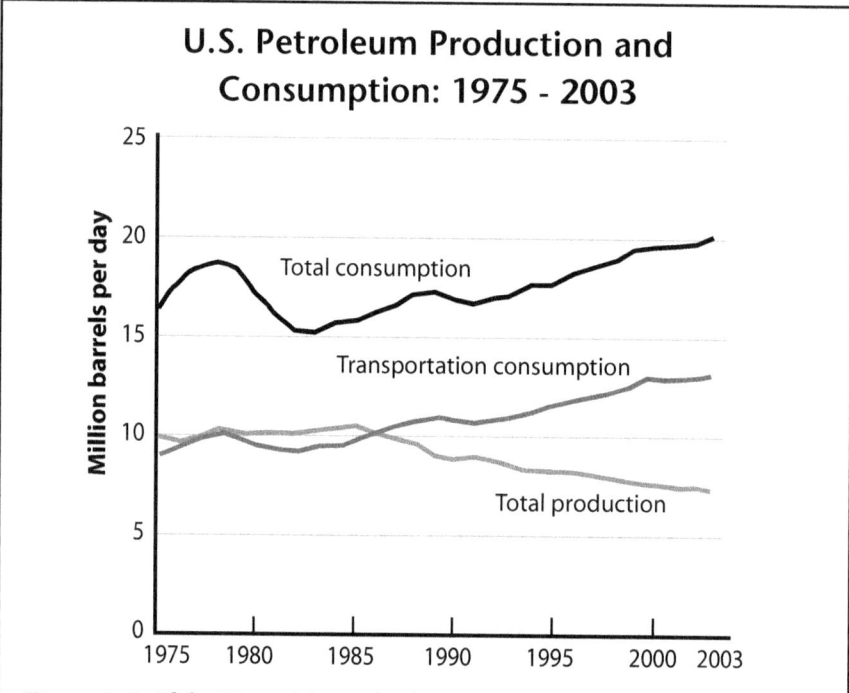

Figure 2-5. If the United States had to rely only on its domestic oil sources at current rates of consumption, it would have less than a three-year supply[2]. Energy-efficiency programs and a retooling of transportation infrastructure are required on an urgent basis. Unfortunately, neither side of the political field seems to want to do anything more than fiddle with inadequate measures. Source: U.S. Department of Transportation, Bureau of Transportation Statistics.

lieving that "bigger is better" and that happiness can only be gained by ever-larger amounts of personal consumption. This is reinforced by politicians such as California Governor Arnold Schwarzenegger, who recently stated that Hummers could be environmentally friendly and was quoted in an April 16, 2007 *Newsweek* article saying that "biofuel is not like some wimpy feminine car, like a hybrid." Even if biofuels were environmentally sustainable (which they are not when considered in the entire lifecycle of transportation), burning two or three times more biofuel than necessary in order to drive a Hummer clearly violates the laws of efficiency, sustainability, and plain old common sense.

Consider that the venerable 1925 Ford Model T should be considered an economy car by today's standards, as it was able to achieve an astounding 25 miles per gallon fuel consumption. After 80 years of technological improvement, the 2005 Ford Expedition (a commuter/shopper *necessity* according to more and more North Americans) gets a mere 12 mpg. Instead of using enhanced vehicle engine technology to improve fuel efficiency the major auto manufacturers have increased

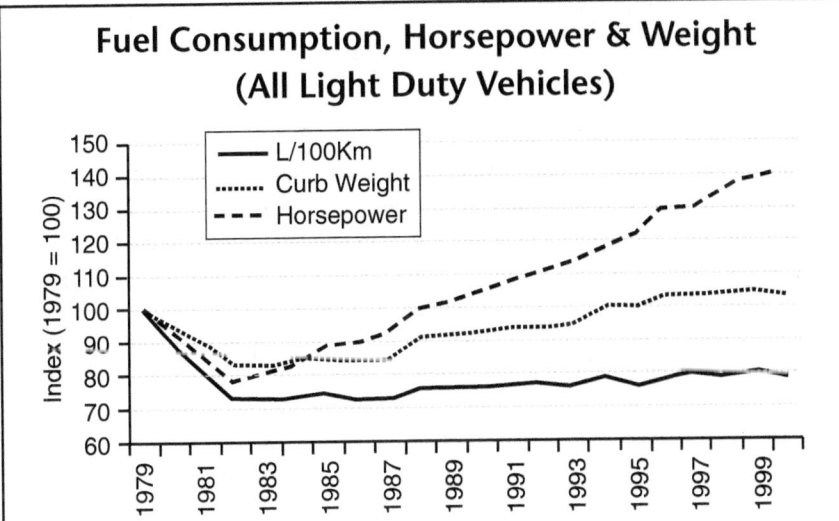

Figure 2-6. Transportation energy consumption continues to rise as vehicle energy efficiency has stalled as a result of lack of political will, aggressive lobbying by the auto industry, and increased passenger-miles of travel. (Source: U.S. Department of Transportation, Bureau of Transportation Statistics, Transportation Statistics Annual Report November 2005).

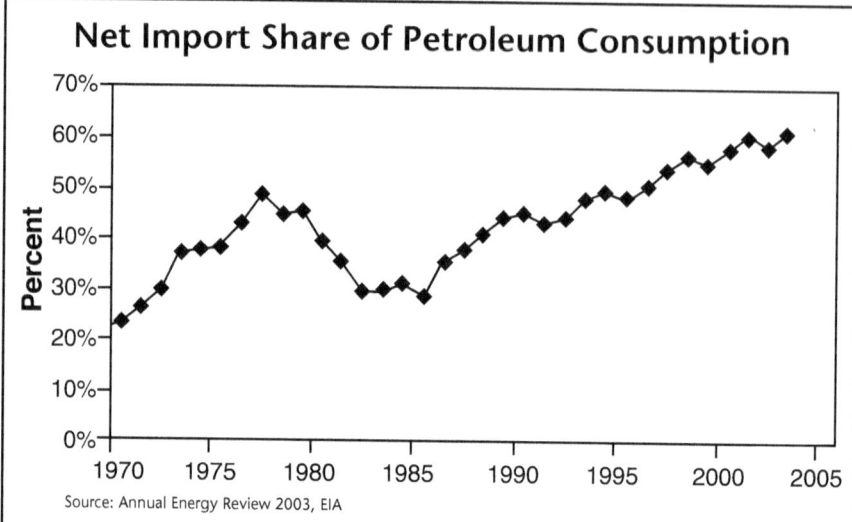

Figure 2-7. Over the past generation, domestic oil production in the United States has dropped to approximately one-third of total daily demand, requiring vast quantities of imported oil to make up the deficit. (Source: U.S. Energy Information Administration, Annual Energy Review 2003)

engine horsepower (providing futile "zoom-zoom" for driving in bumper-to-bumper traffic), adding superfluous weight and power-hungry options that eliminate any potential energy-efficiency improvements. As shown in Figures 2-5 and 2-6, vehicle fuel economy has barely budged in the last quarter century. Add to this increasing vehicle miles driven and rapidly depleting domestic oil reserves and it is no wonder that oil imports into the United States as well as many other countries are skyrocketing.

Consider too that the square footage of the average North American home has more than doubled during the past generation (while family size has halved) and that energy efficiency standards for construction have stalled. The result is that additional energy is required for heating, lighting, and air conditioning in addition to the energy required for commuting to and from home for school, work, and shopping.

Over the same period, domestic oil production in the United States has dropped to approximately one-third of total daily demand, requiring vast quantities of imported oil to make up the deficit. During 2005

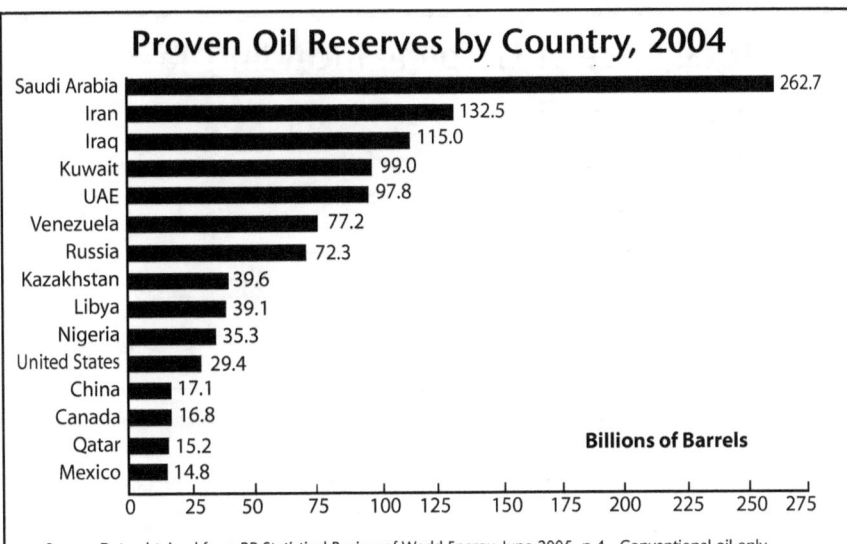

Figure 2-8. The United States reported oil reserves of 21.4 billion barrels of oil as of June, 2005 (U.S. proven oil reserves have declined more than 17% since 1990, with the largest single-year decline (1.6 billion barrels) occurring in 1991); it therefore has less than three years of domestic supply at current rates of consumption. (Source: The Washington Institute for Near East Policy)

the United States consumed 20,800,000 barrels of oil **per day**, of which only 7.5 million barrels (or 36% of total consumption) came from domestic sources. This consumption level equates to over 25 barrels per year for each of the approximately 300 million inhabitants of the United States.

Given that the United States reported oil reserves of 29.4 billion barrels of oil as of June, 2005, (U.S. proven oil reserves have declined more than 17% since 1990, with the largest single-year decline (1.6 billion barrels) occurring in 1991) it has less than four years of *domestic* supply at current rates of consumption[3]. Imports **must** come to the rescue in order to maintain existing levels of use.

Even though the majority of the world's oil is supplied by regions that are politically unstable and often unfriendly to the West, energy dollars are being exported as fast as the U.S. Treasury Department can print them. According to an October 25, 2003 report in *The Economist*,

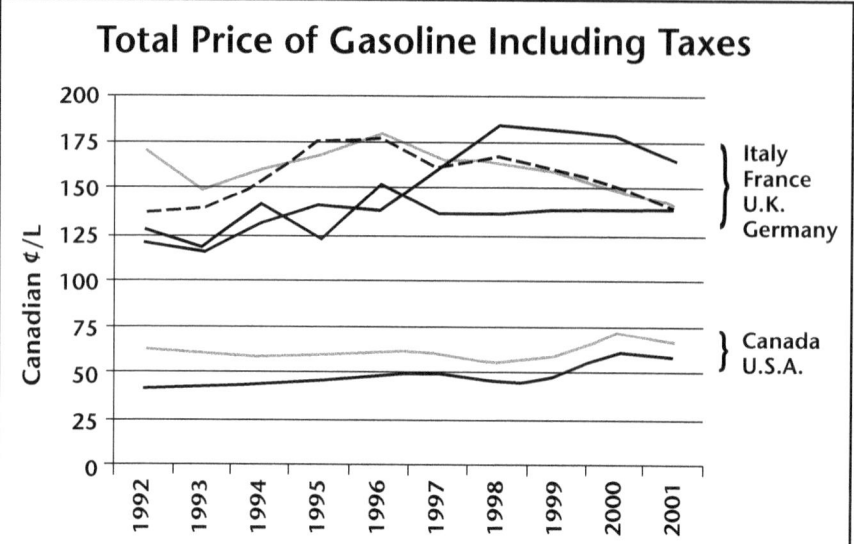

Figure 2-9. The United States has put itself in the very difficult position of having to rely on imported oil to keep its economy running. Unlike European governments, decades of North American administrations have not had the backbone to improve the Corporate Average Fuel Economy (CAFE) laws or to tax gasoline in order to reduce consumption.

OPEC has drained the staggering sum of $7 trillion from American consumers over the past three decades. This massive sum does not even include industry subsidies, environmental and health costs, or the expense of military security required to get the crude safely into North America. If this money had been pumped into energy efficiency, mass transit, and improved building standards, the U.S. trade deficit as well as conflicts in oil-rich regions of the developing world would be greatly reduced or even eliminated.

The list of current beneficiaries of U.S. energy demand may surprise most people. Table 2-1, based on information provided by the U.S. Energy Information Administration, shows the countries providing crude oil imports to the United States for the period 1980 through 2003.

Canada is the major supplier of crude oil to the United States, receiving vast sums of dollars in exchange for a Canadian government-sanctioned rape and pillage of its environment. Although the receipt of this money may be seen by some as a good thing, it is creating massive social and financial strain on the province of Alberta, which is trying

Major Suppliers of U.S. Crude Oil and Petroleum Products
(Thousand barrels per day, average; rank in 2003)

	1980	1985	1990	1995	2000	2003
Canada	455	770	934	1,332	1,807	2,072
Saudi Arabia	1,261	168	1,339	1,344	1,572	1,774
Mexico	533	816	755	1,068	1,373	1,623
Venezuela	481	605	1,025	1,480	1,546	1,376
Nigeria	857	293	800	627	896	867
Iraq	28	46	518	0	620	481
United Kingdom	176	310	189	383	366	440
Algeria	488	187	280	234	225	382
Angola	42	110	237	367	301	371
U.S. Virgin Islands	388	247	282	278	291	288
Norway	144	32	102	273	343	270
Kuwait	27	21	86	218	272	220
Colombia	4	23	182	219	342	195
Total, major suppliers	4,884	3,628	6,729	7,823	9,954	10,359
Total, all U.S. imports	6,909	5,067	8,018	8,835	11,459	12,264

Table 2-1. Major Suppliers of Crude Oil and Petroleum Products to the United States. Source: U.S. Energy Information Administration.

to cope with inadequate services, infrastructure, and housing as well as inflationary problems. More money docs not necessarily mean fewer problems. Furthermore, as a result of the massive increase in tar sands refining, the exported crude oil leaves behind enormous amounts of carbon emissions, contributing to Canada's inability to meet its Kyoto Protocol agreements.

U.S. oil imports equate to a staggering $620 million ***per day*** being drained from the American economy and significantly contribute to the U.S. trade deficit, currently hovering at $725 billion annually (fiscal 2005 figures). While this is good financial news for energy-exporting countries, the United States may have placed its economy on a crash course with bankruptcy.

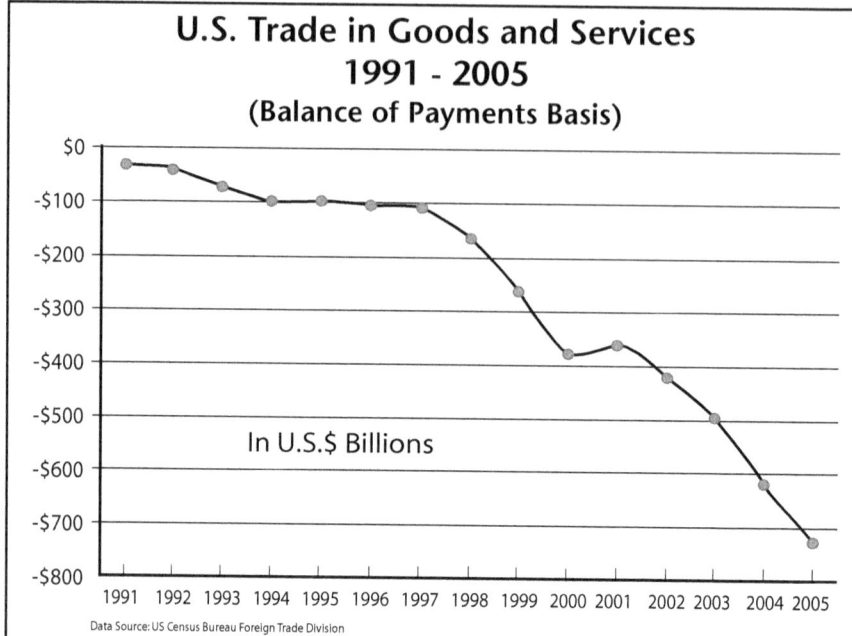

Figure 2-10. The U.S. economy has been operating with enormous trade deficits, partially as a result of massive oil importation payments. This trend has economists worried that it is causing unsustainable national debt ($9 trillion), as well as social security, mortgage, Medicare, and other unfunded liabilities. (Source: U.S. Census Bureau, Foreign Trade Division).

Although some economists believe trade deficits are beneficial as a result of the connection with an increase in GDP and jobs, most claim that long-term deficits generate other more systemic economic problems. The record levels of unfunded government liabilities are believed to be directly attributable to trade deficits. These liabilities are at levels that have never before been tested and include national debt ($9 trillion), Medicare ($30 trillion), Social Security ($12 trillion) as well as immense levels of corporate and personal debt.

Massive consumption funded with debt places an enormous burden on the future generations who must deal with the cleanup of the resulting economic chaos (U.S. dollar collapse, hyperinflation, recession) that inevitably ensues. If oil supplies become restricted or prices exceed a given threshold, the economy will not have sufficient resiliency to withstand the pressure and the U.S. could well find itself in another thirties-style depression.

"Since World War II Americans have invested much of their new-found wealth in suburbia," postulates Barry Silverthorn, producer of the 2004 documentary *The End of Suburbia* (www.endofsuburbia.com). Barry further explained to me that, "The suburbs promised a post-war public a sense of space, affordability and upward mobility that has become iconic in how most North Americans live. But as the new millennium gets underway, the assumptions of cheap fuel, cheap, easy to transport food and upward mobility are all being challenged in the face of peak oil and resource constraints. It would appear that the lexicology of "suburbia" could easily transition towards "slum" as people can no longer afford the fuel and infrastructure costs to keep the dream alive".

Is there a way out of this energy spin cycle? Can North America and the rest of the world continue to find fossil fuels to feed this never-ending thirst? If not, are there any alternatives?

Canada to the Rescue?

The Western Canada Sedimentary Basin (WCSB) underlies, as the name implies, most of the western provinces and has been the main supply of oil production in Canada for the last two generations. Because of the effects of "supply peaking," production has been in steady decline and is expected to fall precipitously within the next decade. Industry is now in a race against the clock to replace conventional light oil supplies with "synthetic" oil from the Alberta tar sands. (The politically correct term is "oil sands," but given that the sticky bitumen is about as liquid as asphalt, I prefer the more visually accurate name).

Extraction of the bitumen from the tar sands is accomplished using massive open-pit mining or steam extraction from underground deposits. Upon extraction, the bitumen is mixed with other hydrocarbons to allow the slurry to flow through pipelines to refineries, where it is upgraded into "synthetic crude oil," leaving behind massive environmental damage from effluent runoff, fresh-water depletion, and greenhouse gas emissions.

Canada reported that proven oil supplies were 178.8 billion barrels in 2005, of which only 8.9 billion were from conventional sources. With current production hovering at around 3.1 million barrels per day, conventional low-cost, easy-access sources are in steady decline and production is only able to increase as a result of escalating output from unconventional tar sands sources[4].

The remaining 169.9 billion barrels of Canadian oil reserves are attributed to tar sands supply. Despite the potential financial windfall of developing the tar sands, there are a considerable number of economic, environmental, and technical hurdles that could hinder project development.

Cost overruns, low world oil prices, lack of fresh water, carbon emissions, and environmental issues, as well as availability of affordable energy to process the bitumen, could easily bankrupt development. Because of the high capital and technology requirements for building the mining and upgrading plants, world oil prices must stay above approximately $30 per barrel and natural gas below $7 per 1,000 cubic feet in order to simply break even. The existing tar sand refineries gulp massive amounts of natural gas, another Canadian commodity that is on the endangered species list and is relied upon to provide heat for steam extraction and bitumen upgrading. Just as worrisome is the question of where the massive quantities of fresh water used in the upgrading of bitumen will come from.

The WCSB provides more than liquid fossil fuels; it is also a main source of natural gas for both Canada and the United States. With the winter 2005/2006 natural gas futures price hovering around $15 per 1,000 cubic feet (up from $2 just a few years ago), it appears that natural gas costs and midterm supply shortages may become the devil-in-the-details that could stall the massive tar sands industry buildup. (There has been considerable talk about circumventing this problem by using nuclear power to provide the thermal energy required for bitumen extraction; however, that's a whole other essay.)

As of January 2007, proven Canadian natural gas reserves were ranked 19th in the world at 57.9 trillion cubic feet (Tcf). (1 trillion is 1,000 billion or 1 million million.) Although this is a staggeringly large number, domestic gas consumption is equally staggering. According to official figures, in 2004 Canada produced 6.5 Tcf of natural gas, consuming 3.4 Tcf and exporting the majority of the balance to the United States to help meet its demand of 22.4 Tcf per year[5].

Given that North American demand for natural gas is expected to rise to 34 Tcf over the next six years, Canadian proven reserves will be exhausted shortly after this period. An increasingly large volume of natural gas is being used to create synthetic oil for the production of gasoline, the majority of which is exported and used without regard to sustainability or efficiency. There is no debate that natural gas supplies in Canada are in steady decline and that it is a necessary fuel required to heat homes (which themselves are energy inefficient and too large). One would therefore assume that natural gas should not be squandered in tar sands refining. At the very least, Canadians (more correctly Albertans) should be aware that they are trading current income dollars for future, midterm energy security while driving domestic home heating and other natural gas-related costs skyward. This is a trade that most will wish they had not made in a few years' time.

While there is no doubt that unconventional gas sources such as coal bed methane or imported liquefied natural gas will begin to supplement conventional sources, even the most optimistic of reports recognizes the folly of assuming they will replace WCSB supplies[6].

A report headed by Rob Woronuk, Senior Analyst, Canadian Gas Potential Committee[7], perhaps sums up the problem of relying on natural gas supplies for "standard" energy-consuming applications:

"It is clear that the 2002 EIA [U.S. Energy Information Administration] forecast of gas imports from Canada is overly optimistic. Supply development from the WCSB [Western Canada Sedimentary Basin] is resource constrained and other supply sources will develop slowly. Both Canada and the United States must endeavour to curtail gas demand, import liquefied natural gas and develop energy alternatives."

With this worrisome outlook for current and future natural gas supplies, it should be clear that synthetic crude will experience production constraints or at the very least become an expensive fuel option, having major financial impact and affecting transportation. If, in desperation, tar sands are relied upon to maintain oil supply levels in the future, the transition will be neither seamless nor painless for the environment and the economies of either Canada or the United States.

Mexico Perhaps?

Mexican oil exports to the United States are a close second to those of Canada, so perhaps Mexico will be able to bridge the growing oil gap? Not so, according to *Oil & Gas Journal*. Proven Mexican oil reserves were 12.9 billion barrels as of January 2006, and depleting quickly. Production at the super-giant Cantarell oil field, which is one of the largest in the world, averaged 2.1 million barrels per day or 63% of Mexico's total oil production during 2005[8]. According to Matthew Simmons, CEO of

Simmons & Company International, a Houston-based investment bank that specializes in the energy industry, "Cantarell, discovered in 1975, was the last oil field found anywhere whose daily production would exceed one million barrels."[9] According to technical data from the U.S. Energy Information Administration and Pemex (Mexico's state-owned, nationalized petroleum company), production will decline by 14% a year, regardless of any additional drilling or expansion undertaken to the oil field[10].

Oil exploration in Mexico is hampered by cost and by difficulties between the company and its political bosses. Major reforms to the economic arrangement between the government and the company will be required before new exploration, drilling, and development projects can begin. Given the fiscal and geotechnical challenges facing Mexico and Pemex, the midterm forecast is a continuing drop in crude oil exports, further aggravating the flow of oil into the North American energy market.

Regardless of Mexico's ability to produce additional crude supplies, it is still a net importer of refined petroleum products, importing over 311,000 barrels per day, of which gasoline comprises 50%. Demand for refined products will continue unless Pemex is able to invest at least $19 billion over the next eight years. Given the immediate requirements for exploration and drilling infrastructure, finding these substantial funds will be difficult at best.

What is perfectly clear is that North American fossil fuel liquids (gasoline, diesel, and heating oil) as well as natural gas are going to be in short supply and increasingly expensive within a very short period of time. Assuming North American oil consumption remains steady at 9.2 billion barrels per year, only 23.1 years of domestic reserves remain, and that's if the tar sands are included in the calculation. If only conventional sources are considered, the depletion time drops to approximately 6.6 years[11].

The Persian Gulf Region

According to the U.S. Energy Information Administration, the Persian Gulf contains 260 billion barrels of proven oil reserves, representing one-fifth of the world's oil supply, and maintains almost all of the world's excess oil production[12]. This excess capacity is required to buffer swings in supply and demand, acting as a mallet to force oil markets in line with Saudi policy.

Forecasters *hope* that production in the Persian Gulf area will reach 12.5 million barrels per day by 2009 and 15 million by 2020. This proposed increase would place Persian Gulf oil capacity at 33% of world total by 2020, up from 28% in 2000[13].

Provided the massive oil supply reserve and production data are correct, many optimists say that imports of crude oil and natural gas from the region should be able to satisfy the North American and world appetite for perhaps another 75 to 100 years. "Not so fast," says Matthew Simmons. "No one has raised a murmur that these growth projections might be mere fantasy." Mr. Simmons' research concludes that these production levels are extremely unlikely and that Saudi Arabia, the world's current largest producer will not be able to maintain, let alone increase, its production capacity to meet rising world demand[14].

Even if Saudi Arabia and the balance of the Persian Gulf states have sufficient reserves and production capacity, supply disruptions present another worry for oil consumers. It is no understatement that the United States economy is extremely vulnerable to Middle East energy shocks. Tight oil supplies from all global energy exporters mean that any sizable disruption in the supply of oil from even one source will cause prices to skyrocket. A recent report on the potential for Middle East energy shocks[15] estimates that oil prices will rise in excess of $7 per barrel for every one million barrels removed from supply lines, with sharp price hikes likely to trigger worldwide recession or at a minimum cause severe hardship to the financially disadvantaged at home and in the developing world. This report has listed several key areas of concern related to attacks on Middle East oil infrastructure:

- Al-Qaeda attacks on Persian Gulf and Iraqi oil facilities which Osama bin Laden has urged on the grounds that they are "the most powerful weapon against the United States";
- an exodus of oil workers occasioned by fear of terrorism and domestic unrest;
- the spread of Iraqi instability into other oil-producing countries;
- a confrontation with Iran over its nuclear program or other "problematic" behavior;
- domestic instability or uncertain political transitions.

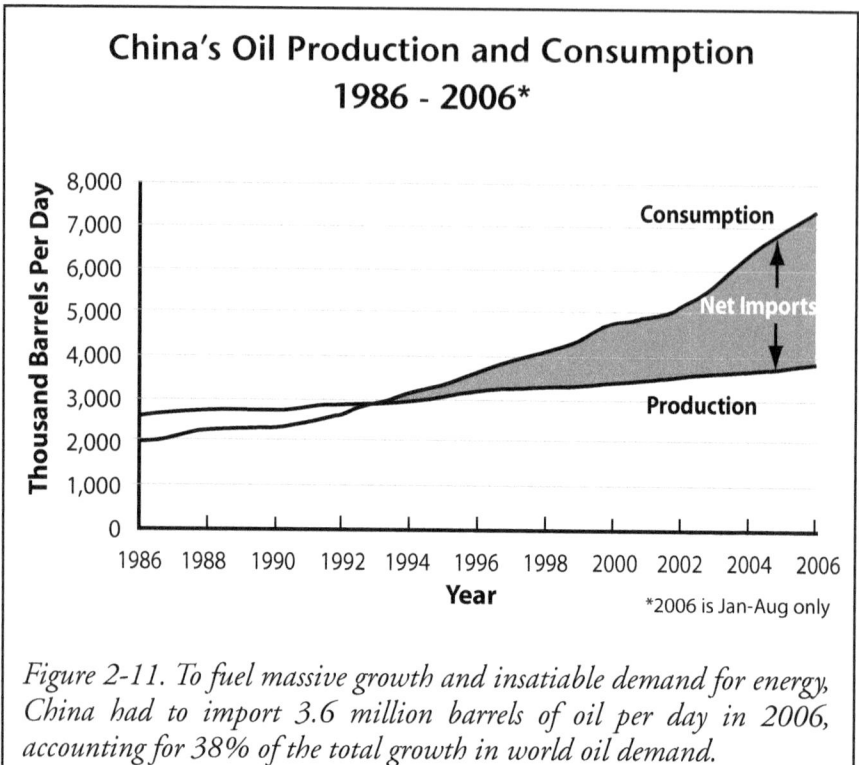

Figure 2-11. To fuel massive growth and insatiable demand for energy, China had to import 3.6 million barrels of oil per day in 2006, accounting for 38% of the total growth in world oil demand.

Because of the lack of transparency within the Saudi regime, it is impossible to understand the true situation regarding oil supply and the stability of the ruling kingdom. Given its importance to all oil-importing countries, the Middle East is a wild card that makes it imperative to find suitable and sustainable alternative resources.

China

China is the world's most populated country with 1.3 billion people, approximately four times that of the United States and Canada combined. It is also a rapidly growing economic powerhouse with a gross domestic product output that the World Bank estimates to be U.S. $ 8.8 trillion based on purchasing power parity (PPP), making it the world's second largest economy after the United States. (PPP is a means of measuring the currencies of different countries based on a basket of similar goods. By way of example, *The Economist* magazine has long used its "Big Mac" index as a slightly anomalous but accurate estimate for judging

PPP in various countries by comparing the price of the ubiquitous Big Mac hamburger as the gold standard financial icon.) Annual economic growth in China is also increasing at an estimated rate of approximately 10%, which will position China as the largest economy within a decade. To put this in perspective, the United States has a gross domestic product of $12.4 trillion, while the number three spot is distantly held by Japan with an economic output of $4 trillion.

To fuel this massive growth, China's demand for energy is also increasing rapidly, accounting for 38% of the total growth in world oil demand[16]. China is estimated to have 18.3 billion barrels of proven oil reserves and produced approximately 3.8 million barrels per day in 2006[17]. China consumed 7.4 million barrels per day in 2006, which equates to 2.7 billion barrels per year, giving China less than seven years of domestic supply. As domestic production for the same year was 3.8

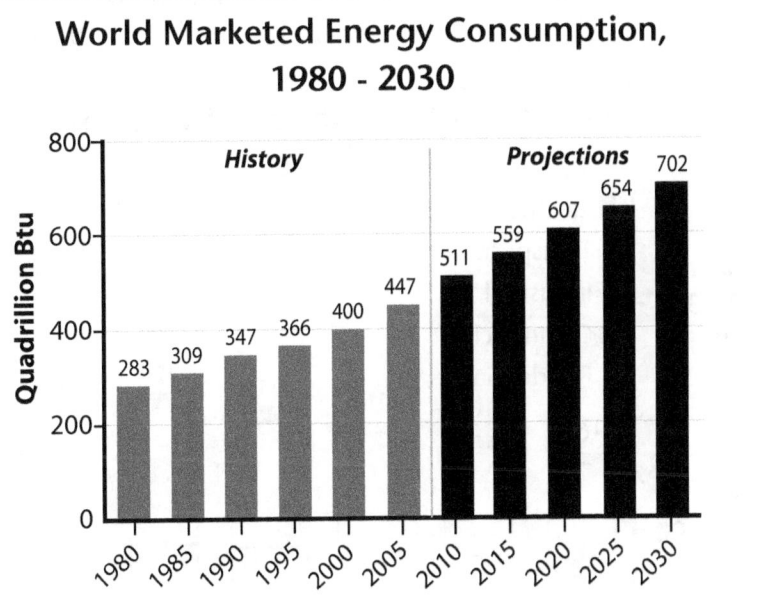

Figure 2-12. In coming years, world energy consumption is forecast to increase dramatically as non-OECD Asia experiences rapid growth, its economies become more industrialized, and personal wealth escalates. The growth in world total energy consumption, including all liquid fuels as well as electricity, natural gas, etc., is projected in this graph. (Source: U.S. Energy Information Administration, International Energy Outlook 2007)

million barrels per day, nearly half (3.6 million barrels per day) was imported to make up the shortfall.

China is both a developed and developing country based on a figurative geographical gulf between the rich coastal area and the poor inland agricultural and resource-based areas of the country. In order to subsidize the vast agricultural areas, cheap energy prices and low energy efficiency have been the norm, resulting in the Chinese government regulating oil prices below international levels. These artificially low energy prices, as well as increased manufacturing and the acquisition of cars and houses by affluent members of society, are creating greater demand and consumption, necessitating oil imports. Accordingly, the Chinese government is looking abroad for oil trading partners and is also attempting to purchase natural resources and oil-producing companies to ensure that supplies can keep up with its growing energy needs.

Naturally enough, the large American and European oil companies are also looking for more reserves, often the same ones as the Chinese.

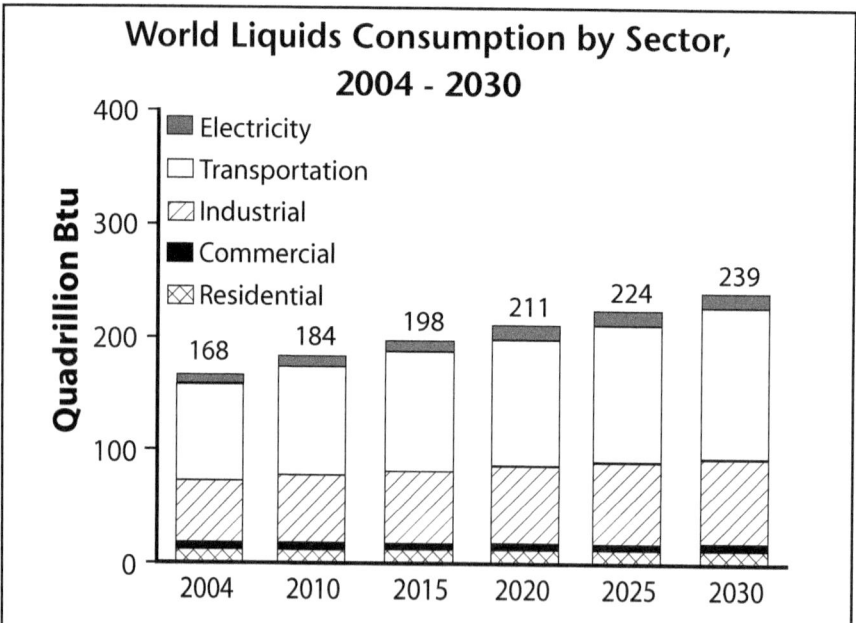

Figure 2-13. The transportation sector will account for the biggest share of the world's energy total as well as the largest growth in liquid fuels energy consumption through the year 2030. (Source: U.S. Energy Information Administration, International Energy Outlook 2007)

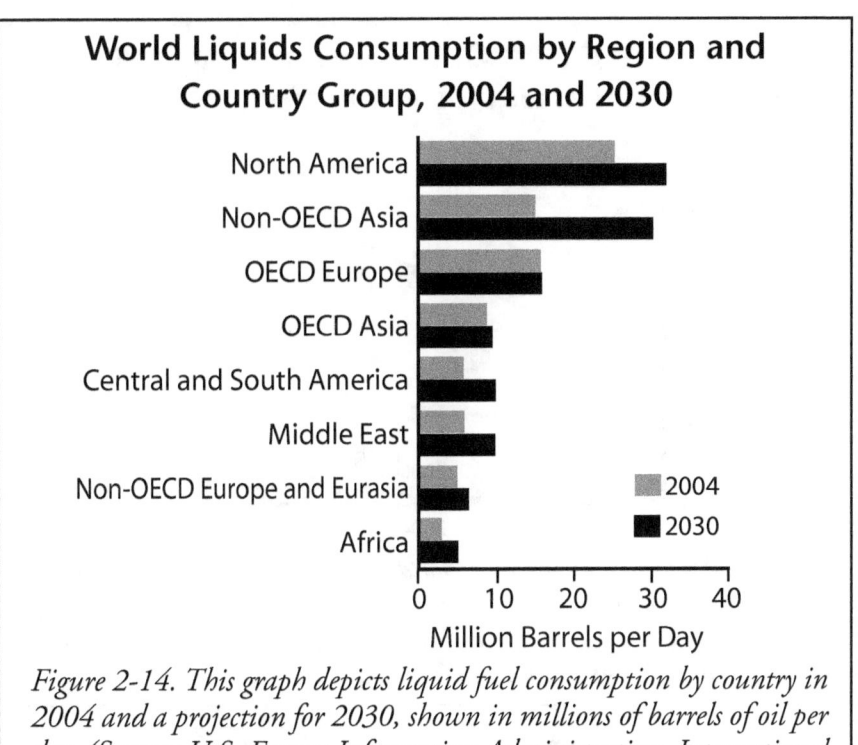

Figure 2-14. This graph depicts liquid fuel consumption by country in 2004 and a projection for 2030, shown in millions of barrels of oil per day. (Source: U.S. Energy Information Administration, International Energy Outlook 2007*)*

This makes me hope that all future competition for supplies will be negotiated with dollars and not guns and bombs.

The World

In the coming years, world energy consumption is forecast to increase dramatically as non-OECD Asia experiences rapid growth, its economies become more industrialized, and personal wealth escalates. Figure 2-12 shows world total energy consumption including all liquid fuels as well as electricity, natural gas, etc. These data are expressed in a common denominator of "quadrillions of Btu."[18]

It is estimated that the transportation sector will account for the biggest share of this energy total as well as the largest growth in liquid fuels consumption through the year 2030 (Figure 2-13)[19]. Estimates also suggest that non-OECD Asia will account for the fastest growth in energy consumption, averaging 2.7% through 2030, while less developed non-OECD countries will account for annual consumption

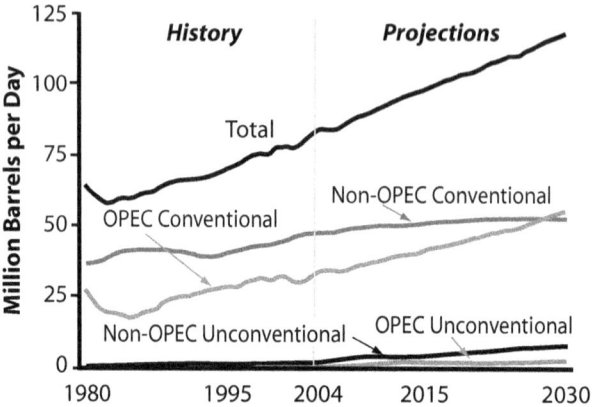

Figure 2-15. Estimates are that world oil extraction will increase by 35 million barrels of oil per day in 2030, with OPEC expected to cover some 65% of this increase. (Source: U.S. Energy Information Administration, International Energy Outlook 2007)

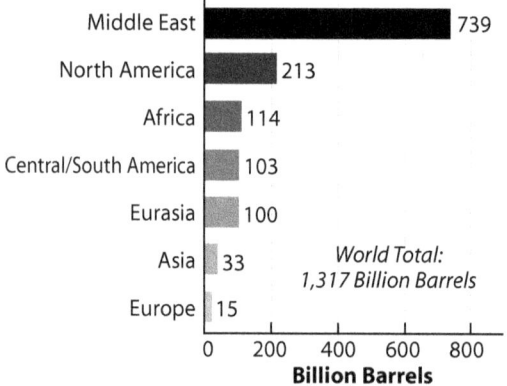

Figure 2-16. World "proven" oil reserves are shown in this graph. These figures are highly speculative, since much of the data is provided by the oil-producing countries themselves and may not be subject to international audit. This is the case for most of the Middle East as well as for nonconventional oil sands reserves. (Source: U.S. Energy Information Administration, International Energy Outlook 2007)

growth of only 1%. Meanwhile, Europe and Eurasia are estimated to see average consumption increase by 2.1%. The Middle East, with its relatively small population base, will see energy consumption growth of 2.3% annually[20].

On the production side of the equation, estimates are that world oil extraction will increase by 35 million barrels of oil per day between 2004 and 2030 (Figure 2-15), with OPEC expected to cover some 65% of this increase with output of 57 million barrels per day (20.8 billion barrels per year) by 2030[21].

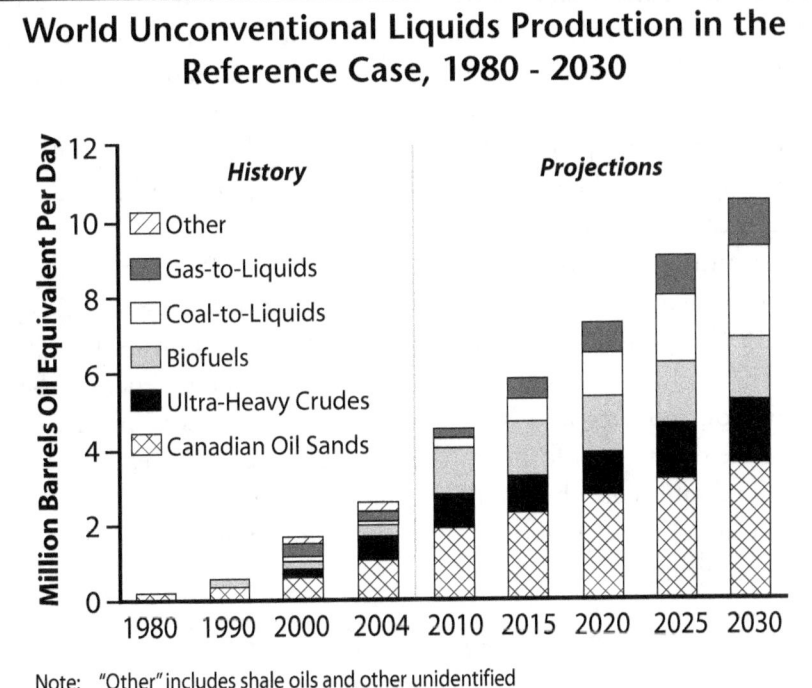

Figure 2-17. Unconventional oil sources can also be tapped to supply growing liquid energy demands. The question that remains is, "What environmental and geopolitical price is society prepared to pay for these fuels?" There are already serious concerns about the supply of fresh water and natural gas that is necessary to convert oil sand bitumen into gasoline, never mind greenhouse gas emissions and environmental degradation. (Source: U.S. Energy Information Administration, International Energy Outlook 2007)

The United States Energy Information Administration[22] calls for increased exploration, cost-reduction programs, improved extraction technologies, and the development of nonconventional resources to provide the balance of this growth. Asking the world's oil supplying nations to pitch in and increase production is one thing; their actually doing it might be quite another.

Proven oil reserve estimates are not as accurate as one might imagine. Much of the data is obtained from countries that have no interest in accurately reflecting diminishing volumes and may not be subject to outside, nonbiased audit. Iran, for one, comes to mind. The current regime of the Holocaust-denying President Mahmoud Ahmadinejad spent considerable time, money, and effort in an attempt to delude International Atomic Energy inspectors about the status of its nuclear ambitions, even as the secretly built heavy-water reactor at Arak and uranium-enriching plant at Natanz were being built. Why would manipulating crude oil reserve data, which can be used to influence world oil politics, be any different?

According to Matthew Simmons, the main OPEC producing nations provided high-quality data regarding their respective oil fields prior to 1982, by submitting production data to well-known sources such as Oil & Gas Journal. When Sheikh Ahmed Zaki Yamani became Saudi Arabia's oil minister in 1982, most OPEC members stopped reporting oil field data or information regarding proven reserves. Given that OPEC members are the dominant oil-producing nations, any absence of data can only be construed as a serious impediment to world energy sector planning.

Peak Oil a.k.a. Production Limits

At a Calgary, Alberta oil patch meeting[23], Mr. Fatih Birol, chief economist of the International Energy Agency (IEA), praised Canada for its efforts in building the tar sands infrastructure and helping to place a ceiling on crude oil prices. However, he went on to say: "It plays a very important role in international oil diplomacy, but only a global conservation effort will keep energy costs from spiraling upward in coming years."

Although Canada has officially reported 178.8 billon barrels of proven oil reserves, second only to Saudi Arabia, over 95% of these reserves are in tar sands deposits and, as explained earlier, are subject to severe technical, energy, and financial constraints. Even if these resources can be fully developed, it is unlikely that Canada will ultimately be able to play the moderating role of either counterbalancing the strength of the Organization of Petroleum Exporting Countries (OPEC) or being able to keep a lid on world oil prices. The reason: peak oil production limits.

When queried about the issue of peak oil, Mr. Birol chuckled, saying that "such concerns rise in step with the cost of oil. It's a fashion. It comes every 10 years, when we have high prices. Four times, we've reached a peak in the last 20 years!"

Mr. Birol's remarks may have been intended as a joke, a means of dealing with the issue, but sadly, they are anything but.

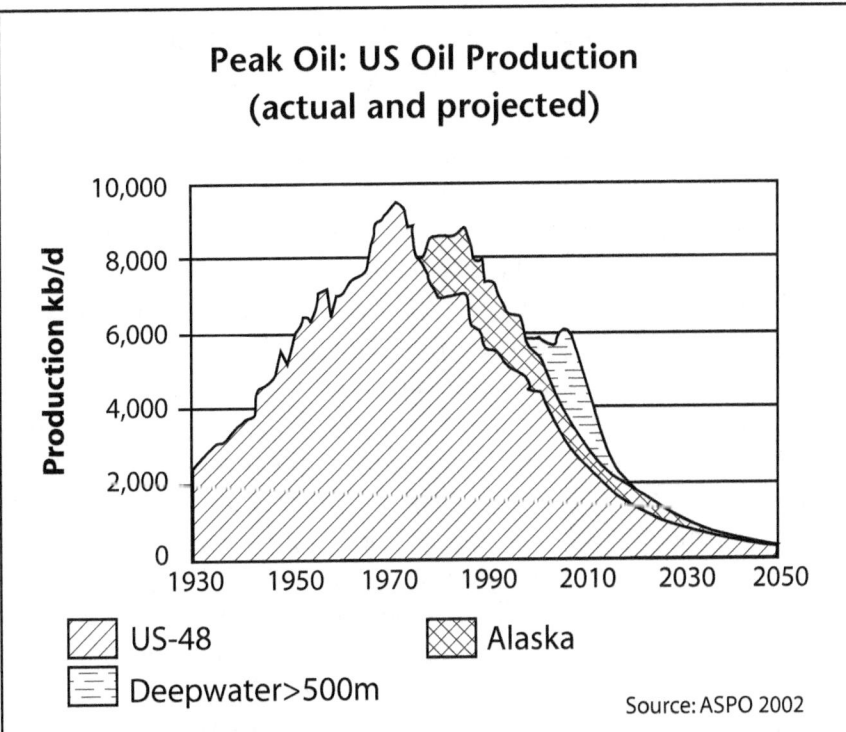

Figure 2-18. Although initially scoffed at, Dr. Hubbert's theory of "peak oil" proved to be amazingly accurate, predicting an oil shortage in the lower 48 states some 20 years before the event actually occurred. (Source: www.peakoil.org)

Defining Peak Oil

Peak oil is a frequently used term which is often taken out of context, so let's start this examination of the theory with the following definition:

The Hubbert peak theory, also known as peak oil, is an influential theory concerning the long-term rate of conventional oil and other fossil fuels extraction and depletion. It predicts that future world oil production will soon reach a peak and then rapidly decline, notwithstanding improvements in oil extraction techniques or the technology used. The actual peak year will only be known after it has passed.

Peak oil theory is named after the late Dr. Marion King Hubbert who worked as a research geophysicist with the Shell Oil Company from 1943 to 1964. He was later a professor of geology and geophysics at Stanford University as well as a research geophysicist with the United States Geological Survey in Washington, D.C.

During his long and distinguished career, Dr. Hubbert developed numerous theoretical concepts regarding the flow of fluids in the earth's crust, which led to the development of important techniques employed to locate oil and natural gas deposits, many of which are still in use today. He is, however, most famously remembered for his postulation that U.S. oil production would peak in the early 1970s and decline thereafter, no matter how much oil extraction technology improved.

Although initially scoffed at, Dr. Hubbert's theory proved to be amazingly accurate, predicting an oil shortage in the lower 48 states some 20 years before the event actually occurred.

Interestingly, despite the data confirming Dr. Hubbert's hypothesis, many politicians, economists, and energy planners do not subscribe to the theory of peak oil. But then again, maybe they do. Mr. Birol's comment about there being four peaks in the last 20 years was intended to refute the peak oil argument, while it unwittingly lends support to it.

In reality, there will not be one but hundreds if not thousands of peaks resulting from the development and extraction of individual oil and natural gas fields located around the world. Further, peak oil does not mean that the world is running out of oil, but rather that there are intrinsic limits on how quickly oil and gas can be extracted from the earth. Indeed, there will be billions of barrels of oil, gas, and oil equivalents in the ground long after the world has reached "The Peak." Long before reserves are depleted, liquid fossil fuel demand will exceed the

Figure 2-19. Oil is not stored in the ground in neat and tidy underground drums, rivers, or lakes. Rather, it is contained within oil-bearing rock and other strata of the earth's crust, often mixed with ground water, complicating both the technological and the financial hurdles of development and extraction.

ability to supply all of the world's collective *needs*, with supply disruptions as well as wars and other territorial skirmishes.

Understanding the Geophysics of Peak Oil

Many people assume that if an oil field is said to hold a specific number of barrels of oil (or natural gas) then those resources can be had by simply extracting them with the appropriate pumping technology. Although this is a reasonable assumption, it is completely incorrect.

Oil is not stored in the ground in neat and tidy underground drums, rivers, or lakes. Rather, it is normally contained within semi-permeable oil-bearing rock and other strata of the earth's crust, often mixed with ground water, which complicates both the technological and the financial hurdles of well development and oil extraction.

Once oil is discovered at a given location, sensing and data acquisition technology is used to determine the gross estimated volume of the field. Upon completion of the gross volume estimation, oil extraction engineering studies and economic analysis must be applied to the reservoir. Oil fields that produce high-quality crude oil with little water

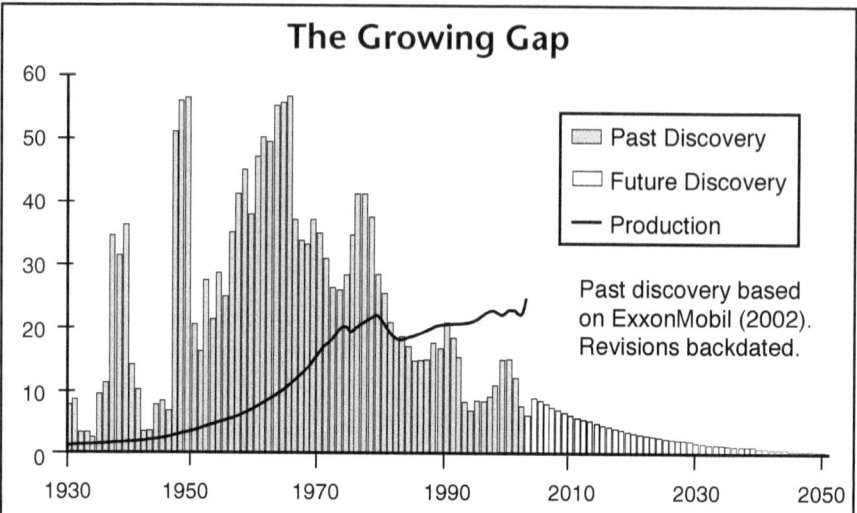

Figure 2-20. "We now find one barrel of oil for every four we consume. The general situation seems so obvious....How can governments be oblivious to the realities of discovery and their implications...given the critical importance [of oil] to our entire economy." *(Source: Dr. Colin Campbell, President of ASPO (Association for the Study of Peak Oil and Gas) in his testimony to the British House of Commons, http://www.oilcrash.com/articles/cooke_03.htm)*

under high natural pressure are considerably less expensive to develop and operate than fields that have high levels of water contamination and low pressure and produce low-grade oil. Application of the technological and economic criteria leads to the "proven reserve" estimate, which will be considerably below the gross volume of the field.

Before the production cycle of a well is initiated, a "depletion curve" is calculated, providing the oil company with the best balance between extraction of crude oil, coproducts, operating costs, and return on investment. According to Matthew Simmons, "The difficulty in determining the optimum rate of extraction is that the rate that maximizes ultimate [oil] recovery will probably not be the rate that maximizes return on investment." Figure 2-21 illustrates the life cycle of a typical water-driven oil well and helps to illustrate Mr. Simmons' point.

When a typical oil well first starts production, natural pressure within the reservoir assists the flow of oil to the well head. This pressure comes from the combined forces of the earth's crust surrounding the

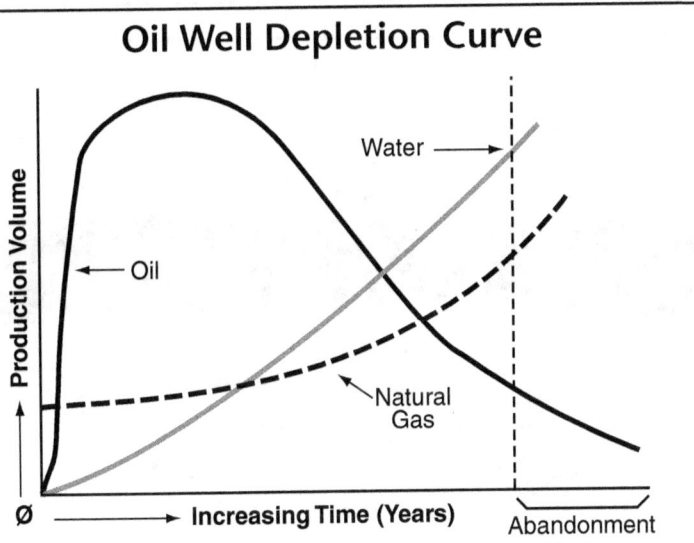

Figure 2-21. Before the production cycle of a well is initiated, a "depletion curve" such as the example shown is calculated, providing the oil company with the best balance between extraction of crude oil, co-products, operating costs, and return on investment. Unfortunately, the difficulty in determining the optimum rate of extraction is that the rate that maximizes ultimate [oil] recovery will probably not be the rate that maximizes return on investment. (Source: Matthew Simmons, Twilight in the Desert)

reservoir, ground water, and dissolved gases in the oil phase.

As pumping continues and the reservoir matures, internal pressure drops, resulting in a peak extraction rate. The void created by oil removal is rapidly filled with additional ground water and the regasification of the dissolved natural gas bubbles. Additional "pumping stations" only accelerate the drop in pressure, further reducing the extraction rate; ironically, the better humans become at extracting oil, the faster the reservoir reaches peak production and the faster we slide down the slippery slope to oil well abandonment.

As reservoir oil extraction and pressure decline continue, a point is reached where natural flow ceases and secondary measures, such as water, steam, or carbon dioxide injection, are required to continue production, albeit at ever-declining rates. Applying economic data such as benchmark oil price and extraction cost will determine where on the life-cycle curve oil well abandonment will occur.

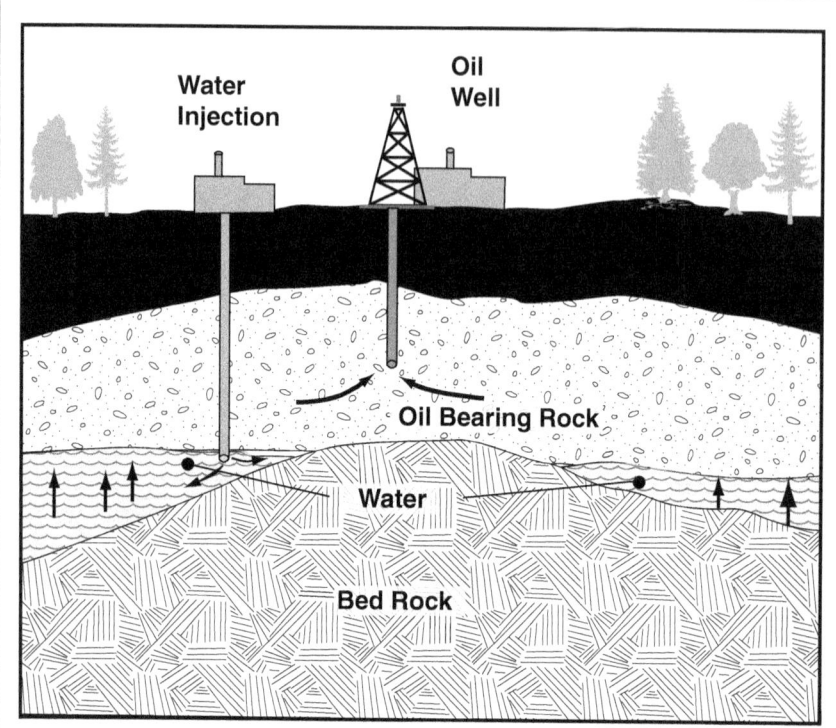

Figure 2-22. As reservoir oil extraction and pressure decline continue, a point is reached where natural flow ceases and secondary measures, such as water, steam, or carbon dioxide injection are required to continue production, albeit at ever-declining rates.

As one can imagine, the "abandonment point" on the reservoir's life-cycle curve will vary considerably, and once reached huge amounts of "unrecoverable" oil will be left behind. It is completely logical to abandon a well one day only to return it to production should favorable economic conditions occur in the future.

One example of abandonment and restart is in the once-prosperous North Sea area, where the major oil and gas companies have sold their positions to secondary oil producers who, with lower overhead costs, can continue to operate wells that are in their twilight years.

A further example of end-of-life-cycle oil extraction comes from the U.S. lower 48 states, where over half a million "marginal" or "stripper" wells produce thousands of barrels of water per day but only a few barrels of oil, offering a perfectly sensible income to individual operators with minimal overhead costs[24].

Even the massive U.S. Strategic Petroleum Reserve (SPR) is subject to the rules of peak oil. In 1975, the Energy Policy and Conservation Act was passed, establishing the SPR to act as a buffer in case of major supply disruptions. To store these vast oil reserves, the government acquired several salt caverns near Lake Charles, Louisiana and Big Hill, Texas and at other locations in the Gulf Coast area.

According to official data, the SPR, which currently has a target capacity of 700 million barrels of oil, is subject to the following maximum drawdown capacities[25].

Days Since Beginning Drawdown	Daily Extraction Rate in Millions of Barrels per Day
0 to 90	4.3
91 to 120	3.2
121 to 150	2.2
151 to 180	1.3

Tinkering with the Data

Given that *proven reserves* represent "dollars in the ground," oil companies and governments have a vested interest in ensuring that the official recorded volumes for these assets are as large as possible. This is a particular problem where a major producing nation does not publish accurate, or any, proven reserve data. Saudi Arabia is a case in point. It is the largest and most important oil-producing nation, controlling nearly all of the world's spare production capacity. The secretive Saudi government does not publish proven reserve data, posing the threat of global economic chaos if production peaks or if it cannot be raised to stay in step with worldwide demand.

Further proof of corporate tinkering with proven reserve data was reported by Reuters in February 2005. Royal Dutch/Shell Group was struggling to rebuild investor confidence after five larger-than-expected restatements of reserves by 1.4 billion barrels of oil equivalent as well as the firing of several members of top management[26].

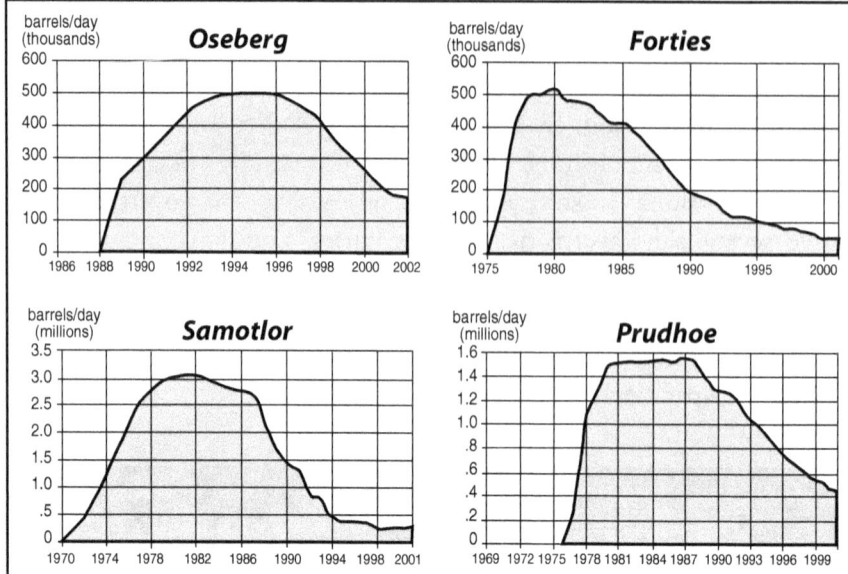

Figure 2-23. These graphs represent production and depletion profiles of four giant or super-giant oilfields in various locations throughout the world. In each case, none of the fields had a plateau, or peak production period, of more than ten years, leaving many industry experts wondering about the true state of Saudi Arabia's "unlimited" oil supplies. It is a very interesting facet of human nature that we agree with the facts in hindsight but are too stubborn to accept that history will repeat itself as we face the same issues in the immediate future. (Source: Matthew Simmons, Twilight in the Desert)

Given the difficulty of accurately determining recoverable oil and gas reserves as well as peak production rates under optimum conditions, padding and misrepresenting or not reporting data only worsens the prospect of the world's ability to keep the supply running.

Without any doubt, peak oil production is a geotechnological speed limit placed on our ability to extract oil and gas from the earth. There may well be billions of barrels of oil in the ground; however, rapidly rising consumption will eventually (some say presently) put the brakes on worldwide economic development.

It is a very interesting facet of human nature that we agree with the facts in hindsight but are too stubborn to accept that history will repeat itself as we face the same issues in the immediate future. Fuel shortages,

Proven Reserves Growth Remains Slow as Capital Costs Rise

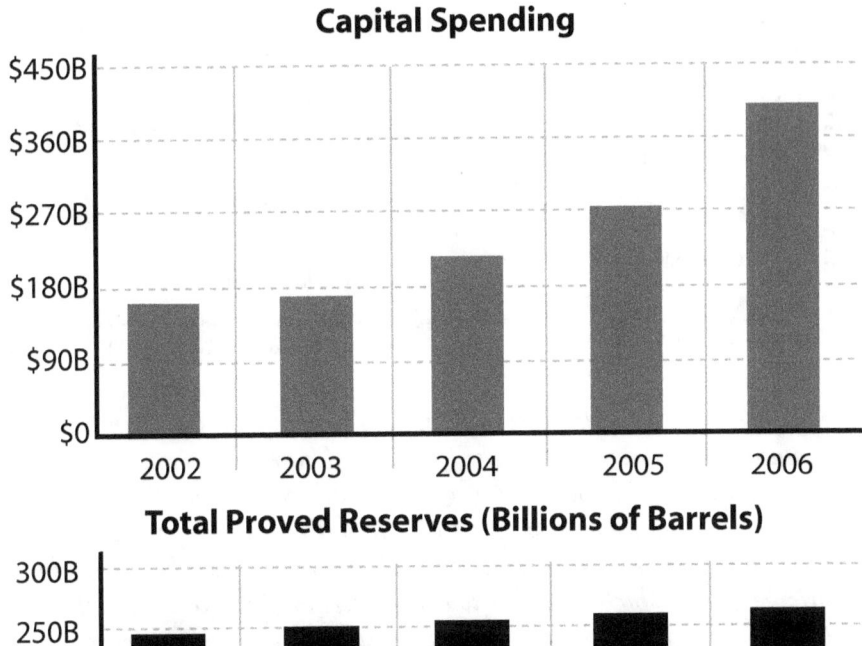

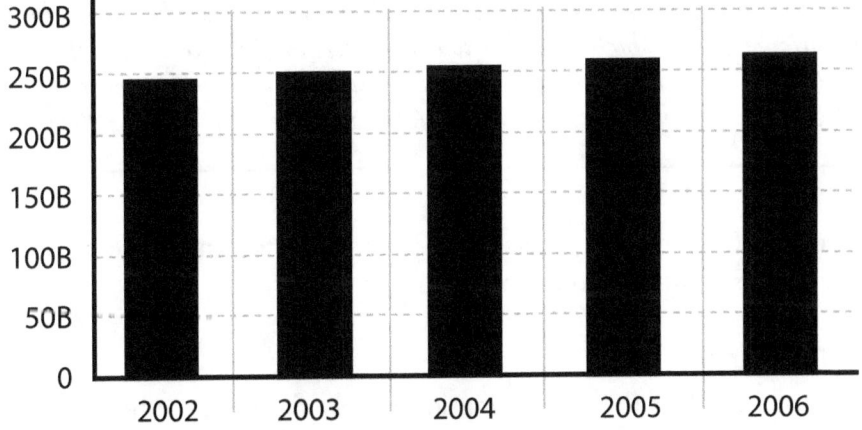

Source: Global Upstream Performance Review National Post

Figure 2-24. We need to look no further than the relationship between world oil reserves and the amount of money spent looking for them to prove that oil supplies and prices are on a collision path. Global reserves would have fallen by 2.4% except for the questionable addition of an increase in Canada's oil sands reserves, despite an increase in exploration spending of 39% in 2006[27]. (Source: National Post/Andrew Barr)

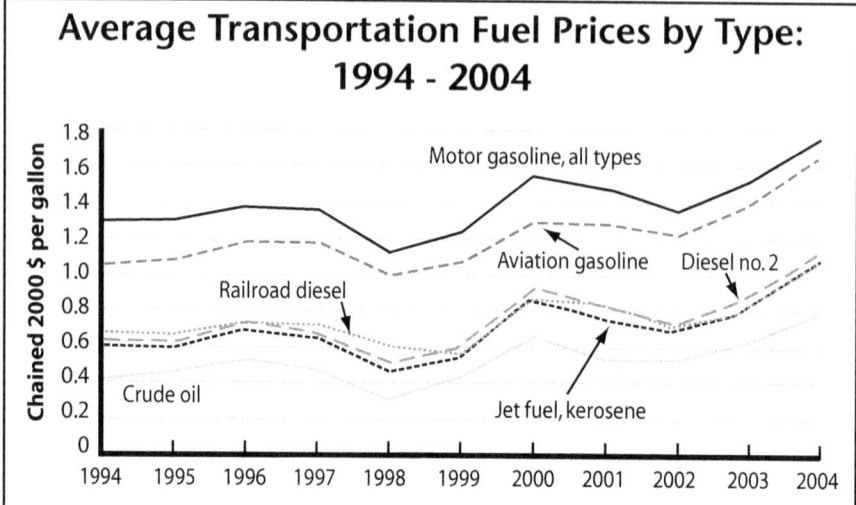

Figure 2-25. The retail price of fossil fuels can play a big part in energy consumption, especially in countries like the United States and Canada. A dramatic rise in the price of gasoline will cause a feedback effect, lowering consumption as people seek alternative means of travel. Over the last decade prices have remained fairly constant, as reflected in this pricing graph which shows various fuel costs in chained 2000 dollars. (Source: U.S. Department of Transportation, Bureau of Transportation Statistics, Transportation Statistics Annual Report, November 2005).

sharply increased energy costs, and warring over remaining resources are the forecast for tomorrow's energy supply.

Summary

Of course, all of these consumption/supply figures are estimates and any number of geotechnical, political, or environmental problems could quickly alter projections. These risks include many unknowns, including when (if ever) Iraq will become politically stable enough to begin production of the 6 million barrels of oil per day it wishes to bring to market and whether geoterrorism will destroy or greatly limit oil production and supply in the many volatile areas of the world.

Economic risk is also a major factor in determining energy consumption. If the United States or other parts of the world enter into an economic slowdown, recession, or worse, energy consumption, particularly in the transportation sector, will immediately be curtailed as fewer goods are purchased and fewer trips are taken for either business or holidays.

Likewise, the retail price of fossil fuels can play a big part in energy consumption, especially in countries like the United States and Canada that have relatively small amounts of taxation built into the price. Low levels of taxation reflect in lower retail gasoline prices, which in turn drive wasteful consumption and the purchase of larger-than-necessary vehicles. Conversely, a dramatic rise in the retail price of gasoline, either through taxation or crude oil prices, will cause a feedback effect, lowering consumption as people seek alternative means of travel, carpool, take fewer frivolous trips, purchase energy-efficient cars, and generally conserve.

Unfortunately, the rules of supply and demand dictate in a free market; therefore, reduced demand will (temporarily at least) cause an increase in fuel supply. If the majority of the retail price is due to crude prices alone, retail prices will drop which will boost demand. Depending on the dynamics that caused the original price increase or price shock, these price oscillations will cause a great deal of uncertainty in the world's oil markets and certainty of supply. It is far better for governments to impose punitive fuel taxation to drive down consumption, rather than allow the free market to create uncertainty through crude oil price swings.

Energy efficiency, conservation, mass transit, better logistics, renewable energy, and well-developed energy policies (including carbon taxation) will go a long way toward reducing the risks surrounding geopolitics and oil; however, it would be fanciful to believe that North Americans and the world economy as a whole will *ever* completely wean themselves from oil. Given the technical and economic hurdles of transitioning away from fossil fuels, particularly supplies from the Middle East, western nations will be well advised to develop policies that allow commerce with this politically unstable region, to improve domestic energy policy, and to prepare for oil supply disruptions if and when they occur.

3
Energy and the Environment

"Enjoy life's journey but leave no tracks."
Aboriginal Commandment

Our current way of life includes the belief that cheap energy is our God-given right. Never mind that the cost of gasoline or imported heating oil does not include the huge subsidies lavished on the oil industry. The price of a gallon of gasoline does not include the cost of the ongoing American military presence in the Middle East, depletion subsidies, cheap access to government land, and monies spent to advance drilling and exploration technologies. And none of these "hidden" costs includes the environmental and health damage caused by the burning of fossil fuels.

Take, for example, a report issued by the Ontario Medical Association (OMA) which shows that smog is taking lives and hurting economies in communities across Ontario, a Canadian province of approximately 12 million people. The OMA report, *The Illness Costs of Air Pollution (ICAP) 2005*[1], shows the negative impact of smog on health and the economy in specific cities across the province.

The ICAP 2005 report reveals that exposure to air pollution will result in almost 5,800 premature deaths in 2005 alone. If nothing is done to further improve the quality of air in Ontario, the number of premature deaths is estimated at 10,000 lives by the year 2026. The combined health care and lost productivity costs are expected to reach well over a billion dollars.

"We hope these numbers will provide the evidence needed for municipal leaders to make the changes necessary to improve the health and economic well-being of Ontario's communities," says Dr. Greg Flynn, President of the OMA. "Smog affects everyone, which is why we need all levels of government to use this new information to push forward with improvements to the quality of air we breathe." A similar picture can likely be painted in most cities and urban areas across North America. And then there is the developing world. A recent summer trip to Beijing

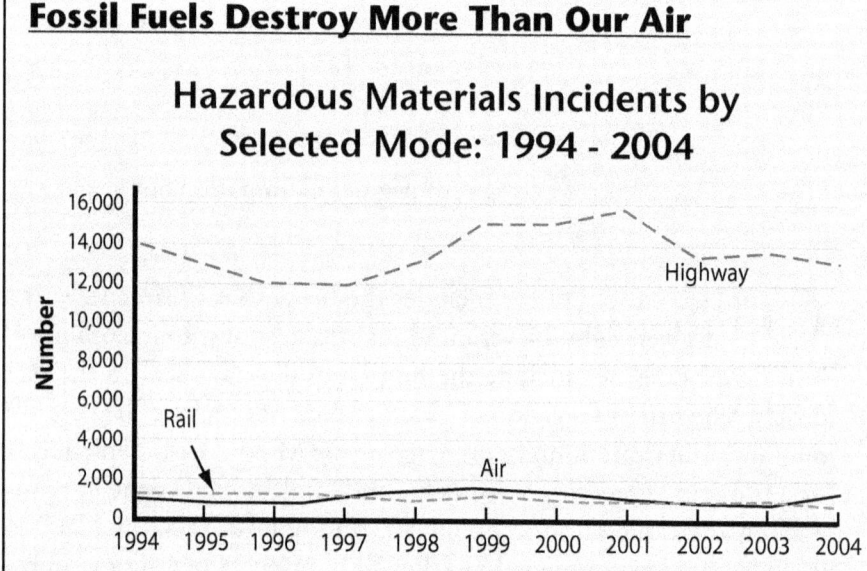

Figure 3-i. There were 14,740 reported transportation-related hazardous material incidents in 2004. Highway vehicles transported slightly more than half of total hazardous materials yet caused a disproportionate number of spillage and other incidents. Reducing liquid fuel consumption and utilizing better rail and logistics systems could reduce these highway-related incidents severalfold. (Source: U.S. Department of Transportation, Transportation Statistics Annual Report (November 2005))

brought me face to face with air that was so thick with smog and particulate matter that my eyes watered and I was left with a very unpleasant taste in my mouth. I wondered how people functioned, walking, bicycling, and going about their lives in this unhealthy chemical soup that was passed off as air.

Fossil fuels don't just power our cars. Home heating and coal- and natural gas-fired electrical power plants rely on fossil fuels as well. With much of the developing world's population connecting to the electrical grid every day, it is obvious that world energy demand will mushroom as more (and larger) appliances are brought online, further exacerbating the problem. In the developed world, middle-class families are demanding larger suburban (and exurban) homes (courtesy of the personal automobile) filled with more and more appliances that were considered luxury items only one generation ago. Central air conditioning, mul-

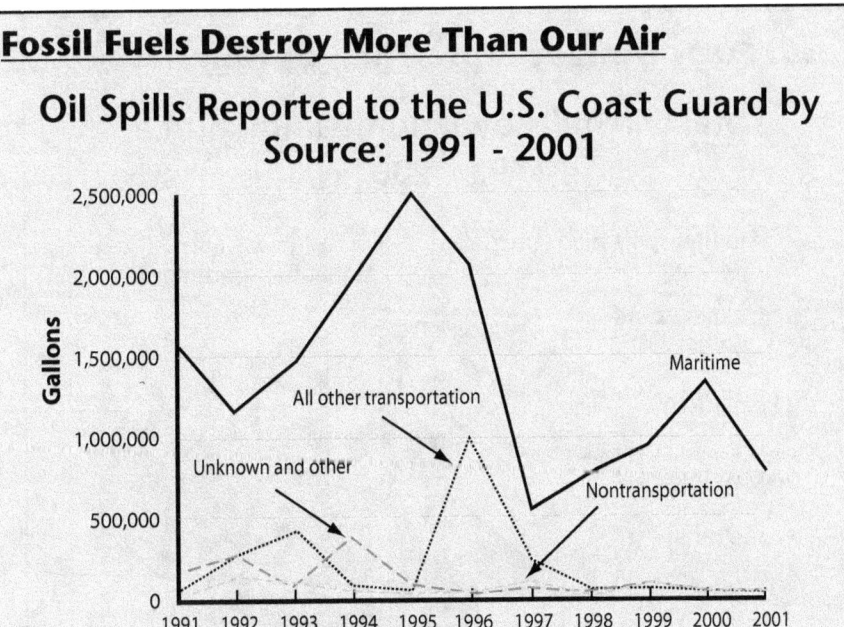

Figure 3-ii. The figures for oil spills into U.S. territorial waters vary dramatically from year to year, as individual spills can account for millions of liters of oil. The vast majority of oil spills were related to the transportation sector, which accounted for between 77% and 97% of all reported spill volumes. (Source: U.S. Department of Transportation, Transportation Statistics Annual Report (November 2005))

tiple refrigerators, computers, chest freezers, hot tubs, and swimming pools, all extravagant or unimaginable items in our parents' day, are now necessities and consume enormous amounts of energy, necessitating the building of yet more coal-fired power stations.

The supposed personal freedom that the automobile gives to society reinforces these entitlements and creates a vicious cycle of energy demand, manufacturing, and consumption, all leading to increased smog and greenhouse gas emissions.

The Chemistry of Smog and Greenhouse Gases

Everyone has heard about them, but what exactly are smog, climate change, and greenhouse gas emissions? Ask anyone who lived in Los Angeles in the early 1980s (or Toronto today) and they will tell you about dirty skies and endless haze. For someone with asthma or other respira-

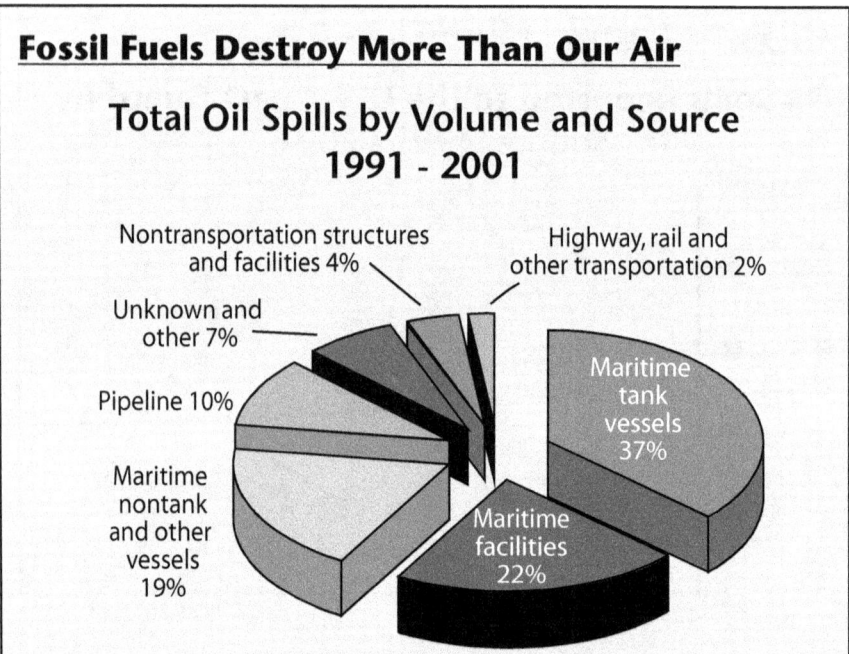

Figure 3-iii. Total oil spills by volume and source are shown in this pie chart, which shows data from 1991 through 2001. There is no information available on oil spilled through the improper disposal of used motor oils or other non-reported sources. (Source: U.S. Department of Transportation, Transportation Statistics Annual Report (November 2005))

tory sensitivities it means staying indoors for countless hours or relying on medical inhalers for each breath. (Numerous reports directly link the burning of fossil fuels to the appalling rise in the number of asthmatics, yet, to the best of my knowledge, the price of "puffers" is not included in the cost of gasoline or coal-fired electricity.)

Steve Ovett wished the air had been cleaner in Los Angeles. Although he won medals at the Olympics in Moscow in 1980, the air in Los Angeles was his undoing. During the 1984 Olympics he collapsed during the 1500-meter race because of smog-induced asthma and spent two nights in hospital.

Smog pollution is created when the emissions from burning fossil fuels combine with atmospheric oxygen and ultraviolet light from the sun. Soot and other fine particulate matter occur because poor engine and emission-system designs spew unburned hydrocarbon fuels directly into the atmosphere. These particulate matter are very tiny, on the order of 10 microns^2 in diameter or less. They float in the atmosphere and are easily drawn deep into people's lungs. Estimates indicate that over half

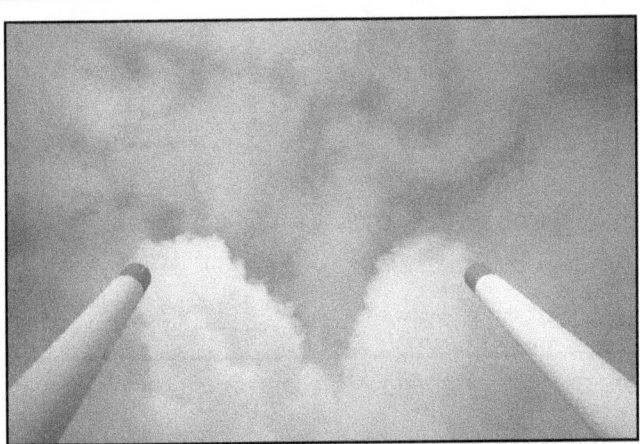

Figure 3-1. Coal-fired power plants are one of the most polluting energy sources in the world, notwithstanding the industry rhetoric about "clean coal technology." This is simply a marketing ploy to help gloss over the facts. Although smoke stack scrubbers and other advanced technological fixes can reduce smog-forming emissions, it is currently impossible to eliminate greenhouse gases from the exhaust plume. Carbon sequestration (a fancy word for pumping CO_2 into the ground) may work in the future, but it is far from proven and not without technical and financial risks.

a million deaths worldwide each year are directly attributed to excessive levels of airborne particulate matter and smog-based pollution.

When smog and soot particles combine with raindrops, acid rain is formed. Clean rain water is neither acidic nor base, meaning that this life-giving element is noncorrosive and restorative to everything it touches. Acid rain, on the other hand, is precisely that—rain water which has become acidic. When it comes into contact with metal, rock, or plants the result is a corrosive action and the decay and destruction of ecosystems as well as man-made objects such as buildings, ironwork, and statues.

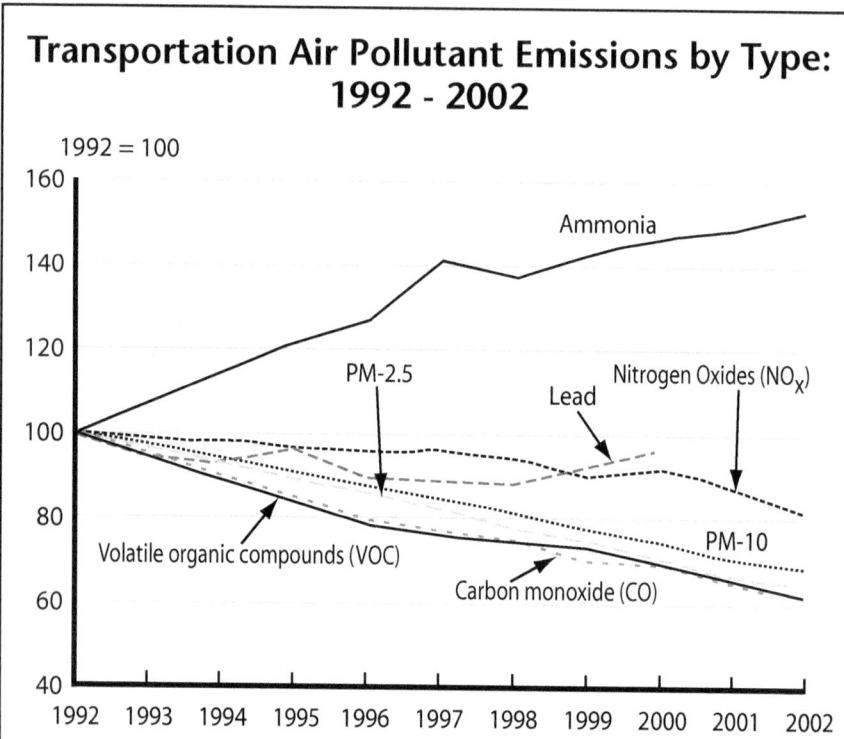

Figure 3-2. Smog has been greatly reduced in the developed world in recent years as a result of decreased sulfur concentrations in gasoline and diesel fuels. Improvements in automotive emission controls and catalytic converters have also gone a long way in reducing airborne pollutants in the transportation sector, with the exception of ammonia, over the last 15 years. Note that PM-2.5 and PM-10 refer to airborne particulate matter that is smaller than 2.5 and 10 microns respectively. (Source: U.S. Department of Transportation, Transportation Statistics Annual Report (November 2005))

Smog has been greatly reduced in the developed world in recent years as a result of decreased sulfur concentrations in gasoline and diesel fuels. Improvements in automotive and coal-fired power station emissions have also gone a long way in reducing smog over the last 20 years. However, the problem is far from over. Increasing demand for coal-fired electrical energy (as the supposed *least-cost* means of generating power) and an ever-increasing number of automobiles on the world's highways, along with lack of emission controls in the developing world, continue to exacerbate the problem. China alone is expected to build another 544 coal-fired power stations to satisfy its insatiable demand for energy.[3]

Carbon dioxide and methane are two of several greenhouse gases that are increasing in concentration in the atmosphere and are directly linked to global warming, although they are not currently regulated in the United States and Canada. It is estimated that coal, oil, and natural gas industries extract approximately 7 billion tons (6.4 billion metric tonnes) of carbon per year[4], all of which makes its way back into the atmosphere.

Carbon dioxide is a byproduct of the burning of any carbon-based fuel: gasoline, wood, oil, coal, or natural gas. With few exceptions, the world's current energy economy is fueled by carbon.

We have to go back half a million years and more to learn how this carbon fuel came to be. All carbon fuel sources began as living things in prehistoric times. Peat growing in bogs absorbed carbon dioxide from the air as part of the photosynthetic process of plant life. The rolling, heaving crust of the earth entombed the dead plant material. Sealed and deprived of oxygen, the plant matter could not rot. Over the millennia, shifting soils and ground heating compressed the organic material into soft coal that we retrieve today from shallow open-pit mines. Allowed to simmer and churn longer, under higher heat and pressure, oil and hard coal located in deep underground mines were created.

Provided these prehistoric fuel sources remain trapped underground there is no net increase in atmospheric carbon dioxide. However, the process of burning any of these fuels reduces the carbon stored in the coal or oil and drives off the trapped CO_2.

Conversely, the burning of *short-term* renewable carbon-based fuels like wood, biomass, and oilseed-based fuels such as biodiesel *may not directly* contribute to net greenhouse gas emissions. These plant materials "recycle" carbon dioxide from the atmosphere into carbon (through the storage of carbohydrates) during the growing cycle. Allowed to die and rot or through a combustion process such as a wood stove or internal combustion engine, these plant materials release the same CO_2 that was originally (and relatively recently) absorbed. Although the burning of prehistoric fossil fuels such as coal and oil also gives back carbon dioxide the atmosphere lost millennia ago, these "new" emissions of "old" material increase atmospheric concentration levels, leading to global warming.

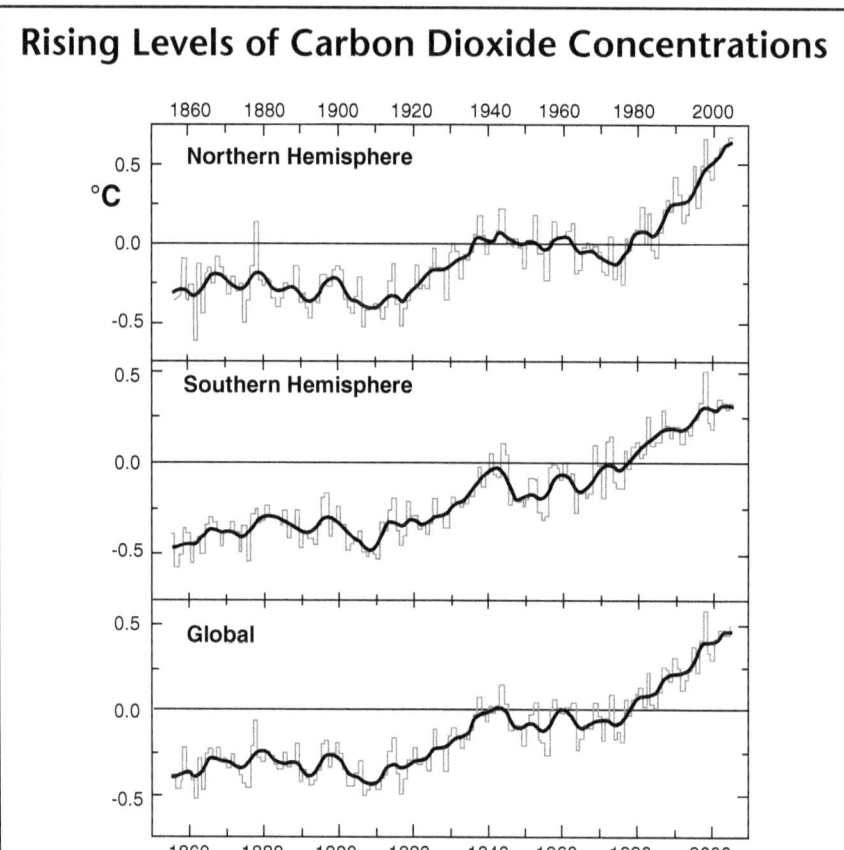

Figure 3-3. Carbon dioxide concentrations are higher than at any time in the last 650,000 years and have directly contributed to a rise of approximately 1°C in the earth's surface temperature over the last century. (Source: Dr. D. Jones, Climatic Research Unit, University of East Anglia)

The Issues of Climate Change

Carbon dioxide concentrations are higher than at any time in the last 650,000 years and have directly contributed to a rise of approximately 1°C in the earth's surface temperature over the last century. It should also be noted that the 1990s were the warmest decade on record, with 1998 the single warmest year of the last 1,000 years. In 2004 the average annual temperature rose by 0.45°C, and the 12 warmest years on global record have all occurred since 1990.[5]

Scientists predict that if the earth's average temperature were to rise by approximately 2°C, numerous devastating effects could occur:
- increased magnitude and frequency of weather events;
- rising ocean levels causing massive flooding and devastation of low-lying areas;
- increased incidents of drought affecting food production;
- rapid spread of non-native diseases.

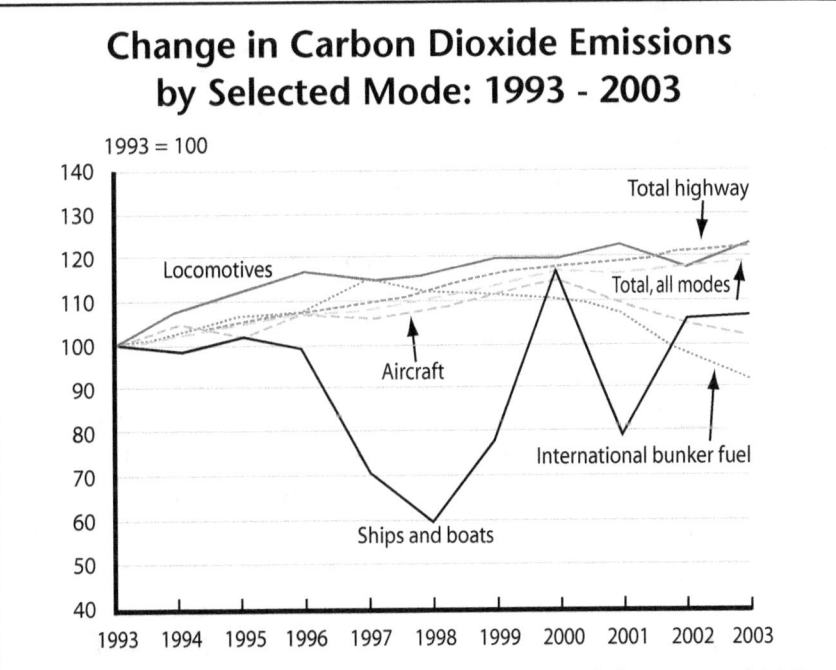

Figure 3-4. The transportation sector accounted for 27% of U.S. greenhouse gas emissions in 2003, an increase of 19% since 1993. (Source: U.S. Department of Transportation, Transportation Statistics Annual Report (November 2005))

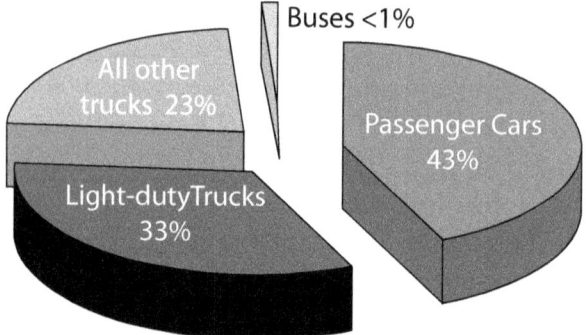

Figure 3-5. Highway vehicles emitted 82% of all transportation-related CO_2 emissions in 2003, a 23% increase from a decade earlier. (Source: U.S. Department of Transportation, Transportation Statistics Annual Report (November 2005))

The IPCC Report

Scientists strongly agree that we are beginning to see the effects of climate change now: melting glaciers, hotter summers, increased droughts, dying polar bears. Consider that the hurricane season of 2005 generated the most named storms ever, the most category 4 and 5 storms, and the record-breaking damage caused by Katrina and Rita.[6] Readers fond of Stephen King novels may wish to consider reading the Intergovernmental Panel on Climate Change report *Climate Change 2007: The Physical Science Basis*.[7] If it weren't for the fact that this report gives factual evidence of an impending worldwide climatic catastrophe, readers might be forgiven for thinking they were enjoying a work of fiction.

Some economists and politicians have gone on record as saying that climate change may actually be a good thing, for example by creating longer growing seasons in Canada or opening arctic sea lanes. Perhaps a longer growing season in Canada is a good thing, but not if it is accompanied by prolonged drought in the prairies, increased hurricanes and tornadoes in the south and midwestern United States. Arctic sea lanes are fine, provided the melting sea ice doesn't flood the coastal areas where the million-dollar penthouses are built or cause Inuit people to lose their ecosystems.

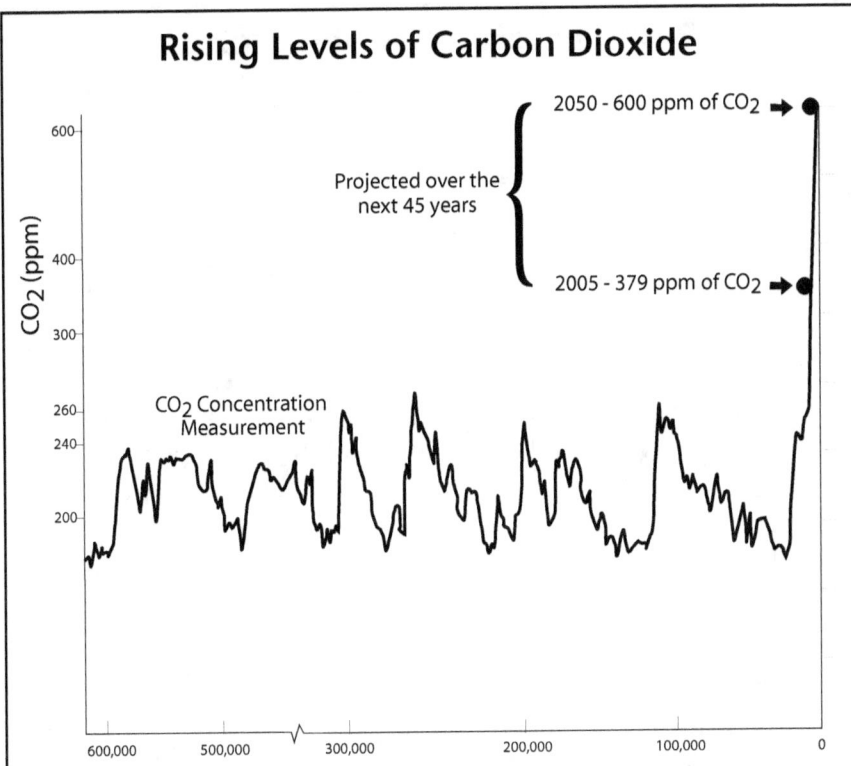

Figure 3-6. Carbon dioxide levels have naturally oscillated around a mean concentration of 230 parts per million in the atmosphere. By 2005, these levels reached 379 ppm as a result of the burning of fossil fuels and other man-made problems and if carbon emissions continue unabated they are expected to reach 500 ppm by 2050. (Source: http://en.wikipedia.org/wiki/Image:Co2-temperature-plot.svg)

It is widely believed that if the carbon dioxide concentrations remain at or continue to climb above today's levels average worldwide temperatures will increase by at least 2°C, with a resulting breakdown of ecosystems that humans will no longer be able to control[8].

Agreeing on the best way to limit the problems of climate change is one thing. Agreeing on what the limits should be is yet another, more important first step. The IPCC report states that, "continued greenhouse gas emissions at or above current rates would cause further warming and induce many changes in the global climate system during the 21st century that would very likely be larger than those observed during the 20th century."

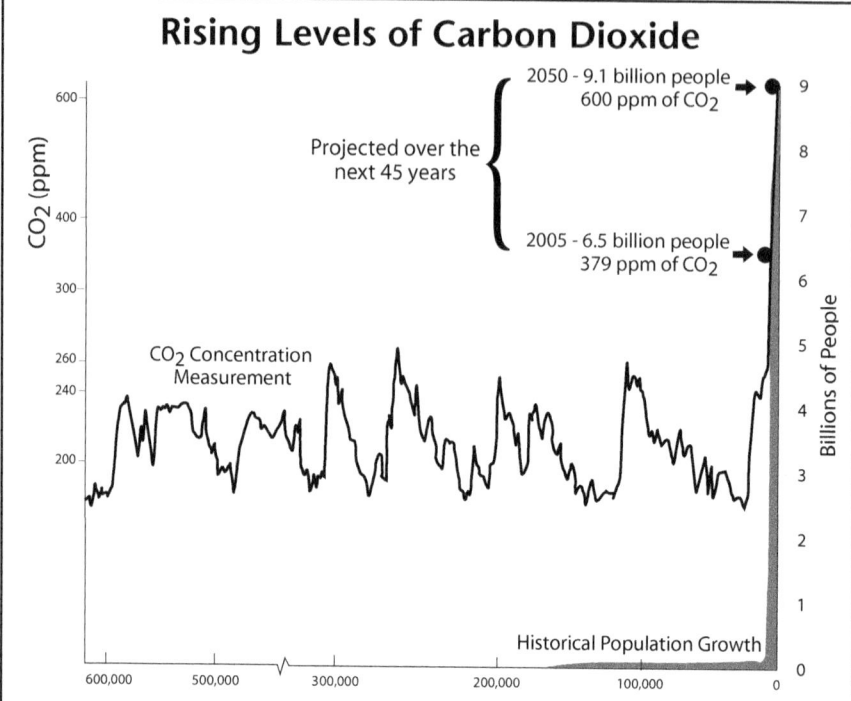

Figure 3-7. Superimposing population growth and CO_2 emissions clearly demonstrates the cause and effect relationship between the two. Although less developed nations like China consume less than 3 tonnes of carbon equivalent per person per year, those of us in North America consume approximately seven times this amount. Obviously, with populations continuing to rise and developing nations continuing to industrialize, decarbonizing society is a top priority for mankind. (Source: http://en.wikipedia.org/wiki/Image:Co2-temperature-plot.svg)

The Kyoto Protocol[9] was negotiated at the United Nations to develop binding means of reducing climate-changing chemical emissions. Although it was hoped that all developed countries of the world would sign and ratify the agreement and follow the modest requirements of the program, the United States and Australia did not sign. Canada signed and ratified the agreement, but the Liberal government spent a lot of time talking about the program and doing nothing. The current Conservative government, arguing that the inaction of the previous government now makes it impossible to comply with Kyoto, has ignored Canada's commitments and has developed a "Made in Canada" program called the Clean Air Act. This cop-out by the oil- and gas-loving Con-

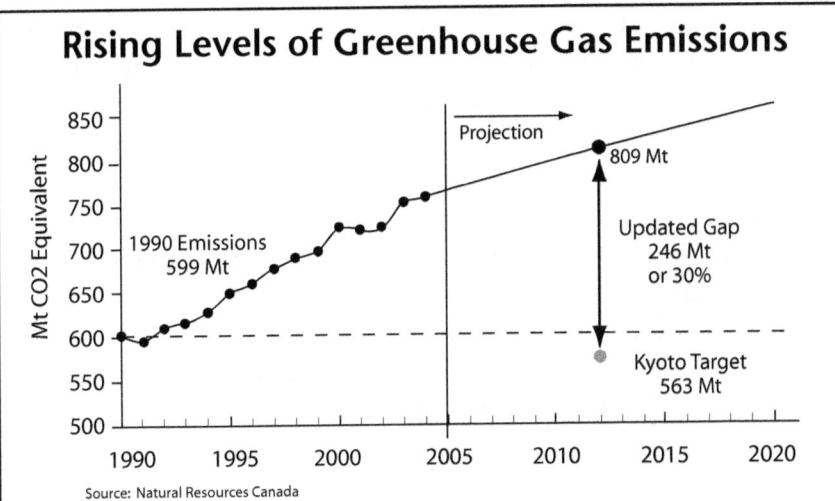

Figure 3-8. Canada's Liberal government signed and ratified the Kyoto Protocol agreement and then spent a lot of time talking about the program and doing nothing. Accordingly, greenhouse gas emissions are continuing to increase, indicating that Canada will be 30% above its binding commitments at the end of the agreement. The current Conservative government now argues that the inaction of the previous government now makes it impossible to comply with Kyoto. (Source: Natural Resources Canada)

servatives confusingly mixes smog and greenhouse gas reductions in the same program and indicates that they do not see climate change as a serious problem.

Thought provoking author and environmentalist George Monbiot quotes numerous sources and reports in his book *Heat: How to Stop the Planet from Burning* that concur with the IPCC report in suggesting that current levels of carbon emissions must be reduced by between 80% and 90%. This means that people in the United States and Canada would have to reduce their emissions by up to 90%, while people living in poverty, say in sub-Saharan Africa, could increase their emissions dramatically to allow a leveling of the playing field. Unfortunately, fair play is not always in the cards of Western society.

Another way of dealing with the issues of climate change is to minimize or deny the problem. With the funding of the BP oil company and the Ford Motor Company, Professors Robert Socolow and Stephen Pacala of the Carbon Mitigation Initiative at Princeton University have proposed that maintaining atmospheric carbon levels below 560 ppm—double pre-industrial levels—is acceptable. According to their plan, emissions would be allowed to rise to 1,200 billion tons annually and thereafter stabilize.[10] In their manifesto, energy efficiency is given a great deal of support, as it should be. However, they make far too many recommendations that make no sense at all. For example, they propose that energy efficiency of cars should double (which is easily doable today) while urging people to drive only half the distance, but they fail to suggest any disciplinary measures to make this happen.

While their action plan calls for a minor carbon taxation of $12 on a barrel of oil (approximately $0.29 per gallon/$0.08 per liter of gasoline at the retail price) they also propose a doubling of automotive fuel economy, which will negate the effects of taxation, lowering operating costs and leading to an increase in mileage driven.

Perhaps because of their funding source, 4 of the 15 recommendations outlined in the report do not propose eliminating or even reducing the number of cars on the world's roads. (There is no mention of

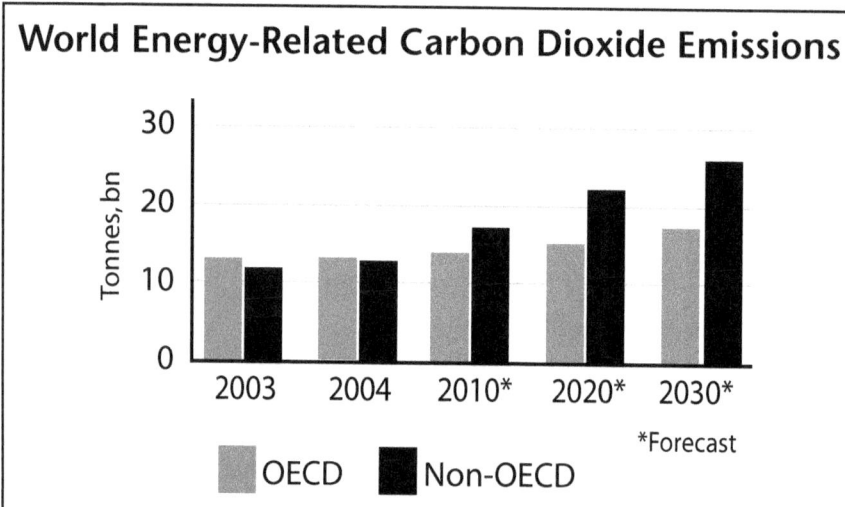

Figure 3-9. Greenhouse gas emissions are a global community problem and the sooner people and governments the world over recognize this and take action with proper regulations, the better. Even now, the greenhouse gas emissions of so-called developing, non-OECD nations equal those of the developed world. There is only one atmosphere and nature does not divide it along political lines. (Source: U.S. Energy Information Administration)

using improved mass transit or freight systems to accomplish emission reductions.) Rather, the recommendations include using modest improvements in fuel economy (from 30 to 60 miles per gallon, which is easily achievable now) and starting a desperate search for alternative fuel sources, using 17% of world cropland to produce ethanol or an 80-fold increase in wind power to make hydrogen for cars. (Never mind the astronomical cost of changing supply infrastructure to fuel these vehicles on hydrogen, the massive cost to maintain roads infrastructure, and the environmental damage from the loss of land use).

Perhaps most misleading of all is the comment in the report that states, "The U.S. share of global emissions can be expected to continue to drop." There is no doubt that a rapidly industrializing world will emit a higher percentage share of carbon dioxide than the United States, even though U.S. per capita emissions will continue to rise and remain the highest of any people in the world. I believe that statements such as this are designed to give the impression that the grotesque consumption of resources undertaken by western society can continue with minimal impacts on our current ways of life as the North American economy moves towards a service and information society and, the emissions that should be attributed to North America are simply exported to other nations. It is well known that China, India, and other countries are the workshops of North American consumption. It may be that all manufacturing moves overseas, with toys, cars, boats, and everything else in everyday life outsourced, but the real consumption, and the greenhouse gas emissions attributed to them, should be reported in the consuming nation as well as that of the producer.

The best change in the environment is no change from historical trends. Greenhouse gas emissions are a global community problem and society has adapted fairly well to the "normal" vagaries of the weather, and introducing the wild card of climate change into the game is hardly an ace up the sleeve. So what then do humans have to do to their transportation systems to reduce atmospheric greenhouse gases at a level that will prevent catastrophic changes in the ecosystem?

Decarbonizing our Energy Supply

Carbon in the form of solid and liquid fuels accounts for approximately 80% of the world's energy supply, and these forms of energy consumption will continue to rise unless politicians and policy makers take cli-

Figure 3-10. Carbon in the form of solid and liquid fuels accounts for approximately 80% of the world's energy supply. Worldwide government carbon-abatement policy is required sooner rather than later in order to move away from the status quo of fossil fuels and limit the damage caused by climate change.

mate change seriously. Although carbon-based fuels are relatively plentiful, packed with high-density energy, and cheap to extract, the legacy of atmospheric carbon left long after their consumption represents a massive ecological and financial liability.

There are really only two points that require consideration here. The first is to recognize that there is no one "silver bullet" fix for the replacement of fossil fuels or the decarbonization of their emissions. The second and perhaps more important point is that sound worldwide government carbon-abatement policy is required sooner rather than later in order to move away from the status quo of fossil fuels and limit the damage caused by climate change.

The technological options are numerous and many make common and fiscal sense even in the absence of government policy. The compact fluorescent lamp (CFL) shown in Figure 3-11 uses 75% less energy and lasts 10 times longer than a comparable incandescent lamp, saving the user approximately 10 times or more the initial cost of the bulb through

*Figure 3-11. The technological options for reducing climate change (and smog) are numerous, and many make common and fiscal sense even in the absence of government policy. The compact fluorescent lamp uses 75% less energy and lasts 10 times longer than a comparable incandescent lamp, saving the user approximately 10 times or more the initial cost of the bulb. As to the environmental impact, there is no comparison. The reduction in energy consumption means that less primary energy is used at the central generating station making the electricity to light your home, with **each** CFL saving at least one wheelbarrow full of coal over its lifetime!*

energy efficiency, providing an after-tax return on investment of 100%, far beyond that of any legal financial instrument. As to the environmental impact, there is no comparison. The reduction in energy consumption means that less *primary energy* is used at the central generating station making the electricity to light your home, with **each** CFL saving about one wheelbarrow full of coal or more over its lifetime.

Yet people do not use them. The complaint is sometimes heard that CFLs cast a pallid, antiseptic light suitable only for hallways. Interestingly, an article in the May, 2007 issue of *Popular Mechanics* describes a double-blind test conducted on CFLs which showed that the expert test subjects, "didn't see any dramatic difference in brightness....and...when it came to overall quality of the light, all of the CFLs scored higher than our incandescent control bulb."

Faced with an overwhelming number of choices, styles, options, and "first cost" of lamps and other appliances, most people are unable or unwilling to become properly informed or simply don't care about the financial and environmental benefits of a given technology. (The same can also be said for companies.) For most people, environmental issues are other people's problems or belong in the realm of governments. Unfortunately, the same apathetic approach to updating other antiquated or inappropriately applied technology is rife throughout the world and may not be corrected unless sound government policy is carefully developed.

Consider the fact that both coal and nuclear energy are often touted as the cheapest or most environmentally benign energy source. Tell that with a straight face to the miner's family after a fatal cave-in, to the tens of thousands of children who require asthma puffers as a result of breathing particulate-laden air, or to people suffering from radiation-induced cancer caused by Three Mile Island or Chernobyl.

There are numerous cases of poorly developed policy on the books. Indeed, many might argue that most policies on the books are poorly developed. I digress and would like to focus on one of the largest policy boondoggles developed in North America: corn-based ethanol applied to transportation.

I have no quibble with the technology per se. Corn-based ethanol makes some of the finest whiskeys and ryes, generating tax dollars and allowing people an escape from the reality of our fast-paced lives, at least for brief periods. But when it comes to putting whiskey in your gas tank, something has gone badly askew somewhere along the policy/technology line.

If governments were simply trying to support corn farmers, I wish they would say so, leave the alcohol and cars out of the picture and develop realistic agricultural support mechanisms. But the argument being made is that somehow corn is good for the environment and the domestic energy business. Really? While there is no doubt that corn-based ethanol is slightly energy- and greenhouse gas-advantaged (and to be fair a number of reports are emerging to the contrary), for the vast sums of money being lavished on this industry I could provide numerous technologies that would provide better GHG-reduction value for the dollars spent. Consider too that it takes the same amount of corn to produce one SUV tank-full of ethanol as it does to feed one person for an entire year.

The problem with pork-based policy is that nobody wins. Instead of precious resources being spent on real greenhouse gas reduction technologies, the money is squandered and the environment loses out. Land is chemically treated and fertilized with chemicals derived from the oil industry. Planting, spraying, harvesting, transporting, processing, and delivering ethanol to market all consume vast amounts of fossil-based energy, all so that consumers can drive around in their SUVs with a clear conscience?

The corn farmer thinks he has died and gone to heaven and spends vast sums of (borrowed) money on land leases and equipment purchases in order to capitalize on these misdirected government ethanol subsidies. The only people who will really make money on this deal are the bankruptcy trustees who auction off the farm assets when the government or tax payers come to their collective senses (or cellulosic ethanol makes its financially viable way to market) and the subsidies stop, as they always do.

Economic theory suggests that governments should not meddle in the markets, nor should they play croupier at the gambling table of technology. Direct subsidization of government-chosen technologies (such as ethanol) rarely works. It would be far better for government to set GHG-emission standards, put a real value on carbon emissions, and let the free markets figure out the rest. With such a system, virtually all Hummers would disappear from the roads; the rest of those belonging to civilians would be heckled off the streets and the armies' stock would quietly rust away, as there would be no need to invade foreign lands to fight over oil.

Corn-based ethanol would be gone in a flash, except of course for those supplies used to make drink. The reason? The true carbon life cycle would be exposed and the financial and business markets would only invest in those technologies that actually make economic/greenhouse gas-emission sense. The market will develop and retail the cheapest commodity fuel, not the one that is best for the environment. If the world was chocking on GHG emissions, fossil fuels were thought to be plentiful and there was no value on carbon, clean, alternative fuels would not be able to compete with oil.

A Carbon-Constraining System

The Europeans were the first out of the gate with a method of controlling carbon emissions by creating a cap on the level of emissions allowed by industry, providing trading allowances, and letting the financial markets take care of the rest. Unfortunately, the system got off to a bad start and has completely stalled. Because emission allowances were given away rather than being auctioned, as is done with similarly abstract radio spectrum allocation, prices have fallen nearly fifty-fold from the desired 50 Euros per tonne.

When carbon is valued at too low a price, the effects of such a program are not punitive, as is the case at the close of 2007. As a result, the offending carbon-emitting industries do little or nothing to reduce or offset emissions.

If society's carbon emissions are to be stabilized or reduced, governments need to consider carbon taxation. It would shift society into a more sustainable mode of operation as well as reduce infrastructure costs, improve urban smog, prevent resource depletion, and improve energy security.

A carbon tax or trading system which operates at $50 per metric tonne of carbon dioxide emissions (approximately $55 per short ton) may provide sufficient financial incentive for large emitters to upgrade their technologies or business practices because of the effect on profit margins. However, small-scale emitters, including home owners, will not be materially affected by these programs and will maintain their same old habits. The *Transportation Statistics Annual Report 2005* published by the U.S. Department of Transportation states that, "While prices of transportation fuels fluctuate over time, vehicle-miles of travel does not appear to be affected." The report points out that over the ten-year period from 1994 through 2004, motor fuel prices rose at a rate higher than general inflation, yet highway vehicle miles driven and aircraft-miles of travel increased markedly. Fuel price volatility may cause people to grumble, but there will not be any significant change in the way society moves about or consumes energy.

The Green Party of Canada has, for example, proposed a European-like pricing scheme on carbon dioxide for retail consumers of gasoline. Under this plan, carbon dioxide would be given a fixed value of C$50 per metric tonne, resulting in a tax of C$0.12 per liter of gasoline.[11]

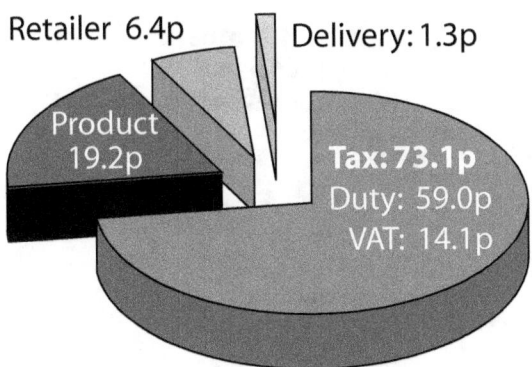

Figure 3-12. For each UK£ (US$2.03/C$2.14), approximately the average retail price for a liter of gasoline in the UK as of August 2007, taxes account for approximately 73%. In North America, taxes represent less than 33%. (Source: <http://news.bbc.co.uk/1/hi/in_depth/world/2000/world_fuel_crisis/933648.stm>http://news.bbc.co.uk/1/hi/in_depth/world/2000/world_fuel_crisis/933648.stm)

(The equivalent calculation applied in the United States would yield a tax of US$0.50 per gallon of gasoline.[12])

Some might howl that this would be a punitive tax and would drive everyone to walk, ride a bicycle, and perhaps even drink all the excess ethanol when people abandoned their cars. Certainly everyone would react and call for a lynching of any politician they could get their hands on, at least in the short term. But there would be nary a dent in driving habits since people realize that price fluctuations like this fall within the "normal" volatility of the market. Gasoline prices have not changed significantly over the last few decades, and it is only recently that prices have nudged upwards as a result of supply constraints caused by the effects of insufficient investment in refineries, terrorism threats, the Iraq war, and other geopolitical events around the world. I would hazard a guess that with a $50 per tonne ($55 per short ton) tax on carbon dioxide emissions, people would complain and then immediately go back to their old driving habits.

I propose a much more aggressive tax on retail fossil fuels. If the plan is to reduce waste, preserve natural resources, stop environmen-

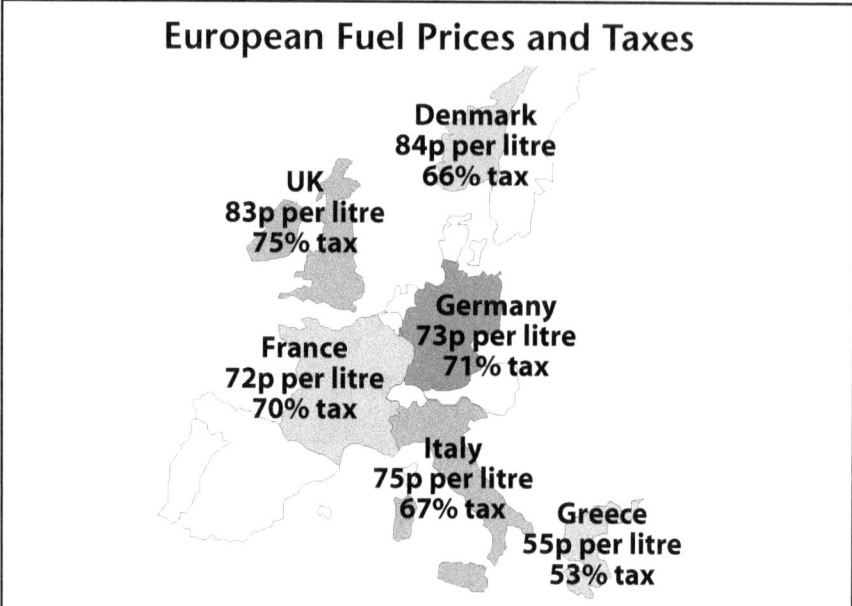

Figure 3-13. Although European gasoline prices are more than double those of North America, Europeans suffer from many of the same personal transportation maladies. A carbon tax of US$500/tonne would raise domestic prices to those of Europe. People would howl in protest, and although more fuel-efficient cars would be developed, total vehicle miles driven would remain the same while suburbs and roads infrastructure would continue to grow. (Source: <http://europe.theoildrum.com/story/2006/5/3/17236/14255>http://europe.theoildrum.com/story/2006/5/3/17236/14255)

tal damage, curtail urban sprawl, improve energy efficiency, and reduce infrastructure costs, then a much more punitive taxation program is required.

At a carbon price of $500 per tonne ($550 per short ton) the effects would be hard-hitting at first but still not sufficient to drive the necessary changes at the societal level. Even though this tax seems severe, it simply would not be punitive enough according to my logic. This tax level would levy a surcharge of US$5.00 per gallon or C$1.20 per liter on sales of gasoline. At the time of writing, this would equate to a retail price for gasoline of US$7.80 per gallon and C$2.15 per liter.

I can almost see people cringing as they read this and thanking their lucky stars that I am not in politics. But however hard this might be for

North Americans to accept, it is simply the cost of day-to-day driving in most parts of Europe.

According to the website www.petrolprices.com, the average price for regular unleaded gasoline on August 9, 2007 was 96.5 pence per liter. At the same-day exchange rate, this equates to a price of US$7.39 per gallon or C$2.06 per liter, which is very close to the current North American retail price for gasoline with a US$550 per ton (C$500 per tonne) carbon dioxide tax.

Anyone who has driven in the UK or any other part of Western Europe will quickly realize that driving is not constrained; in fact the exact opposite seems to be true. After all, it was the City of London that developed the very first congestion tax to help alleviate traffic in the core of the city, the same city where monthly parking fees are nearly $1,200 per month.

At a carbon dioxide taxation level of approximately $500 per metric tonne, the automotive industry would quickly adjust by offering more fuel-efficient, perhaps smaller vehicles. (Given enough time, North America might even begin to model itself along European lines, but unfortunately that time simply isn't available to us.) Drivers, on the other hand, would quickly learn that vehicle fuel costs represent a small percentage of vehicle ownership costs and would continue to burn fuel and drive almost as far and as quickly as before. It takes a long time to knock sense into the heads of a society accustomed to low energy prices and longer still to change personal habits and entitlements. Certainly nothing would really change in society. Suburbs would continue to be built, Wal-Marts would continue to pop up, people would adjust to these fuel prices, and society would almost certainly go back to sleep.

If European drivers continue to increase their commuting miles despite the current level of fuel costs, then perhaps what is required is a carbon taxation policy that is higher still, perhaps as much as $1,000 per tonne of carbon dioxide. At this level of taxation, the price for a gallon of gasoline in the United States would be $12.33 and in Canada $3.35 per liter. At this retail price, most people would have no choice but to change their driving habits. People would demand better access to mass transit, they would move closer to work, workers would begin telecommuting, wasteful or unnecessary travel would be curtailed, people would return to car-pooling, and freight companies would develop better routing and logistics systems. Long-distance trucking would be eliminated, with a resulting decrease in the massive road damage and repair costs it brings.

The Paradigm of Reality Convergence

Several of society's major ills relate either directly or indirectly to the use of carbon-based fuels and the resulting personal entitlements that derive from this:

- Greenhouse gas emissions are rising beyond the atmosphere's ability to absorb them, causing major climatic changes.
- Society cannot afford to build and maintain roads and related infrastructure. A disproportionate amount of state/provincial and municipal taxes is being used for this purpose, with governments straining to justify the expense.
- More roads are producing increased congestion and causing suburban and exurban development, which creates a feedback cycle of increased commuting.
- Oil supplies will become constrained, whether through the effects of peak oil, geopolitical strife, resource depletion, insufficient infrastructure development, or increased worldwide consumption. Supply and demand will dictate higher prices, supply shortages, and rationing.
- The financial system will suffer major instability as energy becomes increasingly more expensive. Inflation in the developed world could easily spin out of control, leading to recession or financial system failure.

Carbon taxation at punitive levels should really be called a "sustainability tax," as it would result in a rethinking of urban development; a boost for the "new urbanization" movement that promotes a more sustainable approach to land use, energy efficiency, and building materials as well as linkages to transit, schools, and shopping.

Consumption of everything would drop if prices reflected the embedded energy, petrochemical, and transportation resources required to produce and deliver those goods. The latest craze toy manufactured in China would disappear. Bottled water from Italy would remain in Italy, where it belongs. Farmers would become rock stars, providing locally produced food to the masses instead of anemic tomatoes trucked in the middle of winter from Mexico to the frozen northern US and Canada.

As a small aside, I was both amused and disgusted this summer when I read an article in a local paper that stated that area farmers could not sell their fresh strawberries to area grocery stores because trucking Mexican strawberries thousands of kilometers was cheaper. If we need any further proof that energy prices are too low and our priorities are skewed, this is a fine example.

In short, carbon taxation is a tool that would allow society to move to a model that is better suited for the future and would ensure that there is a world worth passing on to future generations.

Not all people would approve; indeed most would complain, arguing that their right to the good life would begin to evaporate. Whether we want to believe it or not, we are within a generation or two, at the outside, of being forced to accept such a measure. A retooling today would save an enormous amount of grief in the future.

If a sustainability tax were started early enough, governments could direct the income from the taxation program towards the development of the infrastructure for a sustainable society. Money could be used to develop proper mass transit coach and rail systems, and urban planning rules could be changed to encourage community-based living rather than a commute-to-work society. Road development would be curtailed as fewer and fewer personal vehicles were required. Capital budgets for roads would be directed to maintaining existing roads and infrastructure. Recognizing that new roads *create* sprawl and do not solve it and that existing roads are in deteriorating condition (30% of urban roads were classified as poor in 2003)[13], curtailing new road development is the sensible choice. If you disagree, recall the bridge collapses in Quebec and Minneapolis in 2007 and recognize that there are almost 74,000 more bridges in the United States that are classified as "structurally deficient" according to the U.S. Department of Transportation.

Monies could be used to help industry convert to low- or zero-carbon technologies and to improve energy efficiency. The incentive for industries would be there in any event, as they would want to supply the same or equivalent products without having the burden of a carbon tax levied on their goods or services.

Lastly, governments would directly subsidize those financially marginalized people who currently have little or no access to personal or mass transit systems. Being a democratic and supposedly caring people, we would level the playing field and provide free transit for those who require assistance, as well as upgrade social housing infrastructure, homes, or apartments to ensure that they were as energy efficient as possible. I prefer to see government upgrading the capital stock of social and subsidized housing rather than funding operating costs, as it currently does. Politicians love to blather in the press about how they have compassionately helped the poor pay this year's heating bill. Like putting a Band-Aid on a cancer patient, throwing funds for heating oil at poorly built housing is nothing more than a bonfire for money.

Would the economy suffer? Yes and no. Yes, inefficient businesses would either adapt, leave the country, or go out of business. This would include some of the relics of a past generation that operate today, including the "Big Three" automotive companies, the pulp and paper industry, and a host of other large, inefficient energy users.

In a capitalist society, new industries would emerge to fill the gap, just as has happened in the past. The concept of technological obsolescence has touched us all. The horse and buggy companies died off or transitioned into automotive firms. The recording industry went through the wringer as it transitioned from wax cylinders to records, eight-track tapes, cassettes, CDs, and now virtual music supplied by MP3 and Internet downloading. Coal-fired power plants would be replaced by distributed, "right-scale" generating facilities that are close to centers of demand. Does society really need gasoline-powered leaf blowers, propane mosquito killers, patio area heaters, or three-ton SUVs used just to purchase groceries at the local mall? People have survived for thousands of years without this junk, yet today's suburban homeowner believes these items to be necessities. Perhaps it is time to rethink our priorities, especially when energy-consuming choices will reduce the quality of life of our offspring in the rich world and kill thousands or more in the poor areas of the world.

When General Motors or Wal-Mart fails, a new startup or several will fill the void. This is not hypothesis; this is fact. It is merely a matter of time. A carbon-constrained world does not mean a trip back to pre-industrial times. It does, however, require a change in societal structure, one that should occur as part of planned transition starting now.

Will governments adopt highly punitive carbon or "sustainability" taxes? Not a chance; at least not yet. Governments, politicians, industry, and people in general are in denial about humanity's impact on the environment. The Kyoto Climate Change Protocol is a good start, but many of the world's developing and developed economies (including the biggest per capita emitters, the United States, Canada and Australia) do not subscribe to the program. Unfortunately for climate change naysayers, Mother Nature and the carbon absorption capacity of the atmosphere could care less what politicians say. The destructive forces of nature will continue their relentless march until humans are either beaten into submission or banished to oblivion. Or, as professor and author Thomas Homer-Dixon wryly points out, "we are going to lose some coastline."

With any degree of luck and foresight, mankind will get its act together, enforce *punitive* sustainability or carbon taxation policies on all people collectively, and begin the transition to a carbon-free world.

Figure 3-14. If municipal, state/provincial, or federal governments do not have the political will to develop low- or zero-carbon sustainable energy policies, it is time to move the debate to a United Nations Security Council-like framework to ensure that these programs are developed. The solution is simple, provided society has the will to make it happen. Unfortunately, denial is causing paralysis, which is wasting valuable time and natural resources. Courtesy: United Nations.

4
Transportation Systems in a Carbon-Constrained World

"We want the facts to fit the preconceptions. When they don't, it is easier to ignore the facts than to change the preconception".

Jessamyne West

If you accept the various arguments I have presented earlier in the book, you will no doubt be wondering if I am ready to suggest that society give up automobiles completely and revert to walking or horsepower of the biological kind. In a word, "no"; but as with most issues of this complexity, the complete answer is much more complicated.

Our transportation systems have evolved the way they have for one simple reason: high-energy-density liquid fossil fuel, a.k.a. oil, has dominated all other energy contenders for the last hundred years. Over the millennia, foot and animal power were the primary if not only sources of transportation energy available to the majority of mankind. Wood fires provided heat, light, and warmth. Wind and water power were slowly added to the energy mix, primarily for stationary applications such as the milling and grinding of seeds and grain. Waterborne transportation incorporated the power of the wind, principally for the transport of goods, although a small percentage of people were afforded some degree of early-development mass transit with the advent of the sailing ship.

As the industrial revolution proceeded, solid-fuel wood and coal-fired steam engines allowed the arrival of railroads, resulting in the first true mass transit system, one that was affordable to the people. Some enterprising entrepreneurs even tried to duplicate the technology of the steam train on a personal scale with the development of steam cars and trucks. The Amédée Bolée coal-fired, steam-powered mail and passenger coaches (Figure 4-1) of the early 1880s must have made an incredible impression on the people of France as these outrageous vehicles chugged along the primitive streets delivering people and post. Later, more streamlined steam-powered vehicles including the "Stanley Steamer" (Figure 4-2) and the electrically powered "Milburn" (Figure 4-3) were produced until the 1920s, when the gasoline-powered internal combustion automobile forced the remaining alternative-powered vehicles off the roads and into the history books.

Figure 4-1. The Amédée Bolée coal-fired, steam-powered mail and passenger coaches of the early 1880s must have made an incredible impression on the people of France as these monsters chugged along the primitive streets. This passenger carriage weighed in at five tons and burned coal at the rate of approximately 50 kilograms per hour.

Figure 4-2. The Stanley Steamer, as this car was affectionately known, was built by the Stanley Motor Carriage Company at the turn of the century. Even the steam engine of the "Stanley" was powered with liquid kerosene rather than low-energy-density wood or coal. Sales of steam cars continued until the early 1920s, when the gasoline-powered internal combustion automobile forced the remaining steamers into retirement. (Courtesy: Pat Farrell)

Figure 4-3. As with most early automobiles, the Milburn electric could trace its history back to 1848 as part of the Milburn Wagon Company. Over 7,000 of these electric vehicles were produced until as late as 1927, when lack of driving range and access to electricity, as well as competition from Henry Ford's mass-production factories, dealt a death blow to the firm, with it being purchased by General Motors.

Figure 4-4. Although there has been a resurgence in demand for electric vehicles of late, it is surprising to note that the venerable Milburn still has approximately the same operating range as today's typical electric vehicles, despite nearly a century of technological development. Richard Lane is operating the tiller steering and "accelerator lever" of his restored Milburn.

The development of the internal combustion engine in the late 19th century and the gas turbine engine in the mid-20th century were engineering feats of staggering importance. With these new technologies, man could have enormous amounts of power in a small, lightweight package that would run economically on gasoline, diesel fuel, and kerosene, liquid fuels derived from crude oil.

Early oil explorers were not interested in the gasoline byproduct of the oil extraction process. Rather, their attention focused on a substance known as kerosene, a replacement for diminishing supplies of whale oil, since both of these substances could be used for home lighting. At the time, gasoline was considered of little or no value because of its volatility. Unfortunate folks who tried using it in oil lamps, often ended up with burns and house fires.

It didn't take long to realize that a cupful of either gasoline or kerosene would produce a massive amount of energy, about the same amount as a human doing hard labour for one week. Additionally, liquid fuels were found to be much easier to handle than wood or coal, allowing for automatic feeding into engines with no ash byproduct to dispose of. At the time of its discovery, oil was plentiful and cheap and thought to be relatively clean (at least compared to smoky coal or wood), and it was easily harvested since it would squirt out of the ground almost anywhere wildcatters sunk a drill. The transition from solid to liquid fuel was very fast—it took only a few decades to switch approximately 80% of the world's energy supply to oil. Although supplies are becoming tighter, more expensive, and increasingly difficult to find, switching to alternative fuels and transportation technologies is going to be a very difficult task. Society is going to have a hard time weaning itself from the addictive properties of oil.

Figure 4-5. At the time of its discovery, oil was plentiful and cheap and thought to be relatively clean, squirting out of the ground whereever wildcatters sunk a drill. The transition from solid to liquid fuel was very quick, and it took only a few decades to switch the world to oil. Although supplies are becoming tighter, more expensive, and increasingly difficult to find, switching to alternatives is going to be a very difficult task.

There Is No Substitute for Oil

Hydrogen, ethanol, walking, biodiesel, gas-to-liquids, coal gasification, electricity, or natural gas: Which of these will replace gasoline and diesel fuel for personal transportation in the coming decades? Like the weather person, who cannot predict the weather three days hence, I do not have a crystal ball that can forecast the future. I do, however, believe that society will attempt to use all of the items on the above list and more. As oil supplies diminish or geopolitical issues cause severe supply disruptions or price shocks, society will scramble for whatever energy source it can get its hands on, environmental or physical cost be damned. Humans will simply not let their entitlements and beloved way of life go and will fight to the death trying to maintain the status quo.

Instead of searching for real options to reduce unnecessary travel and congestion, governments will continue their love affair with the personal automobile and continue to subsidize road construction (causing yet more congestion) in an effort not to impinge on people's "right" to freedom of travel. Society will spend countless billions, probably trillions, of dollars on research and development to make whatever sort of high-mileage fossil or even nonfossil-fuel-based, multiwheeled personal vehicle propulsion system possible, including the fuel of the day to go with it. All of this will occur while the very fossil fuels that are in short supply or have become increasingly expensive are needed for more important petrochemical feedstocks for medicines and durable goods as well as to maintain the roads in which society has already invested heavily.

Governments will belatedly reach a point where they can no longer afford to continue building roads (let alone maintain the ones we already have) and people will not be able to afford to drive the sort of distances we currently travel without blinking an eye. It is very likely that society as we know it will grind to halt, as the cost of maintaining current high levels of personal mobility (while trying to reduce carbon emissions and fight the laws of peak oil) will simply swamp the financial and civic infrastructure systems.

When Oil-Based Transportation Comes to an End

As oil-fired transportation systems begin to break down, suburbs and exurbs will either die and become ghost towns or transform themselves into self-sufficient local towns and villages, as of old. Since people will no longer be able to afford to commute by car, towns will (hopefully) redevelop. Towns and cities will be interconnected first by Internet or equivalent ultrabroadband technologies as well by as user-friendly rail and coach. Although people argue that these transitions will not occur, many of the technologies are being developed now, not necessarily because of concerns about energy or climate change but for reasons of cost effectiveness, efficiency, or convenience.

Figure 4-6. As oil-fired transportation systems begin to break down, suburbs and exurbs will either die or transform themselves into self-sufficient local towns. As people will not be able to afford to commute by car, towns will redevelop, providing local shopping and the opportunity to work or telecommute, as well as making education, health care, and social services available within walking or mass-transit distances. Within a decade telecommuting will account for billions of dollars in sales and reduce carbon dioxide emissions through eliminated business travel, not to mention redeye relief for executives with shorter commuting distances. Courtesy: HP Halo

The concept of telecommuting is one example of how business can employ a sustainable model right now, by replacing slow, old-fashioned physical commuting with "telepresence technology." A current sales or engineering call that is conducted person-to-person involves considerable time and the cost of getting to and from a given location. International travel may cause even more delays through the necessity for work visas or immunization shots. The actual work may only require a few hours, but because of travel logistics more time than necessary is used to justify the travel-related expenses.

Telepresence uses virtual reality software for image and sound processing, high-speed Internet links, and realistic, life-size display systems that give the impression that people are actually sitting across from each other having a face-to-face chat (Figure 4-6). Industry analysts suggest that sales of telepresence systems will grow to more than $1.24 billion in sales by 2013, greatly reducing travel-related greenhouse gas emissions at the same time.[1] According to Hewlett Packard, a telepresence-equipment supplier, eliminating one round trip between New York and London saves 3,000 pounds (1,361 kilograms) of carbon dioxide. Assuming that jet fuel were subjected to a carbon tax of even $500 per tonne, the use of telepresence would save $680 in emission costs for the example trip, while increasing the efficiency of the business transaction[2].

Once the telepresence system was up and running, it would be available on a 24-hour-per-day basis, thus allowing multiple meetings to be conducted at the speed of light. Customer-to-client interaction could be increased dramatically and the requirement for physical meetings could be reduced or possibly eliminated. The use of telepresence is one small example of how a new "sustainable" business model could substitute for the old carbon-economy technology of driving or flying from one place to another; an old technology (physical commuting) is removed from the economy, while the new sustainable technology of telepresence is added. The economy doesn't lose capital; it simply adjusts to suit the requirements of the day.

This vision of the future does not sit well with most people today because they are in a collective state of denial, believing that society can and must continue to expand as it has since the beginning of the industrial revolution and with greater acceleration since World War II. Few stop to consider that nature's gift of fossil fuels is finite, that our thirst for these fuels is insatiable, and that the atmosphere cannot absorb the output of fossil-fuel combustion. Something has to give, and our current fascination with personal travel will be one of the first sacred cows to fall.

Moving Ourselves

A quirk of nature, time, and place gave rise to North America's development of vast empty lands prior to the dawn of the industrial age, allowing agriculture and scattered towns and cities to develop. As the industrial age provided the technologies of rail transport, telegraphic communication, and powered machinery, the precursor to today's suburban landscape began to emerge.

Figure 4-7. After the great growth spurt following the Second World War, personal wealth increased beyond anything the world had ever seen, resulting in a myriad of houses, automobiles, consumables, and free time. Unfortunately, urban planning was all but absent in the mix, a problem that continues to this day. Work, shops, churches, schools, community centers, coffee shops, even the local park are all out of realistic walking range, reinforcing the need to drive literally everywhere. (Courtesy Derek Jensen, Tysto.com)

After the great growth spurt following the Second World War, personal wealth increased beyond anything the world had ever seen, resulting in a myriad of houses, automobiles, consumables, and free time. Unfortunately, urban planning was all but absent in the mix, a problem that continues to this day.

I recently went for a drive through a typical, nondescript suburb of Kanata, Ontario, an area that was originally envisioned by architectural designer Bill Teron as a planned "model" city. Mr. Teron's vision for Kanata was what he called a "multi-dimensional" city, one that would provide not only medium-density housing, but areas for living, employment, lakes, parks, golf courses, and of course lots of roads and space for cars. In his plan, green space accounted for 40% of land use, where typical requirements are less than 5%. It sounded like a good idea at the time.

Teron developed the community according to principals that included many good ideas. including the development of a "technology park" to bring high-paying jobs to the area. He was successful to a point, as there are now over 50,000 people employed in the area, where none were employed before.

That was a generation ago. Kanata today has expanded and is nothing like what Mr. Teron initially planned. The original section of Beaverbrook is still there, much as it was in the early 1970s, but what has grown around it is anything but a sustainable community. The Kanata "Town Centre," as it is euphemistically known, is located about two miles from Beaverbrook and is nothing more than a collection of big-box stores, fast-food joints, and a movie theatre. Its location is so far from anywhere that I can only assume the name "Town Centre" to be a joke.

I have driven to a densely populated area of "new Kanata" known as "the Kanata Lake Area" (perhaps another joke, as there are no lakes to be found) about one mile to the west of the Town Centre. Normally, I wouldn't drive this short a distance, preferring to walk or ride my bike and to avoid using my car. However, city planners forgot that people might walk and didn't build any sidewalks; nor did they provide bike paths or bike-locking facilities. In fact, city planners forgot a lot of things. The nearest school cannot be walked to because of the lack of sidewalks, so school buses are mandatory even though, with a bit of planning, walking paths could have been easily implemented. There is no corner store because if people are going to drive anyway, they will just go to the big box a couple of miles away. Work, shops, churches, community centers,

Figure 4-8. How does a supposedly sane society allow the development of such stupid ideas as the drive-through kiosk? Coffee shops prefer to push commuters into a drive-through service that is often packed with 10 cars idling for minutes at a time, wasting precious fuel resources while contributing to urban smog and climate change.

coffee shops, even the "local" park are all out of realistic walking range, reinforcing the need to drive literally everywhere. To make matters worse, the local coffee shop doesn't encourage much of a sense of community either, preferring to push commuters into a drive-through service that is often packed with 10 cars idling for minutes at a time.

These are patterns that occur repeatedly in most areas of North America, and our society will have to change or it will simply vanish when western oil addicts are forced to go cold turkey. Using an automobile to queue for coffee will become a ridiculous phenomenon from another era; using an automobile to drive to work will no longer be a viable option either.

What then are the options to moving people using existing and near-term technologies? Figure 4-9 illustrates the relative efficiency of various current means of travel. It will come as a surprise to no one. Although the use of foot power, buses, and high-efficiency cars are obvious solutions, there are many ways of improving this list of modes of travel and perhaps even changing the dynamic of how people use them. It is interesting to note that the more energy-efficient means of transportation are the least often used. In the United States, 79% of people drive themselves to work alone (the least efficient means), while only 4% use mass transit, and 3% walk (most efficient).

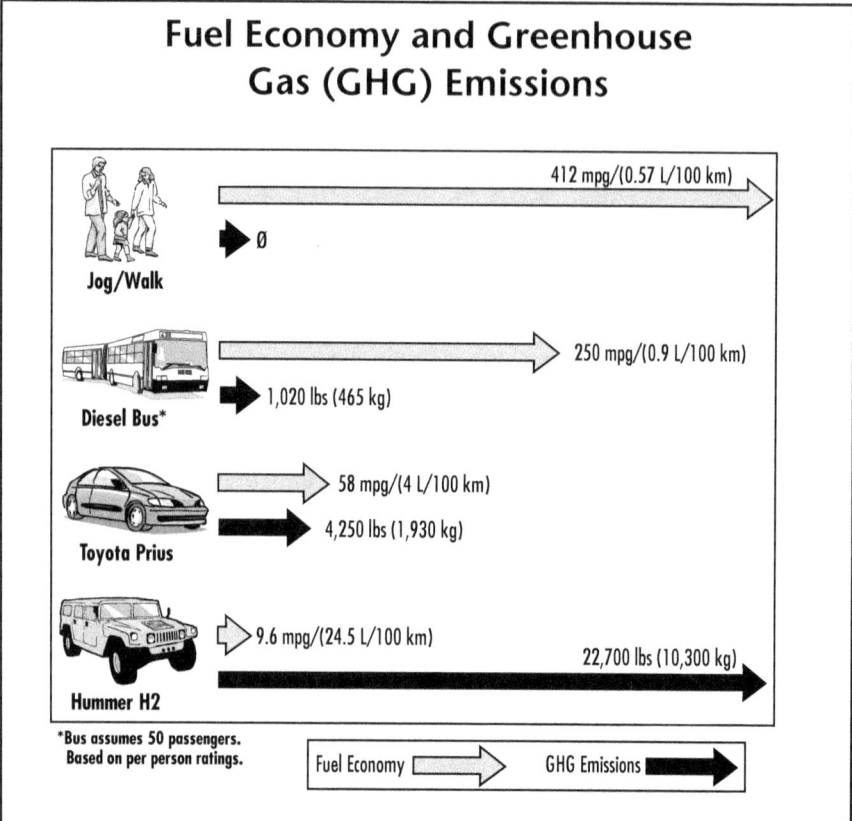

Figure 4-9. This chart compares the fuel consumption and greenhouse gas emissions of four types of urban transportation. Operating costs and greenhouse gas emissions increase in proportion to fuel consumption. It is interesting to note that the more energy efficient a given means of transporting people, the fewer people use it. In the United States, 79% of people drive themselves to work alone (the least efficient means), while only 4% use mass transport, and 3% walk (most efficient).

Why Not Walk or Bicycle?

We have already discussed that the majority of automobile trips are very short, with many falling within the range of access by walking or bicycle, if only town planners were more careful in their road, path, and underground walkway designs. This lack of planning leads to a nation of sedentary car drivers which, when coupled with discretionary income, creates an unhealthy and potentially lethal mix.

Governments complain that the economy will stall if oil prices rise too high, yet they often do not consider all aspects of the economic question. The issue of telepresence discussed above is one example; the rampant rate of diabetes is another. According to the Centers for Disease Control and Prevention, approximately 21 million or 7% of Americans have diabetes and a further 54 million have elevated blood sugar levels, placing them at risk of developing the disease. Obesity and a sedentary lifestyle are known to increase the risk of contracting type 2 diabetes, a situation that is exacerbated by the personally destructive use of the automobile.

Figure 4-11 shows the prevalence of diabetes in the United States in 1990 and 2001 and details a precipitous rise in the incidence of the disease during this decade. The economic costs of the disease were estimated at $132 billion in 2002 and continue to rise unabated.[3] According to Dr. C. Ronald Kahn of the Joslin Diabetes Centre in Boston, "We [U.S. society] really won't be able to afford the amount of health care this is going to cost."[4]

To be sure, adding walking paths to city suburbs will not stop diabetes, but the fact that North Americans are "house and car dwellers" is not contributing to our well-being, and better urban planning with less

Figure 4-10. Electrically assisted bicycles are the rage in the crowded streets of China and could be used here in North America for local commuting. Better civic planning would provide more bicycle paths, shorter and safer routes that would encourage more people to commute by muscle rather than motor.

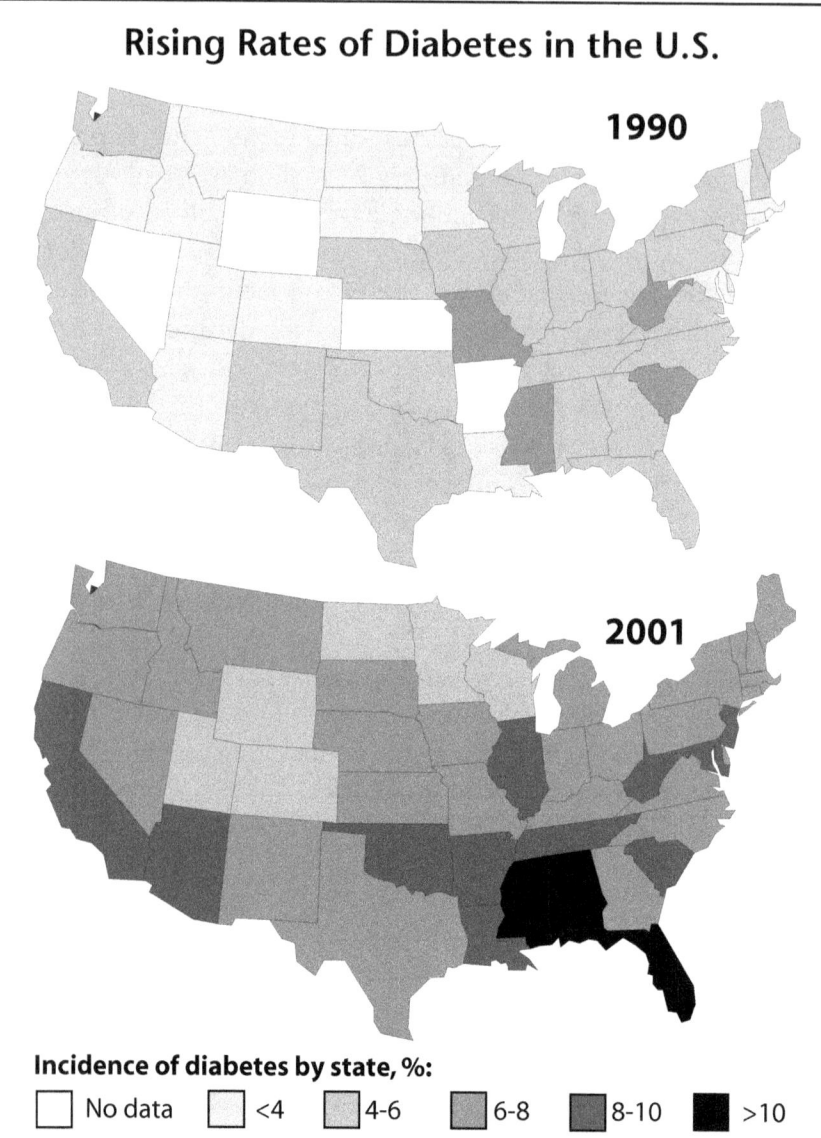

Figure 4-11. These maps detail the incidence of diabetes in the United States in 1990 and 2001, showing a precipitous rise during that decade. The economic costs of the disease are estimated at $132 billion in 2002 and continue to rise. According to Dr. C. Ronald Kahn of the Joslin Diabetes Center in Boston, "We [U.S. society] really won't be able to afford the amount of health care this is going to cost." (Source: Centers for Disease Control and Prevention)

Figure 4-12. This may look like a small motor scooter, but it is actually an electrically assisted bicycle. Since it is equipped with pedals the bike can be driven by minors and people who have not bothered with a driver's licence. After a gruelling day at the office, let the electric motor whisk you home.

emphasis on the car would make an excellent start in the battle against diabetes.

Human- and animal-powered transportation has been the norm since time began, until recently, and it continues to be the most energy efficient and clean method of locomotion. Although walking, jogging, and riding a bicycle may not appear to be an option because of urban sprawl and poor planning, they should not be written off as unrealistic options. If we were to convert the amount of energy contained in a gallon of gasoline into food calories, you could bicycle 400 miles (644 km). Costs for parking, insurance, and depreciation would be virtually eliminated. So too would greenhouse gas emissions. Some people dismiss this idea as completely unworkable, yet millions of people do not own cars, choosing to live, work, and socialize within a relatively small area. For longer trips, mass transit, car sharing, or rentals would be options, all for a fraction of the cost of car ownership.

Inner-City Travel

Improved Public Transit

Properly designed mass transit systems should work wonders in high-population-density urban centers, but they rarely do, as the dismal usage data shown in Figure 1-6 shows. Americans choose public transit for 1.5% of all trips, while personal automobiles account for an astounding 86.5%. As the city sprawls into the suburbs, population density drops, further reducing the effectiveness of transit systems which frequently require multiple transfers from branch to main-line routes. To circumvent this problem some municipalities have developed "park and ride" facilities which allow rural or suburban clients to park their vehicles (often for free or at a reduced rate) next to a main-line hub in order to improve usage and alleviate congestion in the urban core areas. Although this sounds like a great idea, the simple fact is that the majority

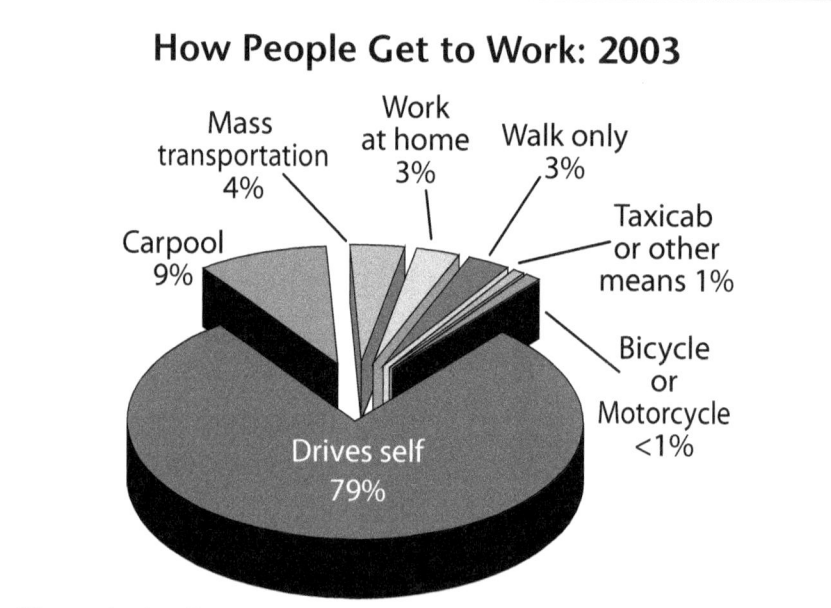

Figure 4-13. For inner-city commuting, there are numerous options available including driving a personal car, walking, car-pooling, or using a motorcycle. Regardless of which form of transit is used, the fact remains that people prefer driving their personal car in congested conditions (a horrible option at best) over using mass transit by a factor of 58 times. (Source: U.S. Dept. of Transportation)

of people don't use mass transit, preferring to use a personal vehicle at greater expense while coping with the daily stress of rush-hour traffic in two directions.

Why does this occur? Why would people subject themselves to the punishment of driving under conditions of heavy congestion when bus, rail or other mass transit options are available? We could argue that people hate mass transit for what it is, a dismal ride packed in crowded vehicles which are less convenient and sometimes slower than the personal car. Granted, people put up with this when they fly because there is no other option. In the case of mass transit, there are plenty of alternative options.

Almost everyone has suffered the indignity of ticket lineups, security screening, and the packing of numerous people into tiny aircraft seats (often the torture-inducing middle one) only to wait for a late departure. Once the flight is airborne, cost-cutting programs and aggressive competition have reduced service levels to below soup kitchen quality.

This is surely similar to using today's bus or rail service on a daily commute to work, yet people put up with this poor service to take a plane trip. Why the discrepancy? The answer is quite simple; there is no other realistic choice to flying vast distances. Unless money is no object and you can afford to hire a private jet, your only option is moving your business from one carbon-copy airline service to another. The benefit is negligible, even if you do pay usurious rates for upgraded "business class" service.

For inner-city commuting, there are numerous options available including driving a personal car, walking, car-pooling, or using a motorcycle. Regardless of which form of transit is used, the fact remains that people prefer driving their personal car in congested conditions (a horrible option at best) over using mass transit by a factor of 58 to 1.[5]

To alleviate underutilization of mass transit a number of policies could be implemented:

- Mass transit must be designed to encourage its use. Comfortable amenities, wireless Internet, cell phone service, and a quality (car-like) atmosphere will be required in order to compete against the personal car.
- System logistics must be planned to ensure travel speed advantage over the personal car. This will require greater use of mathematical

route planning, GPS, and traffic light and dispatch control as communication network capacity is mapped and optimized.
- Dedicated road lanes should be assigned for buses. Light-rail routes should be developed for passenger transit, while road freight is moved to separate freight rail systems.
- Carbon taxation must be punitive enough that people will forego the use of cars in favour of mass transit.
- City congestion taxes and elevated parking fees are required to encourage the use of alternative systems.

In short, if society wants to transition away from the frivolous use of personal transportation, a viable, user-friendly option that offers quantifiable enhancements over the automobile must be developed. This is a far cry from the current horrid conditions of North America's perpetually late bus systems. Government will have to drive the development of user-friendly mass transit by directing resources from carbon taxation income. Mass transit has been allowed to fester and decay, becoming a little-used system for the poor. This trend will have to be reversed quickly to ensure that society still has the physical and financial resources to build a system that actually works.

Figure 4-14. This General Motors hybrid-powered bus delivers up to 60% better fuel economy than conventional diesel systems currently in use. Buses equipped with a hybrid system produce much lower hydrocarbon and carbon monoxide emissions than normal diesel buses, lowering particulate emissions by 90% and nitrogen oxide emissions by up to 50%. (Courtesy: General Motors Corporation)

Regardless of whether the local commute is made by bus, streetcar, subway, or train, the energy-efficiency gains and resulting greenhouse gas emission reductions are achieved by dividing the fuel consumed by the relatively large number of passengers on board. Figure 4-9 indicates that a standard diesel-powered bus with an average fuel economy of 5.0 miles per gallon (47 L/100 km) will achieve the equivalent of 250 miles per gallon (0.9 L/100 km) when averaged over a ridership of 50 passengers. Expressed another way, moving people by bus or train reduces carbon dioxide emissions by a factor of 8 times compared to the automobile.[6]

General Motors has recently developed a hybrid bus technology which increases fuel economy by up to 60% while keeping capital costs within reason. Buses delivered to the Seattle, Washington area are estimated to save 750,000 gallons (2.8 million liters) of diesel fuel per year. According to General Motors, if nine other major metropolitan cities

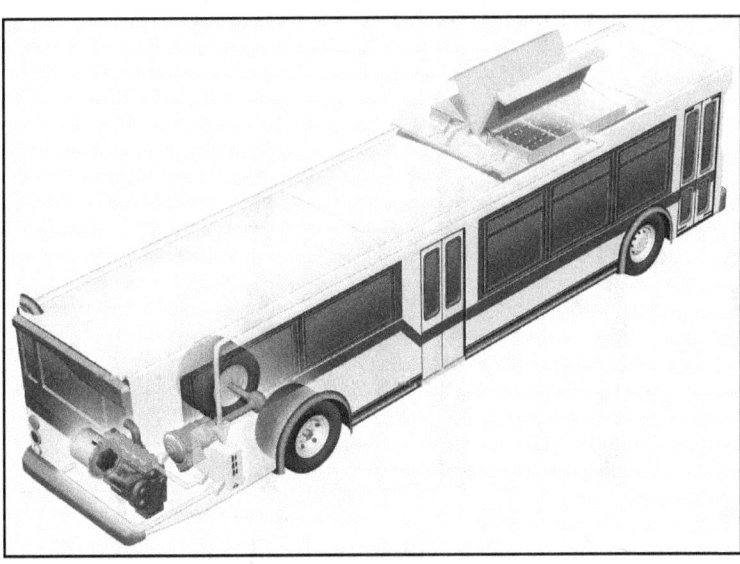

Figure 4-15. Hybrid systems use two sources of power to move the vehicle. Using the parallel hybrid approach, the diesel engine acts as a generator, charging a battery bank located on the roof of the bus. Acceleration is achieved using energy stored in the battery bank, with the diesel engine maintaining speed after the vehicle is moving. (Courtesy: General Motors Corporation)

adopted this technology for their buses, it could mean savings of more than 40 million gallons (151 million liters) of fuel per year.

These buses also improve air quality, producing 90% fewer particulates (tiny pieces of soot and dust which cause lung irritation and asthma) and 60% fewer nitrogen oxides than their standard diesel-powered cousins.

Hybrid systems use two sources of power to move the vehicle. Using the "parallel hybrid" approach, the diesel engine drives the bus and at the same time acts as a generator, charging a battery bank located on the roof of the bus. Acceleration is achieved using energy stored in the battery bank, with the diesel engine maintaining speed after the vehicle is moving. A process known as regenerative braking captures energy normally wasted as brake pad heat and returns it to the vehicle's battery bank. In addition to improving fuel economy, regenerative braking reduces wear and extends brake pad life, further reducing operating costs.

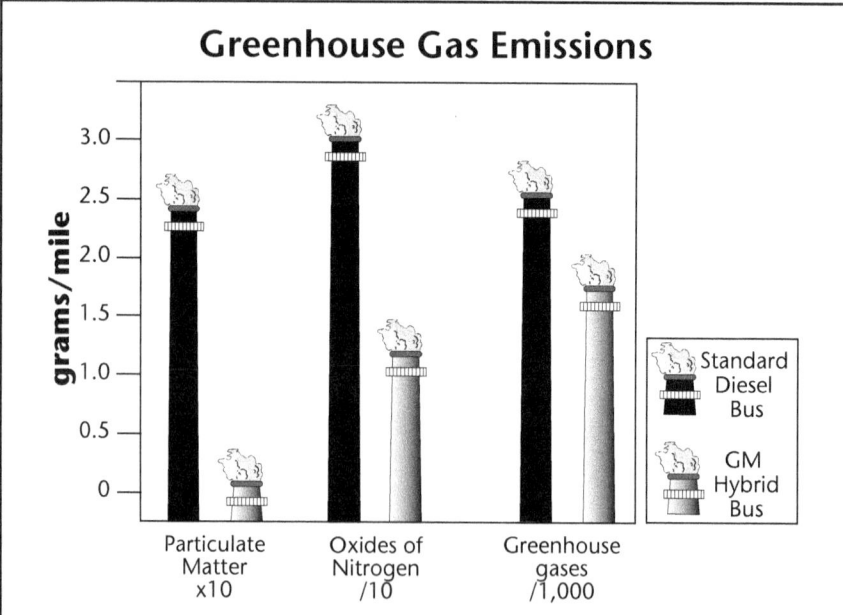

Figure 4-16. Impressive reductions in greenhouse gas emissions, particulate matter, and nitrogen oxides can be achieved with GM's new road-ready hybrid electric bus technology. Fuelling the bus with biodiesel would further improve these gains. (Courtesy: General Motors Corporation)

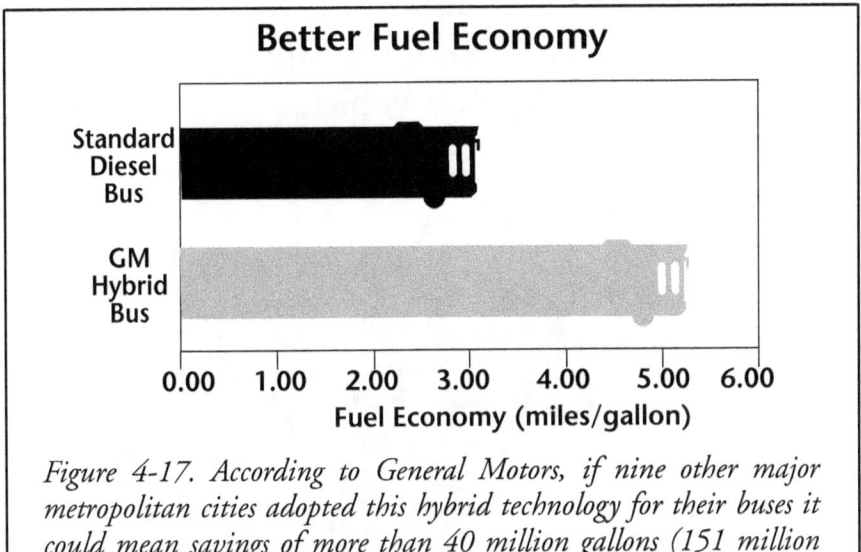

Figure 4-17. According to General Motors, if nine other major metropolitan cities adopted this hybrid technology for their buses it could mean savings of more than 40 million gallons (151 million liters) of fuel per year. (Courtesy: General Motors Corporation)

Ultra-Low Fuel Consumption Vehicles

I must start this section with a word of warning. No matter how excellent the fuel economy of a fossil-fuel-powered vehicle or how much greenery is placed in an automotive ad, there is no such thing as a green car, including all-electric and hydrogen models. My Honda Civic Hybrid is in the same environmental league as the Mercedes Smart™ car, the Toyota Prius or Volkswagen Golf diesel. All of these cars can get excellent fuel economy, although nowhere near what the "official" government figures proclaim. A person with a very light touch on the accelerator and who has no use for air conditioning (my wife for example) can get mileage figures as good as 70 miles per gallon (<4 L/100km) in a Golf diesel. I, on the other hand, can reduce that figure by 25% or more through what I shall politely term "different driving habits."

These numbers may seem illusory to gas-guzzling, SUV-loving Americans, but they are just average figures to western Europeans who have been saddled with higher fuel prices than their North American neighbours. Europeans seem to have discovered what is known as the rebound effect, adapting to high fuel prices by using the higher fuel economy of their vehicles to drive more frequently at a lower per-mile cost than Americans. However, in spite of high fuel prices, European vehicle

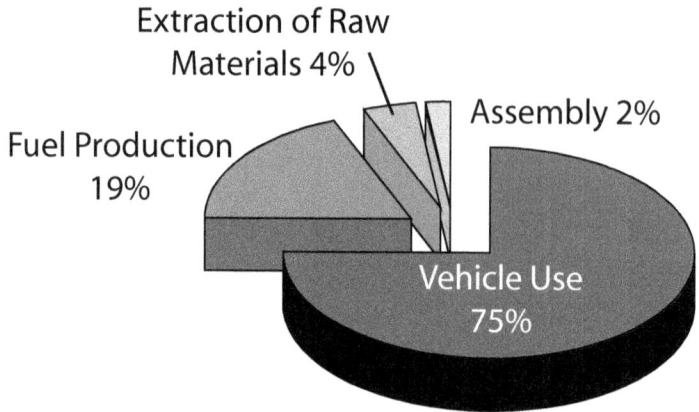

Figure 4-18. This chart[7] illustrates the life cycle carbon dioxide emissions of a typical vehicle with 75% of emissions attributed to vehicle use and fuel consumption and the remaining 25% to manufacturing and refining operations. The picture would change if per-vehicle infrastructure emissions for road and bridge building and maintenance could be properly ascribed. Infrastructure data is simply not considered as part of the total vehicular carbon dioxide emission story and is simply forgotten.

ownership continues to climb. More cars ply the same roads, driving up congestion and creating a demand for more road infrastructure that no government can truly afford, either economically or environmentally.

If every car in the world magically became an economical hybrid or subcompact diesel, fossil fuel usage would decline for a few years but would continue to climb as underdeveloped nations start to industrialize. Add in the embedded energy required in the extraction of natural resources and power to make the car, and you can see that no matter how energy efficient the vehicle you pick, it still has a huge impact on the environment (Figure 4-18).

The only way to make cars green is to stop building and using them, which will occur when carbon taxes rise to penalizing levels or when fuel rationing becomes a daily reality. Until we no longer have any choice in the matter, ultralow-fuel-consuming vehicles are one way of reduc-

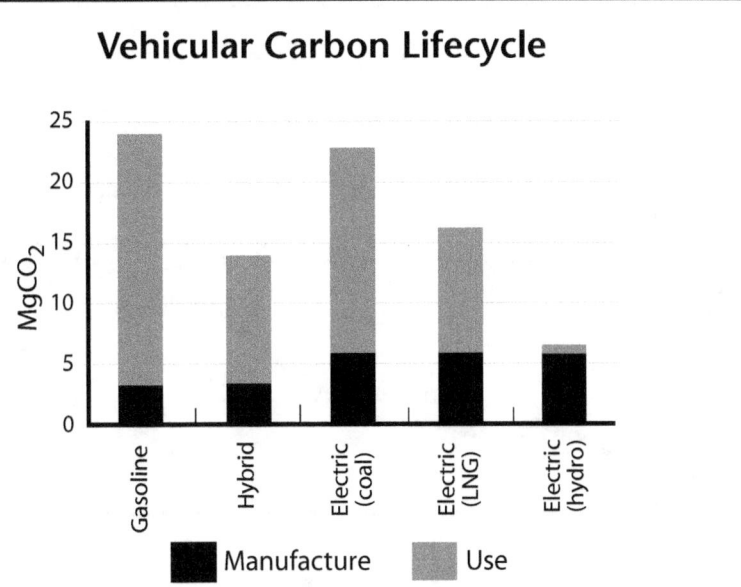

Figure 4-19. Vehicle manufacturers are quick to say how clean electric or hybrid cars are, but this can be very misleading. An all-electric car that derives its energy from a coal-fired power plant has almost the same emission profile as a standard gasoline-powered car. Hybrid cars are better than regular vehicles, but the difference in emission profile will be obliterated by increased vehicle usage over time. (Source: Institute for Lifecycle Environmental Assessment)

ing our personal transportation environmental footprint while saving money at the same time.

For years manufacturers have been offering cars in this automotive class with reduced weight, engine displacement, and vehicle size. One of the pioneers was the Volkswagen Beetle of the 1960s, which you either loved or hated depending on your point of view. The recently resurrected Beetle has been completely redesigned into an automobile that actually works. Coupled with excellent functionality, the gasoline or turbo-diesel versions of this car certainly offer excellent economics and fit the ultralow-fuel vehicle category.

The Smart™ family of automobiles manufactured by the Mercedes-Benz Group is another model that combines modern looks, safety, and economy all in one supercute package. These models use the newly developed CDI (common rail direct injection) diesel engine that greatly

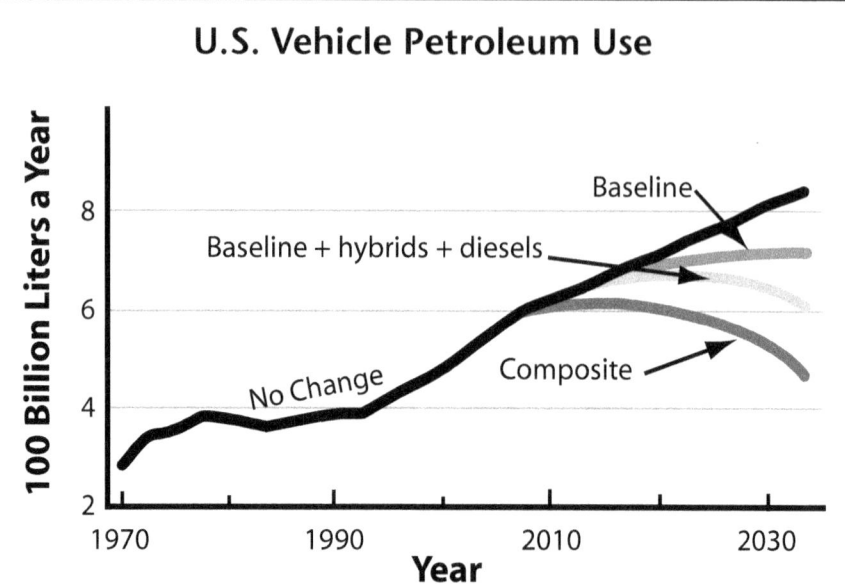

Figure 4-20. This graph is adapted from a report[8] that exemplifies fairytale thinking regarding vehicular energy use. "No change" assumes that fuel consumption remains stagnant, while "baseline" adds evolutionary developments in automotive technology, a difficult proposition to accept considering that manufacturers squander engine-design improvements on increased size, options, and acceleration. The "composite" line assumes a slowing in vehicle sales growth and distances travelled, which is completely at odds with all past data and current thinking. In short, barring any major policy or geopolitical changes, the only realistic trend is "no change," something governments and self-motivated industries are excellent at promoting.

reduces engine noise, smoke, and cold-starting issues. The thrifty little 3-cylinder engine, equipped with turbocharger and after-cooler, weighs in at a mere 152 pounds (69 kg), providing fuel economy of 69 miles per gallon (3.4 L/100km). This low fuel consumption means that CO_2 emissions are a reasonable 5.1 ounces per mile (90 g/km).

Are you concerned about safety? The Smart™ car features a "steel safety cell" which is reinforced with high-strength steel, making the vehicle one of the safest small cars available, according to the manufacturer. Tests conducted by Transport Canada, the country's vehicle safety testing agency, confirm that the car is safe, certifying it for use in

Figure 4-21. The Mercedes Smart™ car is super cute, very safe, and ultra fuel efficient. The 0.8 litre CDI direct injection diesel engine uses the latest engine technology, greatly reducing noise, smoke, and cold-starting issues. (Courtesy: Mercedes-Benz Canada)

Canada. (At the time of writing the Smart™ car is not available in the United States.)

In typical Mercedes fashion, the Smart™ car is equipped with numerous features to help prevent road accidents, including antilock braking, acceleration skid control, engine torque control, and multiple airbags. The car has rear-wheel drive with the engine located immediately above the rear axle, offering excellent weight distribution and good performance in wet, snowy, or icy conditions.

The problem with all of this supposed automotive greenery is that sales volumes of these fuel-friendly vehicles are so low as to have little if any direct impact on total world vehicular greenhouse gas emissions. Furthermore, although they consume little fuel compared with their peers, the sum of their lifetime fuel consumption and embedded energy results in greenhouse gas emissions that are quite significant. Additionally, sales of larger vehicles offset any advantage offered by the fuel-efficient models. Toyota may produce the ultragreen Prius to appease the

Figure 4-22 a, b, c. The Honda Civic Hybrid is an excellent vehicle for people who want safety, reliability, good looks, and good fuel economy. Unfortunately, when gas rationing or carbon taxation finally hits home, even this vehicle will become far too expensive to operate. (Courtesy: Honda Canada Inc.)

movie star and granola set, yet it has no problem offering a Sequoia SUV which is promoted as "Big. Comfortable. Powerful. And loaded with all the 'never quit' features your active lifestyle demands."⁹

It is a free country, almost a free world, and for the time being people can purchase almost any vehicle they want and, with the magic of "no money down" and perpetual leases, purchase vehicles they cannot afford. Governments continue to subsidize personal automobiles with endless road and related infrastructure development, saddling future generations with debts that will become due but remain unpaid as the economy struggles to deal with oil supply and price shocks. Government subsidies for "high-efficiency" vehicles such as hybrids are of little help. Although some people may switch to the subsidized models, there

Figure 4-23. Toyota may produce the ultragreen, energy-efficient Prius to appease the movie star and granola set, yet it has no problem offering a Sequoia SUV which is promoted as "Big. Comfortable. Powerful." Society is not really schizophrenic; it is simply caught at the junction of two paradigms, not sure whether to take the sustainable, sensible low-carbon path or risk it all with endless consumption. (Courtesy: Toyota Canada Inc.)

is no *disincentive* to driving low-efficiency models, which is where the problem started in the first place. The only correct solution is to establish punitive carbon taxes.

In any event, my contrarian views are not aligned with conventional thinking, and industry, media, and governments see no value in disturbing the status quo and delving honestly into this issue. They see no upside to establishing that statements like mine are correct and prefer to ignore the entire issue. As a result, society goes on enjoying its entitlement to personal freedom on the open roads, and all the while the carbon stopwatch continues ticking away.

Neighborhood Vehicles
In my vision of the carbon-constrained transportation system, there will be very little room for individual transport beyond city or town limits. Commuting within town will use much more advanced mass transit than the typical diesel bus trundling down the road, twenty minutes late. I truly believe that human ingenuity will develop a transit system that people will actually want to use. For those times when we cannot use inner-city mass transit, some form of low- or zero-carbon local commuting, beyond walking and the bicycle, must be available.

My mother, who is in her eighties and lives in a town of a few thousand, still drives for groceries, lunches with her friends, and runs local errands. Her little mid-1990s Chrysler Neon is the quintessential "little old lady driven to church on Sunday" car, using a tank of gas every month or so while accumulating almost no mileage. When she wants to travel out of town to plays, concerts, and the like, she either relies on others to do the driving or purchases tour packages from a local travel company that specialises in bus or rail trips to these various events.

For someone like her, and keep in mind that there were 9.3 million US households without a personal vehicle in 2003 (9% of all households), perhaps some form of low-cost, low-carbon, "neighborhood" vehicle may be suitable to augment intercity transit systems.

As the vehicles in figures 4-24 through 4-28 illustrate, electrically powered "zero- emission" vehicles are not only possible but commercially available now. What, then, happened to the mainstream electric automobile and why are there not more of them plying the roads? The electrically powered Impact (later EV1) car manufactured by General

Figure 4-24. The German-made TWIKE is a two-person, three-wheeled, electric/human-powered "car." Although uncommon in North America, this alternate vehicle has been accepted by Europeans for inner-city commuting. It can be powered using an onboard electric motor and integral battery, pedalled, or operated using a combination of the two. Typical range for the 500-lb (227-kg) TWIKE is 25 to 50 miles (40-80 km), with a maximum speed of 53 mph (85 kph). (Courtesy: www.twike.com)

Motors in the 1990s looked as if it might just make the grade. It had great styling and good performance, but GM argued that it was a commercial failure because it was too expensive and couldn't provide the operating range demanded by consumers, even though it could easily meet 90% of a homeowner's local commuting requirements (leaving intercity mass transit for the balance) and would have made an excellent neighborhood vehicle.

The real story behind the development of the EV1 provides a slightly different picture and offers an interesting footnote on the politics between the car industry and environmental regulators.

The California Air Resources Board required that 2% of all vehicles sold in the state were to be of a "zero-emission" designation. The EV1 met this requirement and just over 1,000 of the vehicles were leased prior to GM cancelling the program, even though it had many unfilled

Figure 4-25. There is no way my mother could fold herself into a Twike, but a Zenn (Zero-Emission, No Noise) car might be a viable option. This low-speed, all-electric neighborhood vehicle is limited to a maximum of 25 mph (40 kph) and meets all federal safety requirements. Currently about one-half of all US states allow these fully functional vehicles on the road. They offer no vehicular emissions and feature heating, air conditioning, and other amenities of standard automobiles. (Courtesy: ZENN Motor Company)

Figure 4-26. The General Motors EV1 was the all-electric vehicle that set the gold standard for functionality and range. GM also set the standard for upsetting their car-loving customers by recalling the vehicles from lease and destroying them at about the same time as California withdrew its Zero Emissions Vehicle Mandate. (Courtesy Brad Waddell from www.getmsm.com/ev/EV1/default.htm)

Figure 4-27. The GM EV1 all-electric car used numerous advanced technologies including 3-phase AC drive system, self-sealing and low rolling resistance tires, regenerative braking, and a programmable heating and air conditioning system based on heat pump technology. In many ways, the 1996-era vehicle is still one of the most advanced cars ever produced. (Courtesy Brad Waddell from www.getmsm.com/ ev/EV1/default.htm)

orders for the car waiting on the books and many loyal customers who wanted to purchase the vehicles outright. It is interesting to note that GM and other auto manufacturers sued the State of California because they felt that zero-emission vehicle requirements broke federal laws barring states from introducing legislation that effectively regulated fuel economy. California capitulated to pressure from the oil and auto industry and eliminated the zero-emission vehicle mandate, and GM did the same to the EV1, destroying all but a few of the cars. In essence, GM felt that it was prudent to supply only fossil-fuel-powered vehicles to the market, even though it was a world leader in electric vehicle technology at the time and could have easily improved its green image by selling thousands of the EV1 models. Perhaps if the issue of climate change had been more front and centre with the public, as it is now, things would have developed differently.

Figure 4-28. The Tango electric vehicle will no doubt turn heads, especially if you see George Clooney at the wheel of his. The narrow profile of the Tango makes it appear closer to a Honda Gold Wing motorcycle than a car, although the massive, low-slung weight of the battery bank ensures that the vehicle won't tip over. According to the manufacturer, the Tango has a working range of up to 80 miles (129 km), can accelerate from 0 to 60 mph (0 to 100 kph) in only four seconds, and can park almost anywhere. Sounds like the perfect vehicle for even today's urban traffic. (Courtesy: Commuter Cars)

Who Killed the Electric Car?, a documentary with cult status, theorized that the EV1 electric vehicle program was killed by of the combined efforts of the oil industry and GM itself, as the former would see fuel sales drop and the latter would lose valuable income from normal maintenance service and replacement parts such as oil filters and the like. The reality is probably closer to what GM states is the reason for ending the EV1 program. GM argues that car sales would have been limited to the warmer climates of the southwestern states as a result of battery technology issues and that sales in this area would not have been sufficient to support the vehicle in the long term. It also feels that battery replacement costs would have been unacceptably high.

Figure 4-29. This massive lead-acid battery bank weighs in at a whopping 3,800 pounds (1,724 kg) and contains approximately the same amount of usable energy as a gallon of gasoline, which weighs 7 pounds (3.2 kg). Once you begin to compare the energy density of gasoline to that of electricity and understand the difficulty of storing electrical power, you can see why the internal-combustion-powered automobile overtook electrics early in the 20th century. Advanced battery technologies can improve on this performance, and the ability to charge batteries with zero-carbon electricity at home is also a practical option for many commuters.

The exact truth may never be known, but General Motors made its decision and the EV1 is no longer, although other manufacturers and determined hobbyists have started to fill the void left by the demise of this pioneering vehicle. Neighborhood vehicles are an unusual product today, with only a handful of small niche suppliers such as ZENN Motor Company, but with fuel prices rising and carbon taxation on the horizon, people will gravitate to these products over time.

Intercity Personal Travel and Beyond

According to the United States Census Bureau,[10] US passenger miles of travel (pmt) increased by 27% between 1992 and 2002 to a total of 5 trillion miles (8 trillion kilometers), the equivalent of 17,000 miles (27,000 kilometers) for every man, woman, and child in the country.

Approximately 87% of pmt in 2002 was in a personal vehicle, with air travel absorbing a further 10% of the total. The remaining sliver of approximately 3% was divided between bus, intercity train, and motorcycle (Figure 4-30).[11] The change in passenger miles of travel by mode (Figure 4-31) shows that all forms of transit increased between 1992 and 2002 with the exception of Amtrak train travel. The rapid decline in Amtrak usage between 1992 and 1996 is partially accounted for by a reduction in total track mileage during that period. Although the change in pmt trend shows a modest increase in bus and transit use, the increases start from a very low base and account for less than 2% of total passenger miles of travel.

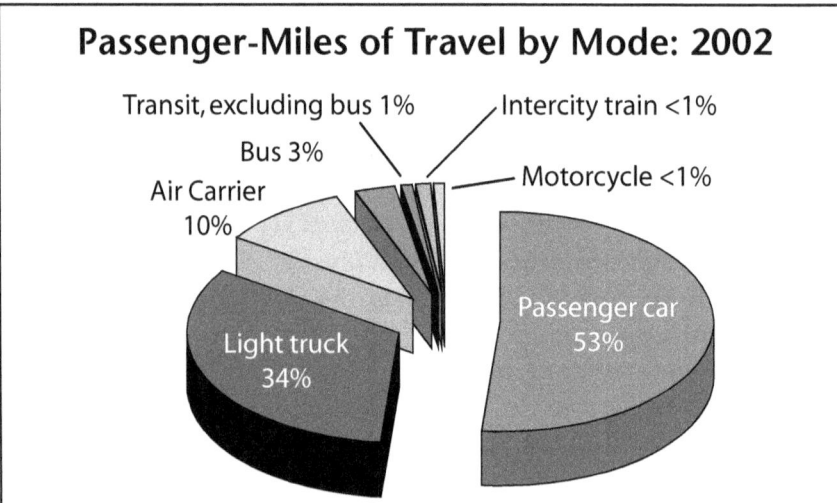

Figure 4-30. Approximately 87% of all passenger miles of travel in 2002 were in a personal vehicle, with air travel absorbing a further 10% of the total. The remaining sliver of approximately 3% was divided between bus, intercity train, and motorcycle. Americans travelled over 5 trillion miles (8 trillion kilometers) in 2002. (Source: US Department of Transportation, Transportation Statistics Annual Report, November 2005)

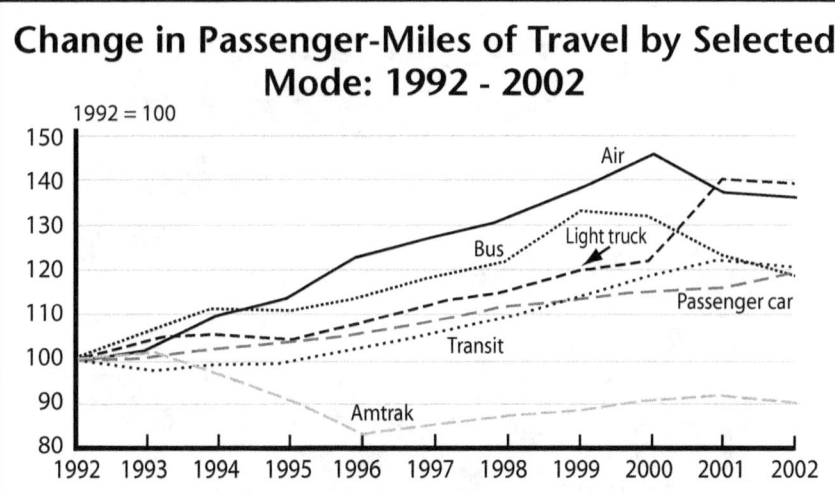

Figure 4-31. Between 1992 and 2002 Americans increased their usage of all transportation modes with the exception of Amtrak train travel. Although the trend shows modest increases in bus and transit use, they start from a very low base and account for less than 2% of total passenger miles of travel. (Source: US Department of Transportation, Transportation Statistics Annual Report, November 2005)

Because of the massive capital outlay on roads and related infrastructure in North America, intelligent planning and investment in intercity transit has been negligible at best, perhaps even negative in real terms. This can best be illustrated by the fact that rail and bus still account for less than 3% of total long-distance-trip miles, while personal vehicles account for 56% and air travel 41%[12]. It seems that Americans simply cannot shake their habit of using an automobile as their most favored means of moving about, regardless of the distance travelled.

Contrast this with Europe, where the combination of road congestion, crowded airports, flight delays, security hassles, and climate change awareness has increased rail ridership and resulted in massive developments in new train technologies, price competition, business models, and high-speed rail lines.

Europe has always been ahead of the curve regarding rail service, partially because of high population density and the relative proximity of cities. The argument has been made that since North Americans are more spread apart, rail service is too slow and other, faster means of

Figure 4-32. Contrast American rail usage with that of Europe, where the combination of road congestion, crowded airports, flight delays, security hassles, and climate change has increased ridership and resulted in massive developments in new train technologies, price competition, and high-speed rail lines.

travel prevail. However, as fuel costs increase and carbon emissions are constrained, flying may not be a viable option, with high-speed rail and "telepresence" technology providing the alternative. The argument that people must fly long distances is further weakened by the fact that 56% of Americans choose a personal automobile rather than flying to distant points[13], which indicates that high-speed rail could easily provide a large amount of long-distance travel.

Of course, if rail offers poor routing, high prices, late departures, insufficient departure selection, and inhospitable service, people will opt to fight traffic and drive or fly. Low rail ridership creates a negative response, as politicians do not want to support or improve an expensive rail system that few are using in the first place. This is especially true in North America, where people have a very hard time prying their hands from their steering wheels and refuse to be convinced that there are better ways to move about.

Here again, the simple economics of carbon taxation could well be the tool required to assist in the transition. As we learned earlier, a typical rail system reduces greenhouse gas emissions by a factor of eight times over the personal automobile, even more if low- or zero-carbon electricity is used to power the train. Obviously, the further one travels the greater the impact of the carbon taxes levied on the personal automobile, while the train could enjoy operating without any penalty at all.

I have always been a bit of a contrarian (which I suppose all environmentalists are), looking for alternative ways of doing things that do not negatively affect or might even improve my quality of life while being cost effective and responsible. Travel is one area which I investigate a great deal, and to my surprise I have found that I do not have to wait for carbon taxation policies to make my point.

As I write this section of the book, I have been retained to go to Toronto, some 250 miles (402 kilometers) away. I could drive this distance, but getting stuck in traffic gridlock and wasting several hours of my time is not a prospect I enjoy. Some suggest that driving is a cheap way to go, costing only $50 or so in gasoline. Of course this is simply not true, as one must consider the cost of ownership and wear and tear, which was 48.5 cents per mile (30 cents per kilometer) at the official US IRS rate as of December 31, 2005.[14] At this cost, a 250 mile (402 kilometer) roundtrip would cost you $242.50. If you do not believe it, check the figures yourself.[15] If you add in the nonproductivity of your driving time, the cost is higher still.

For the sum of C$487.96, I could do what almost everyone else does and purchase a last-minute roundtrip discounted airline ticket (adding in a further $31.80 for guaranteed seat selection so I do not get stuck with that damn middle seat) and save all sorts of time as the plane spews horrendous amounts of carbon dioxide into the atmosphere. Or I could think about it for a few moments and consider alternatives such as the train or bus. At the same day full-fare rate, I could take an economy rail seat and pay C$237.44 or step up to first class for C$377.36, with savings of $250.52 and $110.60 respectively compared to the discounted air fare and not have to worry about cramped seating.

Even though the first-class ticket is cheaper than air, the savings are greater than the simple arithmetical difference. Add the cost of taking a taxi to the airport or parking, approximately $20.00 per day. Rail service

Figure 4-33. Even though a first-class rail ticket is often cheaper than flying steerage class, the savings are greater than the simple mathematical difference. Rail service includes free parking and a meal on the way up and back, not to mention complimentary libations that are unheard of with discount airlines. The value of a coffee served in a real cup or a glass of grape to unwind after a stressful day might be reason enough for some folks to switch travel providers. The massive savings in greenhouse gas emissions are simply icing on the cake.

includes free parking, a decent meal on the way up and back, not to mention complimentary libations that are unheard of with discount airlines. The value of a glass of grape to unwind from a stressful day might be reason enough for some people to switch travel providers. The massive savings in greenhouse gas emissions are simply icing on the cake.

Then there are some of the nonmonetary issues that should be considered. Time is a big issue, with many people assuming that flying is faster. On this heavily travelled corridor, there are typically three flights per hour in each direction during rush hour. People assume a jet moving at several hundred miles per hour must be the quickest way. Not so; I routinely take the train on this same route and handily beat the plane once you factor in traffic congestion and travel to and from the airport, parking time, and queuing for tickets (and possibly checked baggage).

Add in more time for ever-increasing security checks, delayed flights, taxiing traffic, de-icing, circling while waiting for landing space, slow ground support, deplaning, and travel time from the airport. If you are truthful, you will find that rail wins hands down for all but the longest of trips.

In addition to these considerations, the train offers a scenic view and ample room to move about, and you can plug in your laptop and even chat on the phone if you like. You gain a few hours of useful time that can be used to work, sleep, socialize, or read, unlike the bits of low-quality time spent trundling through airports and on planes.

But, you say, there aren't as many trains as planes available and on long hauls the train will take too long. These are very good issues, but they completely miss the point. Poor selection and availability of train or quality bus routing is the result of underdevelopment of these systems over the last 75 years as the personal automobile and road systems have been favored above "freedom-restrictive rail." When cars and roads were first developed, oil was cheap and plentiful and no one had heard of smog, climate change, congestion, or carcinogens. Society bet on the car and all forms of mass transit lost out. In the early days of air travel, many of the same issues applied and once again society played the aircraft hand at the expense of high-energy-efficiency rail and coach.

Now that the proverbial pendulum is swinging in the opposite direction, society has an opportunity, indeed a desperate need, to reinvest in low- or no-carbon intercity rail and coach systems. Society will belatedly come to the conclusion that it cannot afford to build new roads or maintain existing ones, as we have seen earlier. It will take an enormous amount of both capital and political will to invest in rail and logistics systems, ensuring that people (and goods) can be moved when fossil fuel supplies are constrained in the not-too-distant future.

Europe is already on the move, with France showing the rest of the world how it should be done. In the mid 1970s France experienced a huge capacity problem with its rail system linking Paris through Lyon to the Mediterranean region of the French Riviera. Planners knew something had to be done and the thinking at the time was to quadruple the track along this route. Management at French National Railways suggested a bolder scheme: to develop a new-technology, high-speed rail link with aircraft-like speeds of 187 miles per hour (300 kilometers per hour). The outcome was the Train à Grande Vitesse or TGV that runs through the French countryside with few intermediate stations while providing excellent passenger comfort. The service was a huge success and dealt a permanent blow to airline carriers that operated between the cities on this run. This early success led to an expansion of the system and similar versions of the TGV in other European countries.

As national governments and rail operators capitalize on concerns about climate change, airport congestion, and delayed flights, high-speed rail services across Europe are working together to create linkages and partnerships covering routes from the Strait of Gibraltar at the southernmost tip of Spain all the way north to Stockholm, Sweden and west through the Channel Tunnel to London, Manchester, and Glasgow, plying routes that run to several thousands of miles. The London-to-Paris rail route is slightly over two hours, less time than it takes to drive to Heathrow or Gatwick and find your way out of Charles de Gaulle airport at the other end.

Current and future developments include magnetic-levitation trains that "fly" inches above the track using powerful magnetic fields, requiring no wheels and promising superior travel speed and comfort. Technologies such as the maglev Transrapid system developed by a Siemens-backed consortium will continue to advance rail technologies and compete with airlines for mid-distance travel and beyond.

But fast trains are only part of what makes a seamless, intercontinental rail system. A consortium of national railways from France, Germany, Belgium, the Netherlands, Austria, and Switzerland has created linkages with Eurostar and Thalys (both of which already provide cross-border service) in an alliance known as Railteam that competes directly with airlines. Over the coming months Railteam will provide common Web-based ticket and scheduling services anywhere its trains roll. A proper business model that duplicates and improves upon airline customer service and operating efficiencies as well as cooperative linkages with partner companies (for example co-marketing with bus or coach service) are mandatory for the successful deployment of an effective rail service.

Although there are numerous technical challenges to linking European rail systems, for example system voltage, the advantages described previously, coupled with vastly improved environmental performance and the ability to operate on carbon-free or low-carbon electricity, stack

Figure 4-34. Although there are numerous technical challenges to linking European rail systems, the obvious advantages of avoiding crowded airports and endless security checks coupled with vastly improved environmental performance and the ability to operate on carbon-free or low-carbon electricity stacks the odds in favor of a renaissance for intercity, high-speed electric rail service. (Courtesy: Sese Inglostadt, Wikipedia Commons)

the odds in favour of a renaissance for intercity, high-speed electric rail service.

In North America, there would be no technical problem connecting trains anywhere within the United States, Canada, and/or Mexico (most train systems in North America and Europe use standard 56.5-inch (1435-millimeter) gauge track, and power system voltages are easily harmonized), although low population density would require multimodal transit services. A seamless, multimodal transit system would allow low-density towns and villages to be serviced by coach and bus that connected with fast intercity rail hubs using a common ticket and scheduling program. A greater problem is convincing politicians of the need to build these systems and rail infrastructure when current ridership growth trends do not appear to warrant any additional capital spending—a typical chicken-and-egg problem (Figure 4-31).

Flying and Marine Transport

As for flying, the simple answer is it will have to be curtailed. People and goods may still fly, but the financial and environmental penalties associated with flying will be too great for yearly family trips to exotic locations such as Disneyland or Kuala Lumpur. If flying is the only way to a given locale, it may be necessary to go once in a lifetime and stay for an extended period to get the most value from the trip. The likelihood of any ultra-high-speed alternative to carbon-fuel-based flying is pretty remote, although slower-speed dirigible-style airships, which are somewhat more efficient than traditional aircraft, might be a possibility. A more likely alternative would be a new form of marine transportation.

George Monbiot, author of the book *Heat*, suggests that using a ship like the *QEII* to transport 1,790 people at full capacity across the oceans would produce almost 7.6 times as much carbon and resulting greenhouse gases (9.1 tonnes per person) as making the same journey by plane.[16] While this data is no doubt correct, comparing a ship with 920 crew, 7 restaurants, 7 lounges, and a shopping mall with a cramped airplane is to my way of thinking a bit out of order. If people must travel great distances in a carbon-constrained world, there is no reason why marine transport can't be made more efficient. The nearly quarter-mile-long (1,302 foot/397 meter) container ship *Emma Maersk* is one of the world's largest marine vessels and fully loaded weighs in at nearly 2.5 times the mass of the *QEII*.[17] The ship is capable of delivering a ton of cargo 132 times more efficiently than a jumbo jet aircraft.[18] Granted, people are not cargo (or at least do no consider themselves so) and most would not appreciate being stacked in cargo containers like so many mannequins, but better, more humane methods are certainly possible. The marine industry, like many energy-intensive companies in a cheap-energy world, has not worried about fuel costs. As carbon fuels become constrained or taxed, the industry will not wither away; it will innovate and develop alternative technologies. A future passenger transit vessel will not look like the *QEII*; it will be designed for transporting people across oceans in the most environmentally friendly and cost-effective way possible in order to keep ticket prices within reason. (For example, if a carbon tax of $500 per tonne were levied on each passenger of the *QEII*, an additional $4,550 would have to be added to each return ticket.[19]) A passenger transit vessel will dispense with the branch of Harrods and staterooms with butler service and will concentrate on getting you from point A to point B while using the least amount of energy to do it. The ship design and technology already exist. The *Emma Maersk* is equipped with a new environmentally friendly paint that creates a streamlining effect which saves an estimated 1,200 tonnes of bunker fuel per year.[20] Optimized cabins and common areas can easily be designed for business travel, including meeting rooms, telepresence technology, satellite links, cell phones, and computer work stations. Other areas can be fitted to accommodate the travelling family. If engineers can duplicate the same energy efficiency with a marine transit vessel as they have with the *Emma Maersk*, it is quite likely that shipping people by sea will be as efficient as the Train à Grande Vitesse on dry land. As an added bonus, jet lag and red-eye flights will become a thing of the past.

Although it might be heresy to suggest it in an environmental book, it is also possible (indeed likely) that the ship manufacturing sector could take a cue from the military submarine industry and build passenger transit vessels that are nuclear powered. One of the primary disadvantages of nuclear-powered submarines is the large size and mass of the reactor, a problem which disappears with large surface vessels. Properly scaled and applied, it makes just as much sense to use an "emission-free" nuclear reactor on a ship as it does to use one on land. At the same time, all of the arguments for *not* using nuclear power on land are fully applicable at sea, not to mention the additional complications of constantly moving nuclear fuel, a vessel sinking with an operating reactor, and spent reactor fuel handling.

It is very important to note that the capital markets will use the cheapest solutions for a given problem and not the more commonly used solutions, including bunker oil and diesel fuel that people seem to prefer. Given the choice between emission-free (and therefore carbon-tax-free) nuclear-powered transit and using expensive, carbon-based fuels, the markets will choose nuclear power hands down.

It may come as a surprise to many, but over 150 nuclear-powered submarines and surface ships, including civilian vessels, ply the world's waters.[21] Between 1950 and 2003, Russia built 248 submarines and 5 naval surface ships powered by nuclear reactors. The US Navy currently has 11 nuclear-powered aircraft carriers and more than 80 nuclear-powered ships that have logged over 5,500 reactor years of accident-free experience.[22]

Russia has found nuclear-powered surface vessels ideal for operation in the Arctic, where refuelling and power levels required for icebreaking are significant factors. The 61,900-tonne NS *Sevmorput* was put into service in 1988 and serves as a good example of scale, being only 12% smaller than the *QEII*. A ship of this size could easily form the technical model for a passenger transit ship.

The German-built 15,000-tonne nuclear-powered *Otto Hahn* cargo ship sailed some 650,000 nautical miles over a ten-year period without any technical problems, although it was considered to be too expensive to operate and was converted to diesel in 1982.[23] With carbon taxation polices or unreliable fossil-fuel supplies, the reverse might have been true and the vessel might still be operating on nuclear power.

Logistics and the Movement of Goods

While rail is a bit player in the North American movement of people, it is the fastest-growing and largest overall mode of US domestic freight transit. The amount of goods transported within the continental United States is staggering, totalling 4.4 trillion ton-miles (6.42 tonne-kilometers) in 2002, up 18% from 1992.[24]

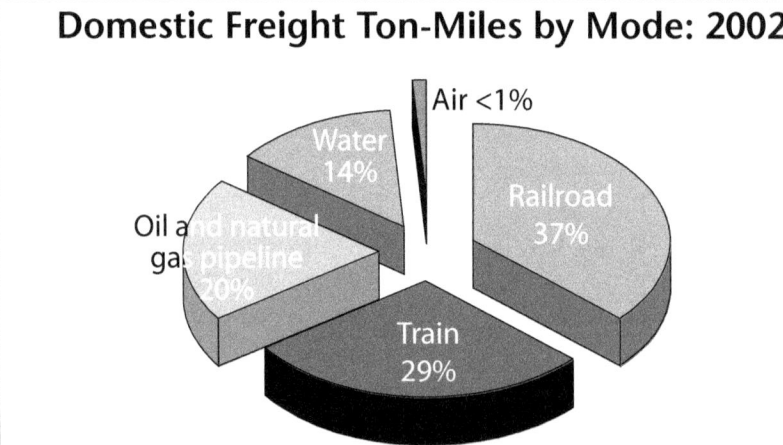

Figure 4-35. According to US Department of Transportation figures for 2002, rail accounted for the largest share of domestic freight-ton miles with 37% of the total. Wholesale and bulk truck transportation took 29% and air accounted for less than 1%, although it still totalled 13.6 billion ton-miles (19.8 tonne-kilometers).

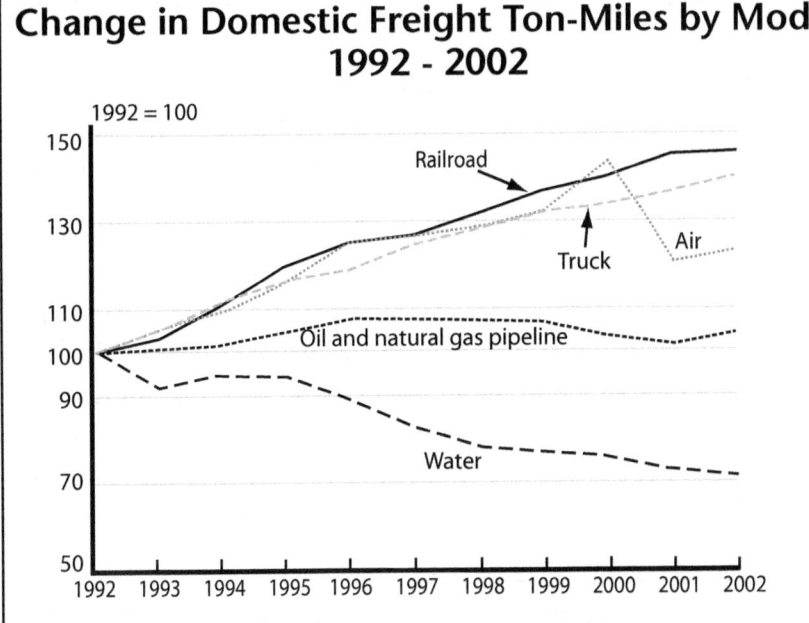

Figure 4-36. *As the domestic economy continues to grow and consumption of "everything tangible" increases, rail ton-miles have increased the fastest (46%), while trucking and air followed with 40% and 29% growth rates respectively between 1992 and 2002. (Source: US Department of Transport, Bureau of Transportation Statistics)*

In 2002, rail accounted for the largest share of domestic freight-ton miles with 37% of the total. Wholesale and bulk truck transportation took 29% and air accounted for less than 1%, although it still accounted for 13.6 billion ton-miles (19.8 tonne-kilometers).(Oil and gas piplines and marine freight transportation made up the balance).[25] This massive amount of freight tends to be averaged over a large number of high-value, low-weight goods shipped by express courier and mail services.

As the US economy continues to grow and consumption of "everything tangible" increases, rail ton-miles have increased the fastest (46%), while trucking and air followed with 40% and 29% growth rates respectively between 1992 and 2002.[26]

Likewise, the number of full rail containers entering the United States from Canada and Mexico increased approximately 350% and 115% respectively between 1994 and 2004[27]. Trains carried approximately 1.5 million and 300,000 full containers from Canada and Mexico respectively in 2004[28].

During the same period, the number of trucks entering the United States from Canada and Mexico increased from 7.7 million in 1994 to 11.4 million in 2004, which accounted for US$453 billion in trade (64% of total trade) in that year[29].

Anyone who has spent any amount of time commuting across North American highways knows that heavy truck transportation takes up an enormous physical share of the roadway. It contributes 96% of total "loading" (Figure 1-23), a disproportionate share of total roadway vehicle volume, causing a vast amount of damage to the road surface and structures.. This leaves cars, SUVs, and light-duty trucks to account for only 4% of roadway loading. If truck transportation could be curtailed, roads would last significantly longer and maintenance costs would drop dramatically.

Figure 4-37. Today it is possible to order just about anything from anywhere and expect it to be delivered within 48 hours. The worldwide courier FedEx Express takes this concept quite seriously, providing air-express shipment of thousands of cases of Beaujolais Nouveau wine from France to Tokyo, Japan, to ensure that the annual uncorking takes place right on time, at midnight on the third Thursday of November. (Courtesy: FedEx)

Naturally enough, current freight transportation systems have been modelled on the concept of cheap and accessible fuel. There is no possible way that our skewed transportation ideals should allow the consumption of Chilean grapes in winter or New Zealand lamb at any time. It simply isn't logical to purchase food or goods from another country that are made available at lower cost because of a transportation system that does not include full-cost accounting which includes the value of nonrenewable-fuel depletion and damage to the environment.

A sustainable freight transportation system still requires the use of roads, rail, aircraft, and shipping, but the manner in which each of these modes of transportation are used varies as a function of energy efficiency and carbon consumption, not as a matter of convenience.

Today, it is possible to order just about anything from anywhere and expect it to be delivered within 48 hours. The worldwide courier FedEx Express takes this concept quite seriously, handling millions of "urgent" packages daily, including the air-expressing of thousands of cases of Beaujolais Nouveau wine from France to Tokyo, Japan, to ensure that the annual uncorking takes place right on time, at midnight on the third Thursday of November.[30] Perhaps I am being a bit unsympathetic, but isn't this an excellent example of stupidity and entitlement, with consumers choosing to burn massive amounts of carbon-based fuels, depleting energy and petrochemical resources that need to be shared with future generations, just to sip a glass of fermented grape juice?

The fact is that there are no regulations against this self-centered behavior, so it continues without anyone blinking an eye.

Over time, questionable shipping practices, especially overnight air delivery, will either become a thing of the past or be so expensive that few companies or people will be able to afford the service. The shipment of urgent drugs or relief supplies are worthy reasons to continue providing fast courier services, but surely people in Tokyo can drink local wine or wait a few weeks for more energy-efficient shipping means to bring them the imported variety.

Of course, shipping goods from one airport to another is only part of a total transportation chain, and a look at the inner workings of FedEx provides a few clues as to how a sustainable transportation and logistics system might be constructed.

Figure 4-38. A look at the inner workings of FedEx provides a few clues as to how a sustainable transportation and logistics system might be constructed. Although FedEx specializes in small packages and light freight, its 547,000 employees handle over 13 million shipments per day to over 220 countries. Implementing an efficient, well-planned business model, designing vast computational and freight logistics systems, and making a substantial investment in infrastructure create a market opportunity and US$29 billion in sales. This same model could be retooled for a carbon-constrained transport system. (Courtesy: FedEx)

According to FedEx, the corporation comprises a number of network companies that specialize in the transportation of goods to over 220 countries and territories. The $29 billion corporation has 547,000 employees, handles approximately 13 million shipments per day, owns 669 aircraft and 109,000 motorized vehicles, and services more than 375 airports worldwide. This is a massive company and one that has fine-tuned its business model around one function: the rapid delivery of small packages and less-than-a-truckload amounts of freight. The company has also fostered a culture of excellence, as evidenced by their receiving for the ninth year in a row the honour of being one of the "100 Best Companies to Work For."

To become this size and be so successful requires an efficient, well-planned business model, vast computational and freight logistics systems, and an investment in infrastructure to ensure that the program works. In short, FedEx has developed a system that can and should be duplicated for moving all freight in a carbon-constrained world, not just the specialized small packages they excel at delivering.

To put this improved carbon-constrained transport system together would require an expansion of the existing FedEx business model to include larger freight shipments. Alternatively, a partnership or amalgamation could be formed with a company with a similar business model that deals in large freight. The A.P. Moller-Maersk Group (MMG) of companies might be a good choice.

Figure 4-39. To put a carbon-constrained transport system together, a small-freight company like FedEx would require an expansion of its existing business model to include larger freight shipments. Alternatively, a partnership or amalgamation could be formed with a company with a similar business model that deals in large freight. The A.P. Moller-Maersk Group is a worldwide organization with about 110,000 employees based in Copenhagen, Denmark that owns one of the world's largest multimodal shipping fleets. The company has a market capitalization of US$51.7 billion (June, 2007) and 2006 sales of US$49.3 billion. (Courtesy: A.P. Moller-Maersk Group)

Figure 4-40. A.P. Moller-Maersk Group has a vast fleet of marine container ships including the Emma Maersk, which is the world's largest container vessel, able to carry over 11,000 20-foot containers which, if they were placed on a single train, would measure over 44 miles (71 kilometers) long. (Courtesy: A.P. Moller-Maersk Group)

MMG is a worldwide organization with about 110,000 employees based in Copenhagen, Denmark that owns one of the world's largest multimodal shipping fleets. The company has a market capitalization of US$51.7 billion (June, 2007) and 2006 sales of US$49.3 billion. The firm has a vast fleet of marine container ships including the *Emma Maersk*, which is the world's largest container vessel, able to carry over 11,000 20-foot containers which, if they were placed on a single train, would measure over 44 miles (71 kilometers) long[31].

MMG moves freight almost everywhere on the planet, with a ship making a stop in China, Taiwan, or Hong Kong every 1.9 hours. It currently operates 1.9 million standard containers and over 200,000 refrigerated units for perishables[32]. Where FedEx moves light freight, quickly, MMG moves the big stuff, slowly. What both companies have in common is an ability to operate logistical systems in many countries, understand local customs and laws, and do their jobs effectively while remaining profitable. They also have a commonality in that neither company currently worries too much about carbon emissions.

Figure 4-41. A.P. Moller-Maersk Group moves freight almost everywhere on the planet, with a ship making a stop in China, Taiwan, or Hong Kong every 1.9 hours. It currently operates 1.9 million standard containers and over 200,000 refrigerated units for perishables. (Courtesy: A.P. Moller-Maersk Group)

A Sustainable Freight System

Times are changing, and with rising fuel costs both firms are starting to look at transportation energy efficiency as a way of improving the bottom line. With the direct link between fossil-fuel energy consumption and carbon emissions, any carbon taxation policies would directly impact profitability. Obviously energy efficiency would improve the public's environmental perception of either company as well as control costs or possibly increase profits. FedEx has already introduced hybrid vehicles into its fleet, while MMG has implemented heat recovery systems and streamlined paints for its shipping armada. These are small steps, but steps in the right direction.

Unfortunately, transportation (and embedded fuel) costs are a relatively small share of the value of many manufactured goods. A 40-foot intermodal shipping container can hold hundreds or perhaps thousands

Figure 4-42. Times are changing, and with rising fuel costs firms such as FedEx and MMG are starting to look at transportation energy efficiency as a way of improving the bottom line. With the direct link between fossil-fuel energy consumption and carbon emissions, any carbon taxation policies would directly impact profitability. Obviously increased corporate energy efficiency would improve the public's environmental perception of either company and control costs or increase profits at the same time. FedEx has already introduced hybrid vehicles into its fleet, while MMG has implemented heat recovery systems and streamlined paints for its shipping fleet. These are small steps, but steps in the right direction. (Courtesy: FedEx)

of high-value electronic or other goods, thus averaging the cost of transportation over many units. At current energy prices, there is little to dissuade businesses or retail consumers from buying things made miles away.

With a punitive carbon taxation program, things would change; air freight would be used only for the most urgent items or, ironically, for those people who could still afford to snub their noses at their environmental responsibilities while continuing to order Beaujolais by overnight delivery.

For the balance of society, the cost of transported goods would reflect most of the environmental damage caused by transportation, giving consumers a true incentive to purchase alternative (perhaps locally produced) items or face the music and pay the fully accounted cost. Of course firms like FedEx and MMG might not survive unless they could adapt to a carbon-constrained business model. Assuming that governments didn't try to dictate how this transition was effected and simply adhered to policy (lowering carbon emissions through punitive carbon taxation), the capitalist markets would adapt as they always do. Consumers might not like it, as they would end up footing the bill, but I for one will be happy when people recognize the real cost of their acts of consumption.

I see firms like FedEx, MMG, and others forming large (shall I say "Grand"?) alliances and business partnerships in the same way that the members of "Railteam" joined forces to create a European-wide consortium to challenge the airlines for business. As freight transportation companies retool, the outcome might fall along these lines:

Local Transport

FedEx and its competitors would alter their business model very slightly at the local or "inner-city" level. Small packages would be picked up by foot, bicycle, and truck couriers as is currently the case. The main difference would be that any transportation at this level would involve low- or zero-carbon fuels, including electricity. The current diesel or gasoline/electric hybrid car or truck of today would be far too inefficient since the carbon taxation on fossil fuel and/or its short supply would increase operating costs. Intelligent competitors would quickly leverage this disadvantage by switching their fleets to lower-cost, low-carbon alternatives.

Pickups would be sorted at the local level and redistributed for delivery to the local client, again through low-carbon means. Shipments destined for distant cities would be aggregated and placed in radio-coded intermodal containers for rail shipment. Likewise, large freight shipments would be aggregated, placed inside intermodal containers, and shipped directly to the rail terminal facility by local truck.

Intercity Transport

At the rail sorting facility, each intermodal container would be scanned and placed on an unmanned, robotic carrier and directed by computer to a loading area, where it would be placed on the desired rail route to the distant city by an automatic crane and gantry system.

Small shipment containers that arrived at the destination city would be opened and their contents sorted and delivered at the local level by courier. Freight containers that arrived at their destination city would be delivered directly to the receiving party via low-carbon trucking.

The key element of the system would be to ensure that trucks and other low-energy-efficiency transportation means were used only at the local level, allowing the trains to do the heavy hauling between cities. Only oversized or specially permitted freight would be allowed to use the nation's highway infrastructure.

Figure 4-43. In a carbon-constrained shipping system, intermodal containers would be uniquely tagged with a radio-coding device, placed on an unmanned, robotic carrier, and directed by computer to a rail car loading area, where the container would be placed on the train by an automatic crane and gantry system. Likewise, large-freight shipments would be aggregated, placed inside intermodal containers where possible, and shipped directly to the rail terminal facility by local (low-carbon-fuelled) truck. If this sounds a bit like science fiction, consider that fully 60% of all US exports to Europe pass through Rotterdam and the impressively automated European Combined Terminals (ECT) using exactly this same technology. (Courtesy: ECT)

Figure 4-44. The ECT is located on the North Sea port of Rotterdam and combines terminal and logistics facilities for sea, river, air, rail, and road services at one location. The Emma Maersk container ship is a frequent customer, delivering up to 11,000 TEUs to the docking facilities. While the ship is still at sea, computer systems onboard transmit freight manifest data to a control system at the ECT. There, computers calculate how to unload the ship as quickly and economically as possible, bearing in mind that each of the 11,000 TEUs will fan out to different locations and upstream modes of transportation. (Courtesy: ECT)

If this sounds a bit like science fiction, consider for a moment that fully 60% of all US exports to Europe pass through Rotterdam and the impressively automated European Combined Terminals (ECT).[33]

The ECT is located on the North Sea port of Rotterdam and combines terminal and logistics facilities for sea, river, air, rail, and road services at one location. The *Emma Maersk* container ship (Figure 4-40) is a frequent visitor, delivering up to 11,000 TEUs to the docking facilities. (A TEU is a twenty-foot equivalent unit. A single 40-foot intermodal container is equal to two TEUs.) While the ship is still at sea, computer systems onboard transmit freight manifest data, including the location of each container, to a control system at the ECT. There, computers immediately start a virtual mathematical dance calculating how to unload the ship as quickly and economically as possible, bearing in mind that each of the loaded TEUs will fan out to different locations and upstream modes of transportation—a daunting exercise to be sure.

When the ship arrives, 83 stacking cranes remove the containers and a fleet of 138 robotic trucks automatically moves them from the ship to stacks on the loading docks. All of these machines are in constant motion, following the unseen electronic maestro's hand located somewhere in the computing system. Coordinating the safe and efficient operation of all this equipment is paramount in meeting customer requirements. ECT says that cargo delivery can be made within 30 minutes of unloading and that client deliveries to ships are possible even during loading operations.

In the proposed land-based system, all freight would be transported between cities in radio-tracked, secure shipping containers which would be passed from one area to the next until they reached their intended destinations. With a suitably designed infrastructure system using many of the advanced logistical, GPS, and automation technologies currently used by FedEx, A.P. Moller-Maersk Group, and European Combined Terminals, it would be possible to remove most truck traffic from the highways and improve delivery of goods while reducing the carbon footprint of the transportation system to near zero.

In order to do this, society would have to massively increase the amount of rail infrastructure and rolling stock to ensure that trains moved between cities as quickly as necessary to meet or better intercity trucking times. The 401 highway running between Windsor, Ontario (just opposite Detroit, Michigan), through Toronto and on towards Ottawa and Montreal is one of the most heavily travelled highways in North America. Truck transportation has risen dramatically on this route, with the number of trucks now appearing to equal that of passenger cars. A high-speed electric rail system that ran in parallel or even replaced some of this highway would allow trains to run back and forth between these cities on a continual basis through the development of a massive "ring line" whereby multiple trains would operate on the line with automated loading and unloading at each city's hub.

The reduction in fuel consumption, carbon emissions, road damage due to truck loading, and congestion would be staggering. As a side benefit, the same rapid rail line could also be used for direct city-to-city passenger coaches that could easily be added to the freight trains as desired.

Marine Transportation

As intermodal containers reached a maritime hub, they would be aggregated in the same manner as is carried out now by ECT in Rotterdam. Containers would be loaded on ships similar to the *Emma Maersk* and delivered to distant receiving ports. One would assume that a ship like the *Emma Maersk* would be very inefficient, but according to her owners the ship travels 41 miles (66 kilometers) for every 1 kilowatt-hour of energy per ton of cargo. As I noted above, a jumbo jet travels only one-third of a mile (500 meters) using the same amount of energy per ton of cargo, making the *Emma Maersk* 132 times more energy efficient. Carbon taxation ratios on freight would follow a similar pattern for both modes of transportation.

However, although a ship like the *Emma Maersk* is much more energy efficient than an aircraft, she still requires massive amounts of bunker fuel and produces equally massive carbon emissions in moving freight from port to port. Although she may be the pinnacle of modern fossil-fuelled marine transportation technology, she is by no means the end of the line for carbon-efficiency improvements. The *Emma Maersk* is one of the largest ships on the planet, requiring a 14-cylinder, 109,000-horsepower (81.3-megawatt) engine, the equivalent of over one thousand family cars. Most industrialized countries have the ability to produce zero-carbon-emission nuclear power reactors of the capacity needed to power a ship like the *Emma Maersk*. Although the public may not be enamoured with the concept, the fact of the matter is that many of the vessels currently in or on the seas are nuclear powered and that our choices for moving goods will be limited in coming years. Nuclear will be part of the equation, so you had better get used to the idea.

Summary

As society moves towards a carbon-constrained world, the movement of people and goods will start to change and many of the personal freedoms we take for granted today will evaporate. People will have to learn to stay where they are planted and live within the local community.

Commuting to work will be a thing of the past unless you are prepared to take rail or coach to distant points. Frequent flights to here and there will be curtailed.

If an immediate imposition of a punitive carbon tax on gasoline, marine bunker fuel, and aviation fuel were implemented, people would demand access to better, lower-cost mass transit and freight transportation options. Oil supply disruptions, repeated terrorist attacks including the use of nuclear materials as a warning to the West to leave the Middle East, or the flooding of Miami or New York City due to rising sea levels might also do the trick, although once these sorts of nightmare scenarios develop the economy might not be able to support any major infrastructure development.

The capitalist economy will continue, and where old transport systems fail new ones will prevail. Rail and bus systems do not need to be left to rot and rust through underuse. With proper competition, improved government policy on carbon, and increased public demand, these systems can and will be enjoyable to use.

It is very hard for people to see this future from the vantage point of a world awash in money, oil, and gasoline. But fossil fuel supplies are finite, geopolitics will play an increasingly important role, and the atmosphere can only ingest so much carbon before we all perish. I just hope that society figures this out before we reach the breaking point and there is no turning back. Even if there is a road to return on, there won't be the fuel to get us there.

5
The Personal Transportation Appliance

Summarizing the Case for the Zero-Carbon Car

The converging issues of climate change, peak oil (resource limits), geopolitical uncertainties, and rising infrastructure costs are conspiring to create a new world reality, the Paradigm of Reality Convergence, where the personal automobile will no longer reign supreme. Recognition of that reality is not here today. Indeed, it may not arrive for a number of decades, when society has used the last vestiges of carbon fuels and financial capital in an effort to hang on to the entitlements we currently take for granted.

The manner in which the transition is made between today and that point in the future when personal vehicles are purged from society will determine how prosperous we will be in the "after-car" years. There is a very high risk that modern industrial society will fail because of the turmoil caused by this transition. Amory Lovins, scientist, environmental thinker, and head of the Rocky Mountain Institute argues that "the global automotive industry is arguably the largest and most complex undertaking in industrial history." There is no question that the economies of most nations have devoted an unprecedented amount of effort and financial aid to the direct support of the jobs and spinoff companies

Figure 5-1. The converging issues of climate change, peak oil (resource limits), geopolitical uncertainties, and rising infrastructure costs are conspiring to create a new world reality, the Paradigm of Reality Convergence, where the personal automobile will no longer reign supreme.

and services that relate to the modern automobile. Support is given for automotive manufacturing, roads, and infrastructure as well as for the production of gasoline and diesel fuel through exploration, refining, and retail distribution. On top of this, we can add the support industries of insurance, service, car rentals, and after-market sales.

For any politician to stand up and condemn or attempt to change this tidy arrangement would be tantamount to political suicide, even if the person genuinely believed that things needed to change. Naturally enough, through a combination of self-interest and the recognition that the oil, gas, and automotive industries are (currently) an unstoppable combination, even the greenest of politicians will do nothing more than tinker with policies that will have no net effect on the many converging issues facing society.

Figure 5-2. The economies of most nations have devoted an unprecedented amount of effort and financial aid to the direct support of the jobs and spinoff companies and services that relate to the modern automobile. Automotive manufacturing, roads, and infrastructure are obvious examples, as is the production of gasoline and diesel fuel through exploration, refining, and retail distribution.

When politicians offer rebates for hybrid cars, tax credits for mass transit, and fuel-efficiency targets for automobiles while exempting SUVs and pickup trucks, or try to grow their way out of imported oil demand by using food-based biofuels, it is clear that they are not taking the problem seriously; the status quo is the political equivalent to easy street. Although the theme of this book is transportation, we must not forget about associated problems. The average size of a North American house has doubled over the last few decades, while family size has decreased by half. Yet homeowners and the construction industry lobby against increased housing energy efficiency, a surcharge on McMansion-sized houses, or the termination of suburban or exurban development which is possible only because of cheap fossil fuels.

Yet society knows deep down that we consume far too many resources and fossil fuels, and some might understand that automobiles have abysmally poor energy-efficiency ratings. The typical automobile uses only 1% of the fuel energy it consumes in actually moving the driver; a vehicle that weighs 2,500 pounds with a 15% energy conversion rate moves a 160-pound driver at less than 1% total conversion efficiency. The numbers are much worse when calculated with heavier vehicles such as the General Motors Hummer.

Something has to give. Public policy demands better fuel efficiency, reduced vehicular traffic density, a movement away from truck-based freight transportation, conservation of all natural capital including raw materials and fossil fuels, and of course a reduction in the effects of climate change.

The previous chapters have attempted to provide an overview of these issues, including the need for a full life cycle carbon taxation policy. Assuming that society started to heed these warnings and did implement a punitive taxation on carbon, the automotive industry would begin to innovate and reshape itself to cope and possibly prosper while dealing with the converging realities.

Figure 5-3. Society knows that we consume far too many resources and fossil fuels, and some people might even understand that automobiles have abysmally poor energy efficiency ratings. The typical automobile uses only 1% of the fuel energy it consumes in actually moving the driver. The numbers are much worse when calculated with heavier vehicles such as the General Motors Hummer.

Figure 5-4. Public policy demands better fuel efficiency, reduced vehicular traffic density, a movement away from truck-based freight transportation, conservation of all natural capital including raw materials and fossil fuels, and of course a reduction in the effects of climate change.

Personal Transportation Technologies and Psychology

In my estimation, personal automobiles will not last much longer than a few decades, even with major environmental advances in automotive technology. While the basic technology of the car has not changed significantly since its inception, there are a number of changes on the horizon that could ensure it lasts until mid-century while improving its hideous environmental footprint. What is required to make this happen are sound worldwide carbon valuation policies and a complete erasing of existing automotive design, with a movement instead to a radical change in technology and usage psychology.

In the West, automotive manufacturers are facing slowing market growth and a public that is becoming increasingly anxious about climate change as well as fuel cost. New sales growth must come from two sources: development of "clean car" policies at home which will foster

Figure 5-5. Assuming society started to heed these warnings and did implement a punitive taxation on carbon, the automotive industry would begin to innovate and reshape itself to cope with the multiple, converging realities.

the growth of domestic transitional technology; and the development of emerging markets in Asia, South America, and Russia through the forging of strategic sales partnerships and joint product development and technology licensing agreements.

Although Western auto manufacturers have no choice but to expand into foreign markets, over time these "partner companies" will convert into sizable competitors as their manufacturing and technology development skills improve. One obvious method of staying ahead of these foreign partner/competitors is to innovate and continue raising the technical superiority bar (something the West excels at) by supporting higher fuel efficiency and carbon emission standards as well as the development of advanced vehicular drive train and chassis technologies—in short, by supporting the development of so-called "green automotive" technologies.

Instead, Western automotive manufacturers are acting in a rather schizophrenic manner, actively lobbying against efficiency standards while showcasing supposedly climate-friendly concept vehicles that run on fuels made from locally grown corn, hydrogen, and propaganda and never make it to market.

A classic example is the Canadian Vehicle Manufacturers' Association, which ran a half-page newspaper ad in 2007 about its supposed "Auto Green Plan," a great ploy for selling more vehicles. The "plan" included the adoption of new technologies, the use of biofuels, government grants to take older vehicles off the road (so manufacturers can sell new cars), government leadership in purchasing new green-technology vehicles, and lastly, changing driver behavior. Interestingly, nowhere did the CVMA address the real issues of reducing the number of vehicles on the roads, developing mandatory fuel efficiency standards, or implementing a carbon tax that would actually reduce driving demand.

No matter how the world automotive markets play out in the coming years, there are several key points that need to be considered in order to better understand which way potential automotive "green plans" may evolve:

1. The average automobile is designed as a "one-size-fits-all" solution, acting as personal fashion/wealth statement, local transportation device, long-distance transportation device, urban/suburban and country vehicle, as well as freight hauler.
2. Most Western households have more than one vehicle.
3. Most people on average drive a vehicle only a very short distance and occasionally extend this to mid- and long-distance travel.
4. Most people currently prefer to commute alone.
5. Automotive companies design vehicles that rot away into nothing long before necessary, wasting materials and embedded energy. (What other major asset would people consider purchasing that has

a 10% to 25% depreciation rate? When people purchase houses and make investments they expect their value to appreciate.)

6. The fuel-efficiency cycle of an automobile is very low, typically less than 15% for the vehicle as a system and approximately 1% where the efficiency is compared to the real work being done, that of moving a passenger.[1]

Since ego and greed are fundamental human characteristics, it is very difficult to create only one "class" of automobile for the market. Keeping up with the Joneses will never go away and the world will not revert to "you can have any color as long as it's black." In addition, many people tire of their cars long before the vehicle is actually worn out.

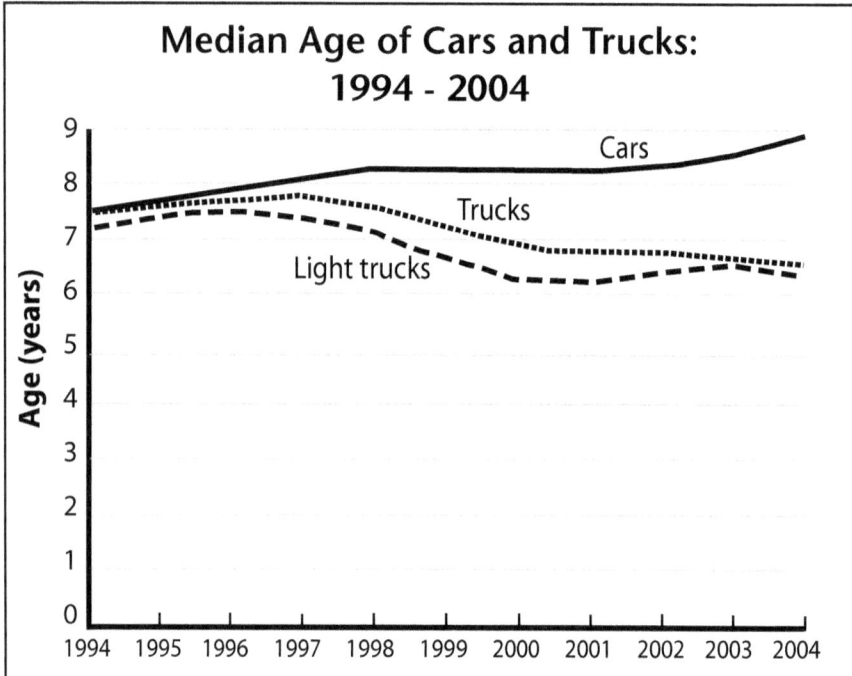

Figure 5-6. The general ten-year trend shows that the median age for vehicles has increased by 19%, indicating that cost of ownership is impacting people's personal finances, causing them to keep the car for longer periods. (As the number of light trucks purchased increased in 1994 the fleet average age has declined over this period). (Source: U.S. Department of Transportation, Transportation Statistics Annual Report, November 2005)

The combination of these phenomena explains the common practice of perpetually financing a car by leasing it for a short term (say three to five years), returning it, and re-leasing a newer model. Since part of the capital worth is returned with the car at the end of the lease, the owner pays interest on the entire value but owns nothing. This dubious practice allows people to operate a more expensive vehicle than they might otherwise be able to afford. This action speaks very poorly of people's understanding of economics (unless they can use lease payments in a tax advantaged way) and their need to appear more affluent than their peers. Although it is a sad situation, it is understood by the marketing departments of all vehicle manufacturers and is heavily exploited in all forms of advertising.

Vehicular energy efficiency, or fuel economy if you prefer, is on everyone's mind as the price of oil reaches new highs, the Iraq war persists, and the US dollar continues its long-term slump. People fight with themselves about the cost of vehicle operation and in particular fuel prices, but the overwhelming desire to own the newest and "best" vehicle they can afford also pushes the public into schizophrenic behavior, with the lust for a nice car or truck generally winning out over the common sense of economy and the environment. (I often listen to people telling me they *need* to own a van or truck, when I know darn well an occasional rental or trailer attached to a reasonably sized car would haul all the stuff to the cottage and back.)

As vehicle ownership costs continue to climb and carbon and other environmental policies take hold, people will need to keep their cars for ever-longer periods. At the same time, they will want to own "new" or at least different models every few years. Perhaps there is a way for automotive companies to balance these competing "needs" with a new type of vehicle that offers a climate-friendly drive train and other technologies including superior energy efficiency, sufficient range, very long life, and the ability to transform or reconfigure function or style. Such a vehicle would, by necessity, use sufficiently complex technologies to thwart copycat competitors and provide the automotive industry with the ability to realistically brand itself as environmentally conscious, perhaps ensuring that it remains financially viable for the foreseeable future.

To understand how these ideas could converge to produce a working vehicle, it is first necessary to examine current and future technologies now being debated and scrutinize them in the light of carbon policies.

Long Life

Earlier we discussed the fact that the embedded energy used in the manufacture of a typical gasoline-powered automobile represents approximately 7% of its total life cycle greenhouse gas emissions (Figure 5-7). As a vehicle's carbon-fuel efficiency improves (for example, an electric automobile charged from hydroelectricity), the portion of carbon emissions attributable to embedded energy rises to 90% of total emissions.

The embedded value of carbon emissions will eventually be calculated, with the end user footing the tab either through higher vehicle cost or through a direct carbon tax tacked on to the selling price at the dealership. In either case, you as a consumer will have to pay the price.

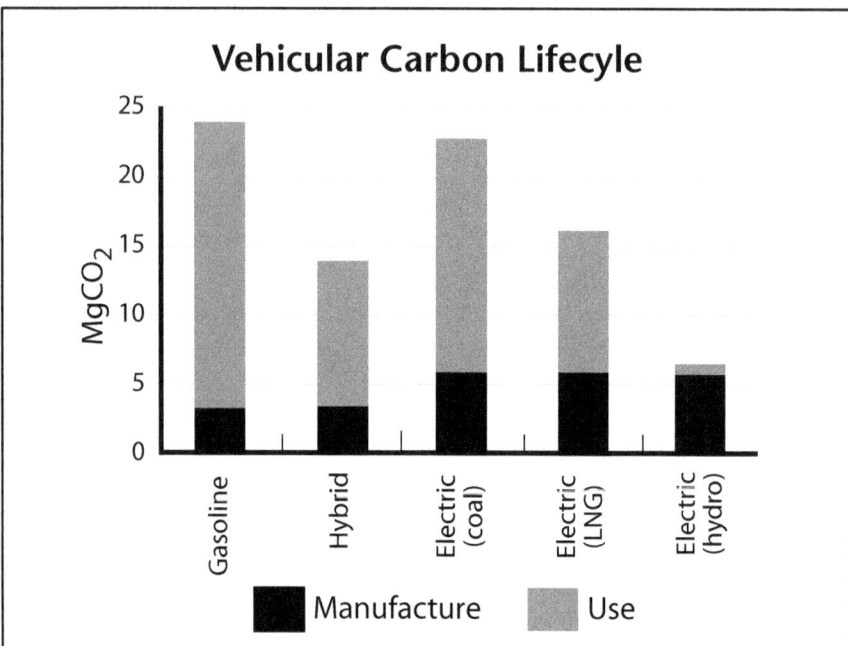

Figure 5-7. As a vehicle's carbon-fuel efficiency improves (for example, an electric automobile charged from hydroelectricity), the portion of carbon emissions attributable to embedded energy rises to 90% of total emissions. The embedded value of carbon emissions will eventually be calculated with the end user footing the tab either through higher vehicle cost or through a direct carbon tax tacked on to the selling price at the dealership. In either case, you as a consumer will have to pay the price. (Source: Institute for Lifecycle Environmental Assessment)

Furthermore, as carbon fuel taxes rise (or the supply of fossil fuels shrinks) the cost of fuelling a vehicle will become prohibitive. Therefore any new-technology vehicle that is energy efficient and operates with low- or zero-carbon fuels will be worth a premium and owners will want a long operating life to ensure a positive return on their vehicular investment.

To keep prices down, auto manufacturers and their associated suppliers will either have to purchase green energy to manufacture their vehicles or find alternative products and processes that reduce or cancel charges related to life cycle greenhouse gas emissions.

Alternatively, consumers will simply have to keep the car longer and amortize these costs over a longer time horizon. If a fixed amount of emissions is amortized into a vehicle with twice the working life of a typical automobile, embedded carbon emissions will be reduced by one half. Simple mathematically, but as we have already discussed, cars depreciate at a faster rate than most people's investments appreciate.

Although the auto industry won't initially like the idea, using electric drive trains (with few if any "scheduled maintenance programs") and lighter running gear will greatly reduce wear and tear. Using regenerative braking (discussed later) extends brake pad and rotor life significantly. Automotive bodies and frames can be produced using materials such as carbon-fiber, lightweight metal alloys and injection-molded panels (using bio-oils derived from waste products through a process known as fast pyrolysis[2]) to extend operating life and reduce weight several fold. Even old-fashioned heavy steel used for stamped body panels can be made "corrosion free" through the use of cathodic rust protection modules.[3] These modules have long been used by the oil and gas pipeline industry to prevent pipes and fittings from failing due to oxidation in salty and/or wet locations. The same technology is now in use in the automotive sector. My Honda Civic Hybrid was fitted with a cathodic protection unit when I purchased the car. Included in the fee of $599 was a lifetime guarantee against rust through and six years' protection against surface rusting and stone chips. This rust protection package was neatly developed using both technology (the cathodic protection module) and a financial product (by way of the warranty program) that mitigates my concerns about premature failure. With a bit more thought, manufacturers could easily extend the total protection of a vehicle to a dozen years or more.

Energy Efficiency (Fuel Economy)

Everyone knows, or at least should know, that trucks and SUVs are very energy inefficient. What may be a surprise is that the same can be said for any fossil-fuel-powered vehicle, no matter how good the fuel economy appears to be. A typical vehicle is enormously inefficient as a result of the cumulative losses incurred during the conversion of the energy contained in gasoline to forward motion (Figure 5-8). Nearly 92% of the potential energy in a gallon of gasoline is lost as a result of engine inefficiencies, idling, driveline friction, and tire rolling resistance. Of

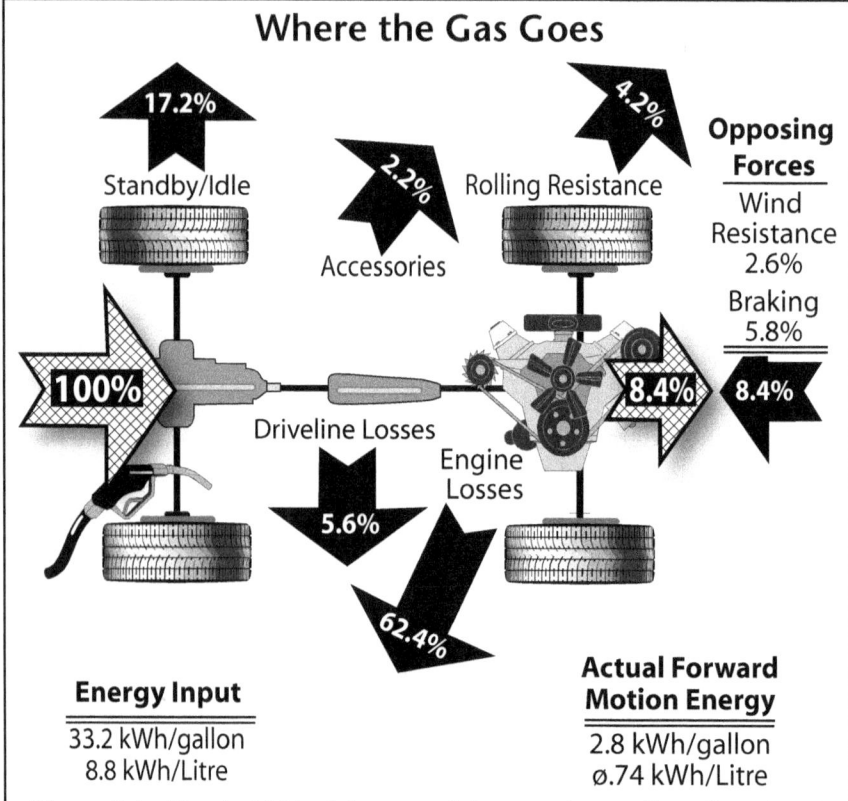

Figure 5-8. Nearly 92% of the potential energy in a gallon of gasoline is lost as a result of engine inefficiencies, idling, driveline friction, and tire rolling resistance. Of the remaining 8.4% of energy that is available for propelling the vehicle forward, 2.6% is required to overcome wind resistance. The remaining 5.8% of energy used to propel the vehicle is then later wasted as heat, bringing the vehicle to a stop. (Source: US Department of Energy)

the remaining 8.4% of energy that is available for propelling the vehicle forward, 2.6% is required to overcome wind resistance. The remaining 5.8% of energy used to propel the vehicle is then later wasted as heat, bringing the vehicle to a stop[4].

To express efficiency in a different manner, consider that one gallon of gasoline contains 33.2 kilowatt-hours (kWh) of energy (8.8 kWh/liter)[5]. A typical current-technology electric car requires approximately 0.25 kWh of energy per mile traveled, (0.155 kWh per kilometer), allowing it to travel 133 miles on the energy equivalent of one gallon of gasoline (47 kilometers per liter).

Another approach to improving mass and energy efficiency in one shot is to consider a neighborhood vehicle such as the ZENN car shown in Figure 4-25. While the low-speed ZENN will not meet all driving requirements, it might be perfect as a second vehicle for performing errands and running around town.

Improving total vehicular fuel efficiency will therefore require a complete change in motive power and drive train technologies in order to reduce these losses to more manageable levels, as outlined in the comparison between internal-combustion-powered vehicles and an all-electric counterpart, as shown in Table 5-1.

Losses From (%)	Internal Combustion Engine	Electric-Drive Vehicle
Engine (Motor)	62.4	5
Idling	17.2	0
Drive Train	5.6	0
Rollling Resistance	4.2	3
Wind Resistance	2.6	2.6
Battery Charging	0	15*
Electronic Drive	0	8
Regenerative Braking	0	-4
TOTAL	92	29.6

*depends on technology

Table 5-1. Comparison of energy losses in an internal combustion vehicle and an electrically powered vehicle.

Vehicle Mass

The energy efficiency calculations noted above assume that moving a 2,500-pound (1.1-tonne) vehicle is actually useful work; but what if all we really wanted to do was move a 160-pound (73-kilogram) person from point A to point B?

Even the fuel-efficient Smart™ car, weighing in at a mere 1610 pounds (730 kilograms), is ten times the mass of the average driver, yielding a total "passenger transport" efficiency of only 0.84%[6]! Vehicles with higher mass-to-cargo ratios (i.e. a Hummer with one passenger) fare much worse. Even if the mass of the average car could be reduced by 50%, vehicle energy efficiency would only improve to approximately 2%, a limit that is simply not acceptable for any energy-consuming device in a carbon- and energy-constrained world.

Reducing vehicular mass while maintaining or improving safety will become a primary concern. Automobiles have been built using the same basic materials, shapes, and processes since the days of Henry Ford, with low-cost, heavy, rust-prone steel being the material of choice. In the carefree days of finned cars, massive engine displacements, and cheap gasoline, vehicle mass was not given a moment's consideration by automotive engineers, and that still seems to be the case today.

Research on the ultra fuel-efficient Hypercar® conducted by the Rocky Mountain Institute[7] suggests that the only way to improve vehicle fuel efficiency is to use a full suite of technologies and construction techniques that includes a massive reduction in vehicle weight and the development of low-aerodynamic-drag bodies.

Unfortunately, many people assume that vehicle mass plays the major role in automotive safety and that driving larger, heavier vehicles is the only means of protection in the event of an accident. The literature does not completely support this view. A report conducted by Monash University Accident Research Centre[8] states that, "the crashworthiness relationship between vehicle mass, size and safety is rather ambiguous." While the report indicates that there is considerable evidence that occupants in larger cars have superior crash protection to those in smaller cars, the precise relationship is not clear. For example, driver habits can play an important part: an unbelted driver in a 4,400-pound (2,000-kilogram) car has the same amount of protection as a belted driver in a 2,500-pound (1140-kilogram) vehicle. Further evidence suggests that

Figure 5-9. While there is considerable evidence that occupants in larger cars have superior crash protection to those in smaller cars, the precise relationship is not clear. Evidence suggests that the passenger space inside the cabin may have a marked influence on the likelihood and severity of injury. Passenger volume does not necessarily require higher vehicle mass, especially if the reduction in weight is accompanied by appropriate engineering and materials as well as active protection systems such as air bags—one of the reasons the diminutive Smart™ car has an excellent crashworthiness rating. (Courtesy: Mercedes-Benz Canada)

the passenger space or volume inside the cabin may have a marked influence on the likelihood and severity of injury. Passenger volume does not necessarily require higher vehicle mass, especially if the reduction in weight is accompanied by appropriate engineering and materials as well as active protection systems such as air bags.

A reduction in vehicle mass and general downsizing will be necessary to improve fuel efficiency and economics in coming years, demanding much more advanced engineering and materials to ensure passenger safety. Anyone who has watched a spectacular Formula One racing crash (Figure 5-9) will no doubt wonder how such a lightweight vehicle operating at 200 miles per hour (320 kilometers per hour), hitting a concrete

Figure 5-10. Fast, lightweight, million-dollar race cars and their equally valuable drivers require very advanced forms of automotive safety technologies. Formula One cars are designed with minimal vehicular mass and use crumple zones and advanced composite materials including carbon-fiber panels in their manufacture. These composite panels could represent a major improvement in the passenger automotive industry, as they are lighter, tougher, and stiffer than most metals. (Photos courtesy of TM Wolf [http://commons.wikimedia.org/wiki/Image:MF1_Collision_28crop29.PNG] and Oom Agent [http://commons.wikimedia.org/wiki/Image:Villeneuve_Germany2006.jpg])

wall, and breaking apart does not normally kill the driver. In a word, "engineering."

Fast, lightweight, million-dollar race cars and their equally valuable drivers require very advanced forms of automotive safety technologies. Formula One cars are designed with minimal vehicular mass and use crumple zones and advanced composite materials including carbon fiber panels in their manufacture. These composite panels could represent a major improvement in the passenger automotive industry, as they are lighter, tougher, and stiffer than most metals. According to Fiberforge[9], a company that develops and manufactures composite materials used in products ranging from skateboards to life-saving military protective equipment and performance-critical aircraft components, composites are highly efficient and a sustainable substitute for traditional materials. Compared to other materials used in automotive manufacturing, composite materials are:

- 500% stiffer than injection molded plastics;
- 60% lighter than steel;
- 600% stiffer than steel;
- 25% lighter than aluminum.

There is also 60% less scrap in the production of composite materials than sheet materials, and wastes can be recycled for use in other processes, reducing embedded energy.[10]

But the potential of carbon fiber has not been fully explored because radically increased fuel economy, reduced greenhouse gas emissions due to lower vehicular mass, and the value of reduced embedded energy in vehicle manufacture are not currently valued by society. Where these issues are important, as with aircraft and race car manufacture and military products, composites are the materials of choice.

To achieve massive increases in vehicular fuel efficiency while reducing mass and improving safety will require the engineering community to make a wholesale switch to computer-designed, low-aerodynamic-drag bodies using lightweight composite materials. Safety engineering will have to take a cue from Formula One racing engineers and incorporate a system-wide process that includes finite element analysis to calculate crash-related stresses in these newly designed vehicles to ensure that the best attributes of the materials are being used.

Virtual Upgrading

If you take a quick look at almost any vehicle—a light-duty pickup truck, a luxury car with snazzy wood and leather interior, or perhaps a zippy sports model with a retractable roof—you will find many commonalities. All of these vehicles have four wheels, a drive train, headlights, a windshield, body work, doors, and seats. Why do so many people own two at the same time?

And why do they trade in their vehicles so often? Most people will freely admit that they become bored with a vehicle long before the unit is ready to be retired to the auto recycler, and in a free world if you have the capital to switch vehicles because of vanity or ego no one is going to stop you. However, unless you are very rich, the ability to switch cars whenever the urge strikes will not continue forever.

As new, advanced vehicular technologies are introduced to the market and a value is placed on the carbon life cycle of the embedded energy required to make the vehicle as well on the energy used to operate it, automotive capital and operating costs will rise dramatically. In addition, I would also expect to see significant "recycling" charges applied to the vehicle for its end-of-life disposal.

Figure 5-11. An interesting possibility presents itself: the design of the vehicle could be more like a multitool, so that the family sedan could be converted into a truck, van, or sports car. General Motors has taken a pretty good stab at this with its Hy-wire hydrogen-powered concept car, originally shown to the public in January 2002. Although the technology behind Hy-wire is stunningly complex, the design concept is very elegant and leads directly to long-term vehicle ownership, with the ability to upgrade on an as-needed or as-wanted basis. (Courtesy: General Motors)

As car owners confront this new reality, the already increasing median age of vehicles (Figure 5-6) will increase dramatically, and people will have no alternative but to own these assets longer. This need for longer life will conflict with the desire to own the newest car on the block, not to mention the "need" to own more than one.

An interesting possibility presents itself: the design of the vehicle could be more like a multitool, so that the family sedan could be converted into a truck, van, or sports car. Provided the power train platform was modular, perhaps Mr. Smith could change vehicle models simply by snapping off the old chassis and replacing it with a new one. With the right mindset and design, this might be possible.

General Motors has taken a pretty good stab at this idea with its Hy-wire hydrogen-powered concept car, originally shown to the public in January 2002. Although the technology behind Hy-wire is stunningly complex, the design concept is very elegant and leads directly to long-

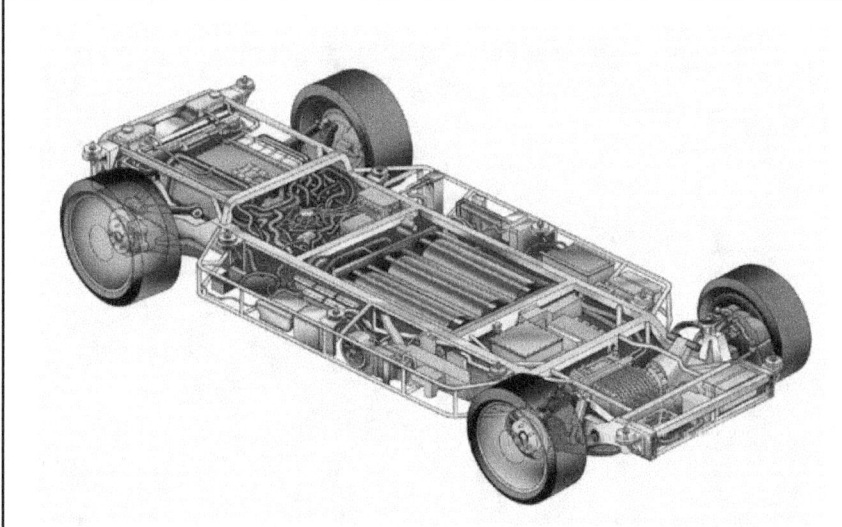

Figure 5-12. The Hy-wire vehicle dispenses with the traditional internal combustion engine, transmission, radiator, steering wheel, and other bulky items required in today's antiquated vehicles. GM engineers created a skateboard-like power module that contains an electric drive system, hydrogen storage tanks, and a fuel cell in place of the mishmash of mechanical stuff currently used. (Courtesy: General Motors)

term vehicle ownership, with the ability to upgrade body styles on an as-needed or as-wanted basis while retaining the working bits of the power module for the long term.

The Hy-wire vehicle dispenses with the traditional internal combustion engine, transmission, radiator, steering wheel, and other bulky items required in today's antiquated vehicles. Instead, GM engineers created a skateboard-like power module (Figure 5-12) that contains an electric drive system, hydrogen storage tanks, and fuel cell in place of the mishmash of mechanical stuff currently used. Of course, many different power systems and energy options could be used in place of hydrogen, but the important thing is make the power module very robust, ensuring ultralong life to amortize its value over a time horizon of at least 12 to 15 years, possibly longer.

The streamlined power module would allow designers to create a variety of bolt-on body styles to suit the owner. The vehicle could be

Figure 5-13. The Hy-wire vehicle dispenses with many of the common elements found in today's automobile. Instead of a gas pedal, the driver twists the hand grips of the steering wheel, rear-view mirrors have been replaced with miniature TV screens. Although these items may seem avant-garde they are the way of the future. (Courtesy: General Motors)

shipped from the factory with a BMW-like facade for day-to-day use and then driven to a facility where the body could be quickly removed and placed in storage while a truck or alternate body could be attached. This process could even be automated, featuring a drive-in, drive-out service that made the chassis exchange.

Because the power platform provides only power and data feed to the vehicle's chassis, automotive "stylists" are given free reign as to what the vehicle looks like. Taking this concept just a few more steps into the future, it might also be possible to unite the vehicle's chassis with a semi-custom body made with composite materials. Using a computer-aided manufacturing process similar to the Fiberforge Relay™ station, a chassis designer could produce near net-shape panels in minutes. Although this technology cannot currently complete the entire process of making body panels from a computerized model and carbon thread, the idea is very tantalizing and provides a glimpse of what is to come.

The Myth of Zero-Emission Vehicles

Many people refer to hydrogen and electric cars as zero-emission vehicles, a statement which can be very misleading. As shown in Figure 5-7, an electrically powered vehicle that is recharged from electricity produced from coal has a similar carbon dioxide emission profile as a typical gasoline-powered vehicle. By comparison, the same electric vehicle charged with hydroelectricity has an emission profile four times lower than that of the gasoline model.

Not all fuels are created equal, and with the race to find the replacement for oil, statements are being made about emissions that are not always correct or are at best misleading. The common statement that hydrogen-powered vehicles are zero emission depends, for example, on where the hydrogen comes from. There are currently approximately 9 million tons of hydrogen produced annually, the vast majority of which results from a process known as the steam reformation of methane (natural gas). On an energy balance basis, for each unit of methane consumed, 0.66 units of hydrogen are produced, giving a negative energy conversion factor.[11]

In order to convert methane (CH_4) to hydrogen (H_2), the carbon molecules must be knocked free from the hydrogen in the form of carbon dioxide (CO_2), a greenhouse gas. Currently, there is no commercial use for the by-product CO_2, nor is there any means of permanently capturing the gas; therefore it is allowed to escape into the atmosphere. Tellingly, for each kilogram of hydrogen produced during the reformation process, 11.88 kilograms of CO_2 are produced[12]. By way of comparison, each kilogram of gasoline burned produces 3.2 kilograms of CO_2. Emissions related to hydrogen are therefore 3.7 times *greater* than those created by burning gasoline. Unless the hydrogen is produced from zero-carbon sources, the vehicle will be anything but zero emission, and you might just as well burn the natural gas in a standard vehicle and forget about hydrogen completely.

Chapters six, eight and nine will delve into the specifics of zero-carbon energy sources, including hydrogen, electricity and liquid fuels. For now, bear in mind that the *zero-emission vehicle* moniker tells only half the story; in order to understand the full carbon emission profile, it is necessary to investigate the "well-to-wheels" lifecycle of the input energy sources.

Summary

Tinkering with hundred-year-old internal combustion technology to increase fuel efficiency or comply with regulatory requirements of the day might improve things incrementally, a few percentage points at a time. But even if vehicular weight could be reduced fourfold and fuel economy doubled, overall energy efficiency would only improve to a pathetic four or five percent, hardly worth the bother.

Radical improvements will only be made by wiping the slate clean and starting over with a new industry-wide competitive mindset. New, lightweight composite materials and light alloy metals must be used to fabricate aerodynamically shaped vehicles. All-electric and electric/hybrid drive systems using the latest in computer technology must replace the burning of carbon-based fuels as the motive power system.

We have the means to make the transition happen, and many companies and researchers are already beginning to see the fruits of their labors, as we will see in the next chapter.

6
A Closer Look at Advanced Automotive Technologies

In the previous chapter, I reviewed some of the physical problems facing automobiles that rely on antiquated technologies such as the internal combustion engine. Automotive engineers and manufacturers are well aware of these limitations, but have little incentive to move away from the technologies they have refined over the last one hundred years. Any further improvements will be incremental at best.

This chapter will take a closer look at some of the advanced vehicle technologies that are in the early stages of development, sales, or marketing. I will not bother to dwell on technologies that simply fiddle with the status quo; variable valve timing, direct-injection gasoline engines, and 6-speed automatic transmissions are fine engineering advancements to existing systems, but it is the radical, disruptive technologies that will provide the major environmental improvements that will be considered here.

Governments and automotive manufacturers are, for the most part, in agreement about the need to research and develop these advanced vehicles. In January 2002, United States Energy Secretary Spencer

Abraham and officials from Chrysler LLC, Ford Motor Company, and General Motors Corporation announced the formation of a partnership known as the FreedomCAR (Freedom Cooperative Automotive Research) and Fuel Partnership Plan. The program was designed to examine and advance the precompetitive, high-risk research needed to develop and produce affordable vehicles and the necessary fueling infrastructure. A sub-theme of the program is to wean Americans off of their need for imported oil and to minimize smog and carbon emissions—all laudable goals.

Where the program goes off the rails is with the final bit of patriotic prattle about the "freedom for Americans to choose the kind of vehicle they want to drive, and to drive where they want, when the want; and to obtain fuel affordably and conveniently."[1]

Cheap oil and the ancient technology of internal combustion power have been well refined over the last century and are well past their pin-

Figure 6-1. The Toyota Prius hybrid has become the icon of the "green automobile." Using a combination of a small gasoline-powered engine, advanced drive train technology, and electric motor assist system, the Prius has improved fuel efficiency figures by 30% over a similarly equipped car without a hybrid system. Although this is an impressive achievement, it is only the start of the advanced-vehicle journey. (Courtesy Toyota Canada Inc.)

nacle. To meet the goals of weaning the United States (or any country) off oil, especially the imported variety, will require major sacrifices related to personal mobility and cost. To assume that any new technology will allow all Americans to drive affordably is simply wishful thinking; all that will happen is that ever more people will be marginalized and left on the side of the road. Advanced vehicle development is fine, but advanced mass transit and freight infrastructure and logistics must also have their own "Freedom" program.

Hybrid Automobiles

Today's Hybrid Power Systems

The sale of hybrid cars and light-duty trucks reached nearly a quarter of a million units in 2006, spread over twelve models throughout North America. These sales volumes did not happen by accident and are a tes-

Figure 6-2. The 2007 Toyota hybrid "synergy-drive" power train produces more power than that of previous models, giving the Prius performance comparable to that of non-hybrid four-cylinder midsize automobiles while offering an impressive 58 mpg (4 L/100 km) fuel consumption. (Courtesy Toyota Canada Inc.)

tament to the public's desire to improve fuel economy and lower their environmental impact while demonstrating their willingness to pay a premium over a similarly equipped car without hybrid technology.

The Toyota Prius is the first midsize hybrid electric vehicle (HEV) to enter mass production. This model employs Toyota's so-called hybrid "synergy drive," its third-generation gas/electric power train technology which improves on earlier generations released in 1997 and 2000. The synergy-drive power train produces more power than that of previous models, giving the Prius performance comparable to that of non-hybrid, four-cylinder midsize automobiles while offering an impressive 58 mpg (4 L/100 km) fuel consumption. Although these efficiency improvements are impressive, they are not sufficient for the long term and cannot be called truly disruptive technology. However, today's hybrid is really the stepping stone into the automotive future, providing a building block for substantial improvements to come.

Figure 6-3. The hybrid system is, for the most part, seamlessly integrated into operation of the vehicle and hidden from the driver of the vehicle. To provide a little "sizzle with the steak," Toyota has included an LCD screen in the Prius that provides radio and cabinet comfort levels as well as a real-time display of hybrid system parameters. (Courtesy: Toyota Canada Inc.)

The Nuts and Bolts of Hybrid Technology

Automotive hybrid power systems are available in two major design configurations and numerous alternate combinations. The two configurations are known as parallel and series systems, each of which utilizes the same building blocks: a small fossil-fuel engine, an electric motor/generator, an inverter control unit, and a rechargeable battery bank. In the simplest of terms, a hybrid electric vehicle uses a combination of internal combustion engine and electric motor/control/battery to move a vehicle. Energy supplied by the electric motor assists the internal combustion engine, thereby reducing fuel consumption relative to that of a similar nonhybrid vehicle. The battery is charged by the engine when excess energy is available (for example during light-load driving) or during a process known as "regenerative braking," which uses the inertial energy of the slowing vehicle to generate electrical energy and charge the battery bank.

Parallel hybrid configurations (Figure 6-4) connect both the fossil-fuel engine and the electric motor to the drive wheels through a power-transfer unit, allowing energy from either or both sources to move the vehicle, depending on driving conditions. While coasting or decelerating, the electric drive motor is actually being driven by the motion of the vehicle and thereby acts as a generator which charges the batteries.

Series hybrid systems (Figure 6-5) are configured so that the fossil-fuel engine drives an electric generator, charging the battery bank and simultaneously powering an electric motor to drive the wheels of the vehicle. The advantage of the series system is that a relatively small fossil-fuel engine can operate at its maximum efficiency level, simultaneously powering the car and charging the battery bank. The vehicle is always powered by the electric motor(s), thus dispensing with a complex mechanical transmission system.

Series hybrids have been used for decades in diesel/electric locomotive systems, where the diesel engine drives an electrical generator. Energy from the electrical generator is supplied to electric motors mounted in the "wheel truck" or "drive wheel" assemblies of the train.

In either a parallel or a series hybrid system, the battery bank is connected to an inverter control unit which in turn is coupled to the electric drive motor. The inverter control unit is configured to direct the flow of electrical energy through the system, depending on driving and operat-

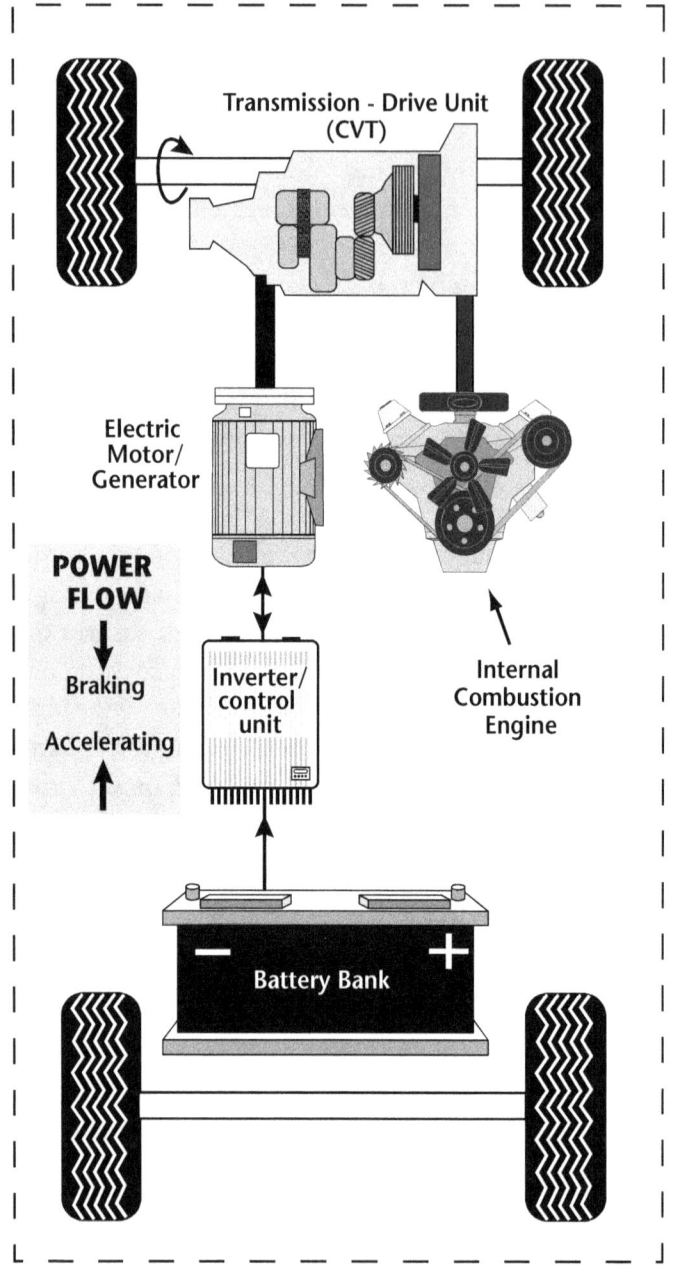

Figure 6-4. Parallel hybrid configurations connect both the fossil-fuel engine and the electric motor to the drive wheels, allowing energy from either or both sources to move the vehicle, depending on driving conditions.

Series Hybrid Technology

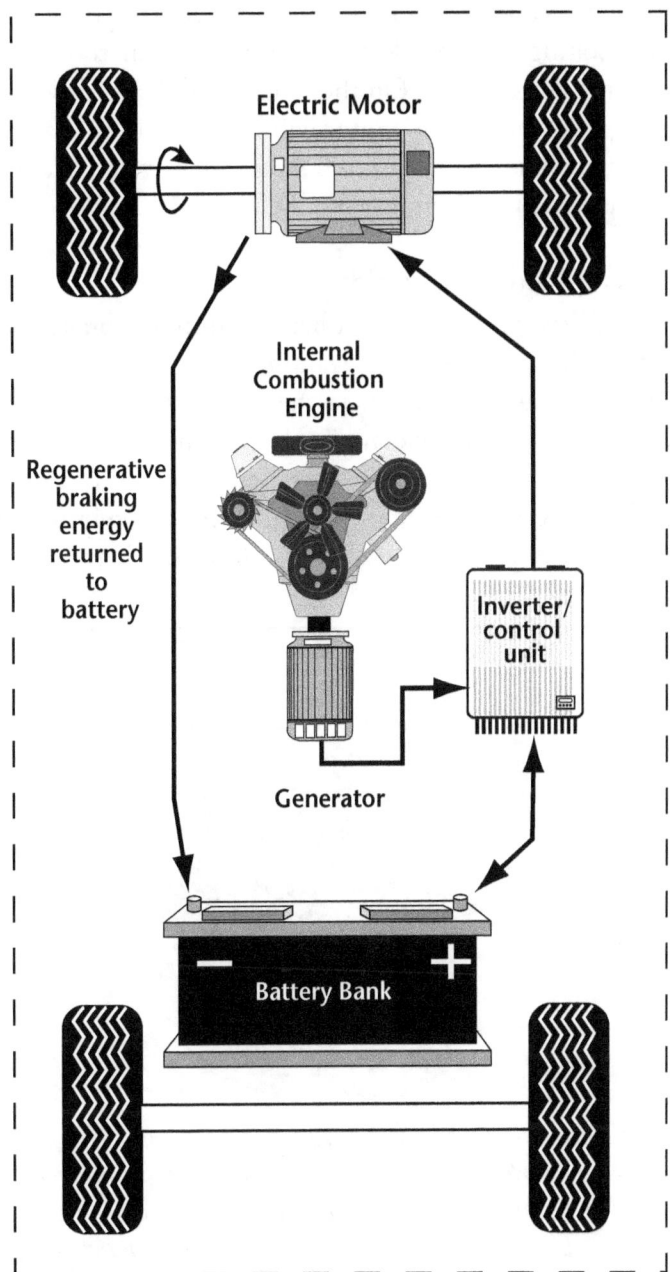

Figure 6-5. Series hybrid systems are configured so that the fossil-fuel engine drives an electric generator, charging the battery bank and simultaneously powering an electric motor to drive the wheels of the vehicle.

ing conditions. For example, if the driver *gently* presses the accelerator pedal, energy stored in the battery bank is directed to power the electric motor, accelerating the vehicle to driving speed. In this instance the gasoline engine remains off. On the other hand, if the driver wishes to accelerate rapidly, the electric motor will be driven in the same manner as above but the gasoline engine will start instantly and provide power to assist the electric motor in moving the vehicle.

"Weak" Versus "Strong" Hybrids

As soon as the driver removes his or her foot from the brake of a Honda Civic Hybrid, the gasoline engine is immediately started, allowing both gas and electric power to accelerate the vehicle. The Toyota Prius, on the other hand, is able to operate in electric-only mode for short distances before the gasoline engine is started. The ability to operate in electric-only mode indicates a "strong" hybrid configuration, whereas the Honda Civic is a "weak" hybrid. The difference between the two is

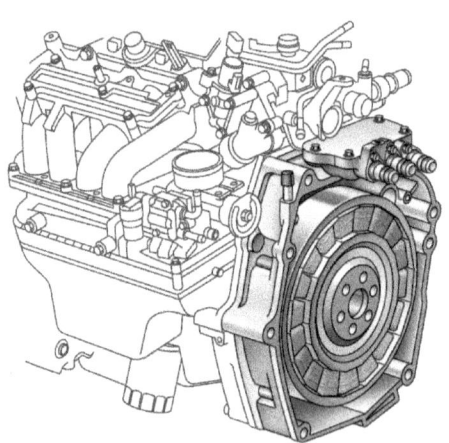

Figure 6-6. The Honda Civic hybrid's gasoline engine (left) is mated to an electric motor/generator which can provide additional muscle on demand. The phantom view of the car details the layout of the various components. The gasoline engine with its electric motor is mounted transversely above the front wheels. A power cable connects the electric motor to the battery bank and control unit located below the rear deck of the back seats. (Courtesy Honda Motor Co.)

of little practical importance to car buyers today as both vehicles have superior fuel economy compared to their nonhybrid cousins, with most buyers simply choosing the vehicle style they prefer. However, the picture changes dramatically when we discuss plug-in hybrids later in this chapter.

Energy that is used to accelerate a vehicle must be dissipated in order to slow the vehicle down or bring it to a stop. A typical automobile will dissipate this energy as heat in the brake pad and rotor assemblies. During braking, the hybrid system uses the inertia of the moving vehicle to reverse the flow of energy through the drive motor, in effect converting it from a motor into an electric generator and charging the battery bank. This is the regenerative braking process. Because the vehicle is expected

Figure 6-7. The internal combustion engine has a narrow operating speed range where performance and energy efficiency are at their highest levels. Because a vehicle is always changing speed, the engine will not always be within its preferred operating band and a speed transmission system is required. In most vehicles, the transmission offers sequential or "stepped" gearing which causes the engine to operate over a wide speed range. The continuously variable transmission (CVT) shown here provides nearly infinite "gear ratios," better adapting engine performance to road speeds.

Figure 6-8. The CVT dispenses with gears and replaces them with variable-ratio pulleys and a drive belt or chain as shown in this model. Moving the "sides" of the pulleys inwards or outwards, changes the inner diameter and thus the ratio. Using this process provides essentially infinite ratios, within the design limits of the CVT.

to stop frequently, the battery bank is never fully charged, leaving some spare capacity to store the energy recovered as a result of regenerative braking.

Hybrid Performance

Hybrids receive a bad rap in the press, with many people complaining that they are not worth the extra cost or that they do not get the advertised fuel economy in the real world. While it is true that hybrids cost more to produce than conventional vehicles of the same make and model, the return on investment data is skewed because of artificially low fuel costs. Add in a proper carbon tax and most everyone will gravitate to some form of high-gas-mileage vehicle, hybrids included.

Even without carbon taxation, hybrids provide a positive financial benefit when the additional capital costs are amortized over the true life cycle expenses of the vehicle. The numbers would be even better if owners kept their vehicles longer rather than "flipping" to the latest model.

The US and Canadian governments both use test programs that do not necessarily provide real-world fuel economy figures. A heavy foot on the accelerator, air conditioning, windy roads, and deep snow negatively impact the surreal mileage ratings, but this is true for all types of vehicles and not just hybrids. Secondly, if the difference between fuel economy figures in the lab and in the real world is, for example, 10%, the increase in fuel consumed will be greater for the nonhybrid model. Studies show that a full-size SUV HEV can reduce smog precursor emissions by up to 20% and petroleum consumption and CO_2 emissions by 30% under similar driving conditions[2].

Large Hybrid Vehicles

The General Motors near-term development program revolves around improving today's automotive technologies without making anyone "sacrifice" performance or size of vehicle. Naturally enough, these improvements include relatively minor incremental advances such as electric power steering, more efficient alternators, clean diesel engines, and a variable displacement engine control system known as Displacement on Demand. The company is also developing three different hybrid systems on several of its most popular models of cars and light- and heavy-duty trucks.

In 2003 a flywheel alternator/starter hybrid system went into production for installation in 2004- model trucks. The truck's hybrid system links a 5.3-liter engine with a compact 19 hp (14 kW) electric motor and battery bank. A regenerative braking system recovers energy that would otherwise be lost during braking. An idle-stop feature shuts off the engine at stop lights in order to save fuel. The large battery bank can also supply power to a 120-volt inverter which can be used to operate power tools and other accessories. Based on third-party testing, fuel efficiency is 15% better than that of standard models.

The General Motors 2006 Malibu and some Saturn models will combine a belt alternator starter system with their Ecotec four-cylinder engine. Under idling conditions, the engine shuts down and the battery

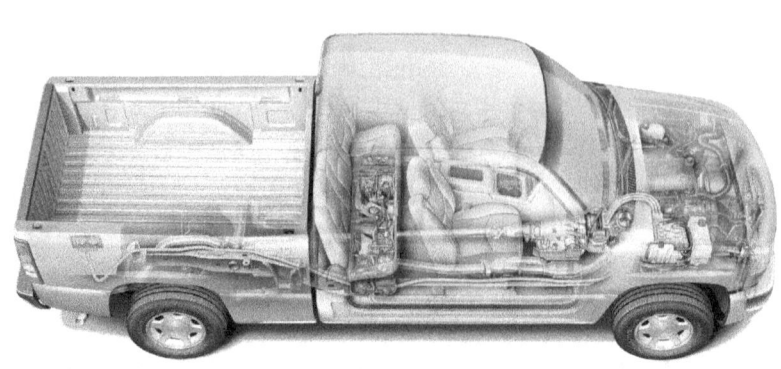

Figure 6-9. The General Motors philosophy is to offer advanced "near-term" hybrid and other technologies to create "no-compromise" vehicles designed for North American driving habits. The flywheel alternator starter hybrid system introduced on GMC Sierra and Chevy Silverado trucks in 2003 is one of the first products of this philosophy. Based on third-party testing, fuel efficiency for these vehicles has improved by 15%. (Courtesy General Motors Corporation)

bank powers accessories and air conditioning equipment, providing a 12% increase in fuel economy.

Using expertise developed by GM's Allison Division for hybrid buses (described earlier), full-size SUVs using this technology will soon become a reality. Beginning in 2007 the Chevrolet Tahoe and GMC Yukon will achieve fuel consumption savings of up to 35% compared with today's models. These are acceptable incremental achievements, for today, but will be completely unacceptable in the long-term.

Hybrid Car Summary

After spending a lifetime in a fossil-fuel-powered car, hearing the gasoline engine stop at just about every traffic light brings about a sense of panic and an inclination to push the car off the road and call for help. Fortunately this reaction dissipates after the first few hours of driving as you realize that the engine will instantly restart as soon as your foot leaves the brake.

The next most remarkable feature is what you don't feel. Considering the vast amount of computational power and dozens of new technologies used in making the hybrid car operate, everything works normally, just like a real car. The seamless integration of software and machine blends the power from the gasoline engine and electric motor. The driver could be in any typical mid-size sedan.

In an effort to further optimize efficiency, the industry standard four-speed automatic transmission has been replaced with an electronically controlled, infinitely variable model (Figures 6-7 and 6-8). This arrangement allows the internal combustion engine and electric drive motor to operate in their maximum-efficiency zone regardless of road/wheel speed.

Figure 6-10. Dual-powered vehicles have been around a long time. In the depression era of the 1930s, the "haybrid" vehicle was seen in operation since gasoline was scarce and biological horsepower was still available. As we approach peak oil and deal with increasing geopolitical tensions in coming years, this sight might well be repeated. (Courtesy: University of Saskatchewan Archives)

The hybrid gasoline/electric vehicle is an important first step on the road to more energy-efficient advanced vehicle technologies. Using the combination of advanced electric drive systems, rechargeable batteries, and integrated computer control provides the building blocks necessary for the vehicles we will explore next.

A Battery Electric Vehicle Primer

At the end of the horse-and-buggy era in the early 1900s, the race to determine which technology would supercede biological power was quickly decided with the discovery of cheap and plentiful oil. This, coupled with the long range of the internal combustion engine and Henry Ford's low-cost manufacturing of the Model T, meant there was no looking back.

During the transition period, steam and battery electric vehicles (BEVs or EVs) directly competed with the early gasoline-powered vehicles of the time. Early in the race, it was not clear which technology would win, as each vehicle type had its pros and cons. The Milburn Electric Car (Figures 6-11 and 6-12) started production in 1914 and offered a vehicle that was simple to operate, clean, and noise free. Perhaps most importantly, drivers didn't have to crank the engine to start the Milburn, as they did with gas-powered vehicles. This feature caused the Milburn and other electric cars and trucks to be favored by women

Figure 6-11. During the transition period between horse and car power, steam and battery electric vehicles (BEVs) directly competed with the early gasoline-powered vehicles of the time. Early in the race, it was not clear which technology would win, as each vehicle type had its pros and cons.

Figure 6-12. The Milburn had a driving range of 60 miles (100 kilometers), which was plenty in its day. However, the lack of access to electric power and the need for driving range in rural areas, coupled with electric starters and plentiful, cheap gasoline, spelled the death of the early electrics. A "season mileage" and trip odometer along with a clock were important features of the Milburn Electric.

drivers and in-town delivery services. The driving range of 60 miles (100 kilometers) was plenty in urban settings. However, in rural areas the limited range and lack of electric power, coupled with electric starters and plentiful, cheap gasoline, spelled the death of the early electrics. Interestingly, General Motors purchased the Milburn Automobile Company in 1923 and continued to produce electric cars until the 1930s.

After this, BEV technology was limited to niche applications such as forklift trucks and wheelchairs. It was not until the release of the General Motors EV1, the Honda EV Plus, and the Toyota RAV4 EV that BEV technology came briefly back to life, at least in the eyes of the general public. After the auto industry successfully forced the State of California to repeal its Zero Emissions Vehicle Mandate, all major car manufacturers have likewise killed off their BEV production, leaving only small, specialty companies to fill the void.

The Home-Built BEV

One group that didn't want to see an end to BEV technology is the thousands of people who keep on looking for alternatives to fossil-fuel vehicles. A quick Google search for "electric car" brings up nearly two million hits, so some people are clearly interested in the technology.

One of the people interested in BEV technology is a quiet, sprightly man by the name of Fred Green. Fred has been involved with electric vehicles since the early 1970s. At 91 years of age, Fred still enjoys driving around town in his newly rebuilt 1990 Volkswagen Jetta BEV. "My current car has just been updated because my '85 Jetta rusted out," Fred explains. "The 1990 model was in better shape and came with an automatic transmission. We stripped the electric drive parts from the old car, popped them into the newer body, and away we went."

Figure 6-13. Fred Green has been involved with electric vehicles since the early 1970s. At 91 years of age, Fred still enjoys driving around town in one of a string of vehicles, his newly rebuilt 1990 Volkswagen Jetta BEV. Fred Green and author Kemp chat about Fred's long involvement with BEV technology.

Figure 6-14. "My current car has just been updated because my '85 Jetta rusted out," Fred explains. "The 1990 model was in better shape and came with an automatic transmission. We stripped the electric drive parts from the old car, popped them into the newer body, and away we went."

Fred explains that he drives his BEV around town all year, even on the coldest of days. "My 96-volt battery system operates an electric cabin heater and I get almost immediate heat," says Fred. "With 23 horsepower and all the torque I need to accelerate, I can get away at the lights just as quick as anyone."

Growing up on a farm in pre-depression Saskatchewan, Fred was one of four boys. Although he describes himself as the sickly one he had a knack for technical details, and after finishing high school in Moose Jaw he went on to teachers college and graduated in the mid '30s. After a brief stint teaching, he went on and earned his Masters degree at the University of Toronto and finished his education with a Ph.D.

"I worked developing radar and showing how radar was better than optical sighting equipment," Fred explains. "After the war, I moved to Ottawa [Canada], joined the Defense Research Establishment Ottawa (DREO), and worked there until I retired. After retirement, I taught a course at the local community college on the fundamentals of electric vehicles. We had an awful lot of fun driving around in the cars we built!"

Figure 6-15. Fred explains that he drives his BEV around town all year, even on the coldest of days. "My 96-volt battery system operates an electric cabin heater and I get almost immediate heat," says Fred. "With 23 horsepower and all the torque I need to accelerate, I can get away at the lights just a quick as anyone."

Figure 6-16. The interior of Fred's 1990 VW Jetta looks like a typical vehicle with automatic transmission. What gives it away as a BEV are voltage and current meters that are used to replace the gas gauge of a gasoline version. Fred points out that he can easily travel 37 miles (60 kilometers) on a full charge, which is plenty for him.

Fred went on to explain to me that he got his start in electric vehicles during his work at DREO. "I would often have to sign out a station wagon to take my equipment out to the radio test labs and field. It was an awful lot of trouble and a huge waste of time having to do that. One day, someone brought an electric golf cart over to me and I was hooked. It saved me loads of time, and I was able to move my equipment around when I needed to. I used the old buggy for years."

"I was introduced to an early electric car from a Florida company called Sebring-Vanguard that made a little wedge-shaped vehicle called the Citicar[3]. It was the neatest thing, and so I bought one. Since then, I have had several electric vehicles that I have converted, including a 1993 Fiero which is still on the road today and of course my 1990 Jetta.."

Today, Fred is still in demand for speaking engagements, which he drives to, naturally enough, in his electric car. "I really like these cars. I plug one into a regular wall socket and it's always ready to go. The only time I stop at a gas station is to fill up the tires with air!"

As I get ready to leave, I ask Fred if he has any other projects on his obviously active mind. "Well," he says, "I think my next project should be to convert a Ford Escape hybrid into a plug-in hybrid. I have to have something to keep me busy!"

A Few Details about Fred Green's Car

Fred's converted VW Jetta is typical of thousands of standard-production cars that have been eviscerated, their internal combustion bits removed by hobbyists, schools, and electric car clubs around the world. The reason for taking a functional vehicle and converting it differs from person to person, but the urge to convert can be put down to the sheer challenge of the task, wanting to save the environment, or even "sticking it to big oil." Whatever the motivation people have for converting a gasoline-powered vehicle to electricity, they tend to follow the same path:

1. The internal combustion components are removed, including the engine, exhaust and fuel systems, and many of the accessory components such as the air conditioning compressor, alternator, and water pump in an effort to reduce power consumption and vehicle weight.
2. After removal of these items, a direct current electric motor is fitted

onto the face of the transmission/clutch assembly using a custom-machined adapter plate. The motor is sized to provide sufficient performance according to the vehicle weight, range requirements, and battery voltage.
3. If power steering is desired, a drive pulley is fitted from the front shaft of the electric motor to the power steering pump.
4. Insulated battery boxes are then installed in the empty engine compartment and trunk area. Deep cycle batteries are installed and wired in series to form a string of batteries at the desired voltage, typically between 96 and 144 volts.
5. The fully loaded vehicle is checked for suspension height, and stiffer "booster" springs are added to ensure that the typically heavier vehicle maintains the same wheel/ground clearance as the stock car.
6. An inverter/controller is wired between the batteries and the electric motor. A "pot box" is connected through a control cable to the accelerator pedal. (Can't call it a gas pedal anymore.) The output signal from the pot box is fed to the inverter, which provides motor speed signals according to pedal pressure.
7. A voltage and current meter are installed in the passenger compartment and connected to the battery bank. The meters allow the driver to judge battery condition in order to determine state of charge and estimated driving range.
8. A battery charger is (usually) installed on the vehicle. The power from the charger supplies the batteries, while a heavy-duty extension cord connects the charger to the house power supply.

In this simplified overview, the converted BEV is now ready to go. Stepping on the accelerator will move the control cable, actuating the pot box. This signal is fed to the motor control inverter, which in turn supplies power from the battery bank to the drive motor in proportion to the degree of accelerator actuation pressure.

As with a gasoline-powered vehicle, the harder you press the accelerator, the more power is supplied to the motor, and away you go. Driving the car is pretty much the same as driving a normal vehicle, except for the silence. Moving along smooth pavement, it is eerie how quietly BEVs operate, not unlike driving a regular car with the engine turned off.

Battery Electric Vehicle Overview

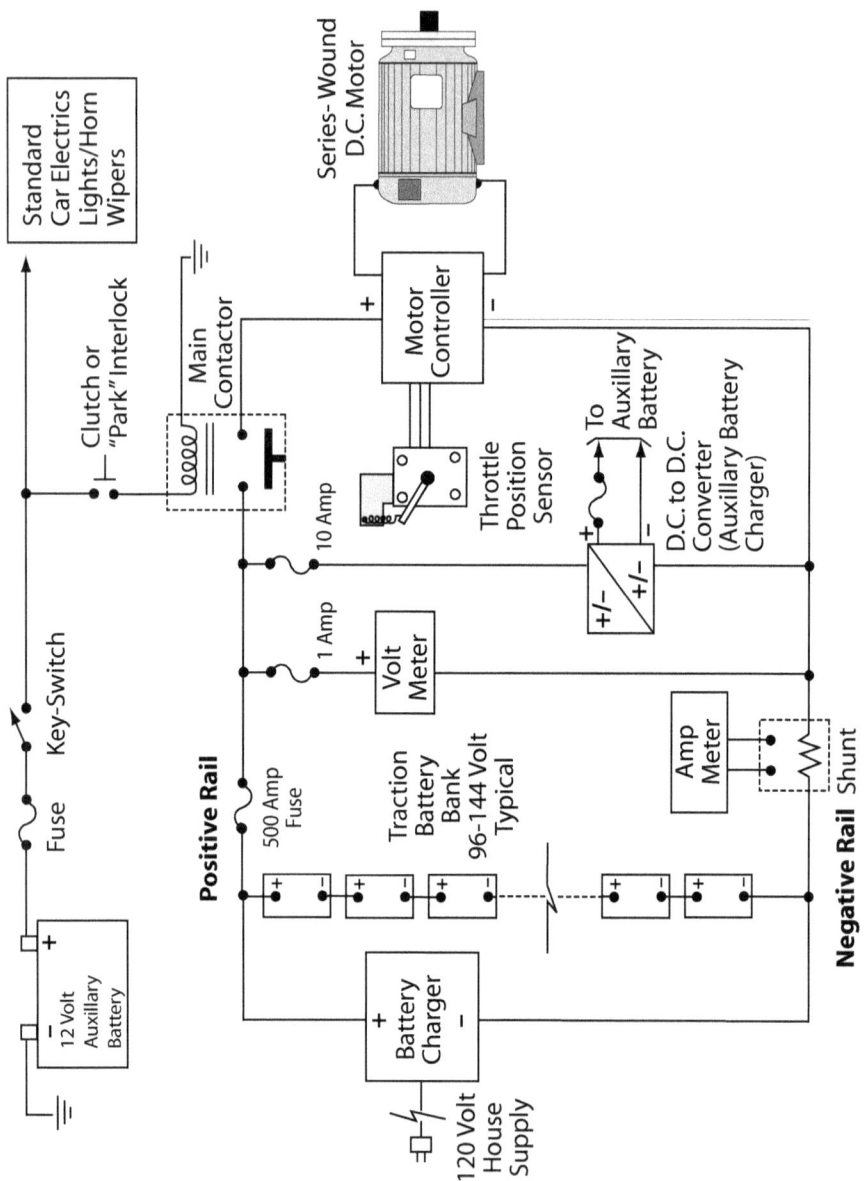

Figure 6-17. This schematic diagram provides an overview of the classic series-wound direct current battery electric vehicle. (continued on next page)

The typical BEV integrates new power storage and control systems into the electrical system of an existing vehicle. In this drawing, the 12-volt auxiliary battery is supplied with the vehicle and is activated via the key switch, supplying power to the lights, horn, wipers, and radio. The activation of the key switch and clutch depression, engages a main contactor which enables the electric traction system.

In most BEVs, the battery bank consists of a series of deep-cycle lead-acid batteries (Chapter 7) that are arranged in a 96- to 144-volt string. Higher operating voltages ensure better overall performance. Battery power is fed along two "rails" which conduct energy through the main contactor into the motor controller. The motor controller monitors the angular position of the accelerator pedal and provides a variable voltage/current to the traction motor, allowing the vehicle to accelerate/decelerate as required.

Current and voltage meters monitor system performance and act as the "fuel gauge" for the battery bank.

The DC to DC converter lowers the traction battery voltage to a level sufficient to maintain a charge on the auxiliary battery, as there is no longer an alternator to fulfill that function.

A battery charger on-board the vehicle replenishes the electrical energy consumed during vehicle operation. An average charge requires six to eight hours to complete.

Comparing BEVs to Internal-Combustion-Powered Vehicles

In many ways, the home conversion of a production vehicle to BEV allows the driver all of the advantages of the donor vehicle, including style, safety, acceleration, speed, maneuverability, operating cost, and comfort. Interestingly, even the average home-built BEV has approximately double the energy efficiency, or miles per gallon equivalent (MPGe), of its internal-combustion-powered predecessor, owing to the inherently higher conversion of energy with the chemical battery/electric motor than the internal combustion engine (Table 5-1)[4].

Although many of the features of the BEV are equivalent to the gasoline-powered version, the main benefits that are given up are also two of the most important: driving range and recharging time.

BEV Range

Many proponents argue that BEVs are not meant to be operated for long ranges, and since most households have two vehicles, the BEV should be used for short-haul commuting and local errands within the typical 50-mile (80-kilometer) range. Furthermore, since most people typically drive only short distances, a rental car or mass transit could be used for longer trips.

The range of a BEV depends on a number of factors including battery type, vehicle weight/design, rolling resistance, and driving style. A gasoline-powered vehicle looks at fuel consumption (in miles per gallon) as the popular measure of its efficiency, although this gives a very poor assessment of the actual energy used, as discussed earlier. A BEV uses the amount of net energy delivered by the battery and consumed by the electric motor to define its efficiency; a typical BEV may consume 240 watt-hours (Wh) per mile (155 Wh per kilometer).

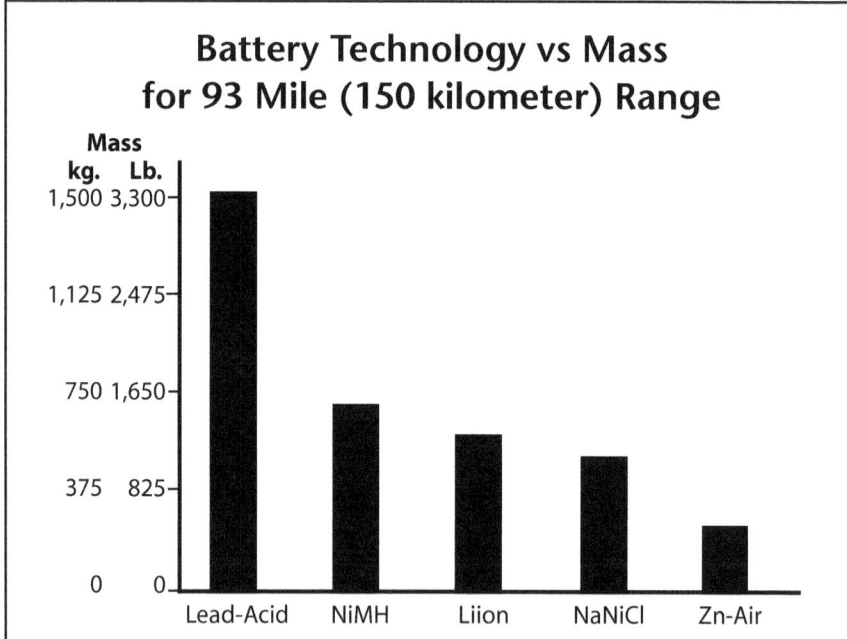

Figure 6-18. This chart shows the relative mass and range of various battery technologies used to power a BEV over a range of 93 miles (150 kilometers). Because of the mass of lead-acid batteries, it is clear that they are suited only for short-range vehicles[5].

It stands to reason that the greater the "volume" of energy stored in the battery bank, the further the vehicle will be able to travel. A car that has a gas tank twice as large as that of a similar vehicle should be able to drive approximately twice the distance on its larger tank of fuel. The same theory does not apply to BEVs as a result of the massive increase in weight and volume required by an expanding battery bank. For example, a BEV with a 600-pound (272-kilogram) battery bank might have a range of 20 miles (32 kilometers), but even if there were the physical space to hold a 1,200-pound battery bank, the extra weight would potentially mean a maximum range of only 30 miles (48 kilometers).

In order to increase BEV driving range, it is better to install higher-efficiency batteries than to use more of the same type:

- Lead-acid are the cheapest and most commonly used in conversion and early-stage production vehicles. Expected range is between 20 and 50 miles (30 to 80 kilometers).
- Nickel-metal hydride (NiMH) batteries can hold more energy per unit of volume and mass (energy density) than lead-acid batteries and can be expected to provide a range of up to 120 miles (200 kilometers).
- Lithium ion battery formulations (having one of the highest energy densities) can provide up to 300 miles (500 kilometers) of range as witnessed by early-stage production vehicles now entering the market.
- Advanced formulations including sodium nickel chloride and zinc-air may provide even greater range as these technologies reach the production floor.

Some enterprising folks have even created a solution to the need for greater range by developing trailers with gasoline-powered electric generators, additional battery banks, or "pusher engines." JB Straubel, the engineering genius behind the Tesla Roadster electric supercar (discussed below), started quenching his thirst for electric speed by converting a 1984 Porsche 944 into a BEV. The 850-pound battery bank comprising twenty 12-volt batteries gave the sports car a range of 20 miles (32 kilometers), which was not sufficient to get to electric-vehicle racing events favored by JB. To improve this range, JB lopped off the back half of a front-wheel-drive Volkswagen, turning it into a power unit/trailer that could push the electric Porsche as far as required (Figure 6-19). It might look a bit odd, but the combination performed as intended.

Figure 6-19. JB Straubel, the engineering genius behind the Tesla Roadster electric supercar, started his rush for electric speed by converting a 1984 Porsche 944 into a BEV. To improve the range of his vehicle, JB lopped off the back half of a front-wheel-drive Volkswagen, turning it into a power unit/trailer that could push the electric Porsche as far as required. It might look a bit odd, but the combination performed as intended. (Courtesy: JB Straubel)

Figure 6-20. The Solar Electric Company has developed a Toyota Prius with a 215-watt photovoltaic panel on the roof of the vehicle to assist in battery charging. Although the concept is interesting, the effective increase in range for a BEV or modified hybrid under real-world conditions is so slight that this option is not much more than a novelty[6]. (Courtesy: Solar Electric Vehicles)

Recharging Time

Proponents also argue that the "at work" or overnight charging time of the BEV is generally not a problem except where the driving range has been extended beyond the ability of the vehicle to make a return trip. If this does occur, plugging the car in to any "friendly" power outlet along the way for even a short period will generally provide enough juice to get home.

Although it is possible to provide a quick, partial charge of some battery technologies, it is still far too slow for the average person to accept as a limitation of a primary vehicle. We are a society on the run, and even filling up with gasoline at the local service station takes too long for many people.

Some experts have proposed the development of "battery-swapping stations", where a car would drive in to a service bay and an automated system would remove the depleted battery bank and exchange it for a fully charged one. If the exchange labor could be completed quickly enough and at a competitive cost, this concept might have some merit, although there are no commercial facilities offering this service to date.

A further improvement on battery recharging is the possibility of refilling or refueling the unit by replacing either spent electrolyte liquid or a spent electrode. The aluminum air battery, for example, is recharged by replacing the used negative electrode. In theory this concept could offer a more rapid "recharging" program than plugging in to a wall socket; however, it remains to be seen if it will be a suitable solution in view of cost and infrastructure issues.

Operating Cost

Operating costs for BEVs are generally thought to be lower than those of comparable internal-combustion-powered vehicles. This is due to increased operating efficiency and the lower cost of electricity relative to that of gasoline, a trend that may well continue in the future, especially for low-carbon sources of electrical power.

As well as the issue of operating cost resulting from energy consumption, there are a number of additional elements that must be factored into the economic equation before the complete picture becomes clear:
- Electricity costs from fossil sources will increase dramatically as carbon taxation is added to the base price of energy.

- Battery life cycle depreciation must be considered. Lead-acid batteries that are constantly depleted and recharged will reach the end of their working life very quickly. As any battery technology ages, its capacity and therefore vehicle range and performance suffer, further accelerating the need to replace.
- Electric vehicles do not require the same level of routine maintenance as traditional internal combustion models. There are no spark plugs, lubricating filters, oil, air cleaners, or other items that require changing. Some battery technologies are virtually maintenance free.
- BEVs that include regenerative braking in their design will have dramatically reduced brake pad and rotor wear.

A Collage of Battery Electric Vehicles

Figure 6-21. The ZENN neighborhood car is a commercial version of Fred Green's BEV Volkswagen Jetta, with the exception of speed. Low-speed vehicles such as the ZENN are limited by local regulations requiring them to operate at a maximum of 25 miles per hour (40 kilometers per hour) although they are fully capable of operating at highway speeds. Variations in crash-worthiness regulations around the world determine speed limits. Maximum published range for the ZENN car is "up to 35 miles" (56 kilometers) with an 80% recharge time of four hours, and a complete charge in eight hours. (Courtesy: ZENN Motor Company)

Figure 6-22. The General Motors EV1 was the most advanced BEV when it was released in 1996. The vehicle improved upon the series-wound direct current motor design described above by installing a 3-phase induction motor and frequency inverter as the drive system. This feature, as well as regenerative braking, gave the car better electrical efficiency. Early units were supplied with lead-acid batteries with a capacity of 53 amp-hours, while later models were fitted with nickel-metal hydride batteries. Lead-acid battery cars were reported to get between 55 and 75 miles (90 to 120 kilometers) range per charge, while cars with ECD Ovonic Company nickel-metal hydride batteries got between 75 and 150 miles (120 to 240 kilometers) range per charge. Recharging took up to eight hours.

The lead-acid configuration consisted of twenty-six 12-volt batteries holding 18.7 kWh of energy, while the nickel-metal hydride batteries held 26.4 kWh of energy. This resulted in average energy consumption of 287 watt-hours per mile (178 Wh per kilometer) for the lead-acid model and 234 watt-hours per mile (145 Wh per kilometer) for the nickel-metal hydride configuration. (Courtesy Brad Waddell from www.getmsm.com/ev/EV1/default.htm)

Figure 6-23. The Myers Motors NmG (No more Gas) is a comic-book-like vehicle that uses the same basic electrical configuration as Fred Green's BEV and the ZENN Car. The NmG is not classified as a car but as a three-wheeled motorcycle, allowing the unit to operate at speeds in excess of 70 miles per hour (113 kilometers per hour). (Courtesy Myers Motors)

Figure 6-24. The NmG has an effective range of between 25 and 35 miles (40 to 56 kilometers). It uses thirteen 12-volt sealed lead-acid batteries arranged in a 156-volt string to give the vehicle highway performance. (Courtesy Myers Motors)

Plug-In Hybrid Vehicle Technology

Figure 6-25. The Tango by Commuter Cars Corporation is a narrow-bodied electric vehicle weighing in at a massive 3,057 pounds (1,387 kilograms), of which approximately 1,000 pounds (454 kilograms) are battery. A total of twenty-five 12-volt batteries are series wired to provide a nominal 300-volt drive system. The vehicle is equipped with two drive motors, one for each rear wheel. (Courtesy: Commuter Cars Corporation)

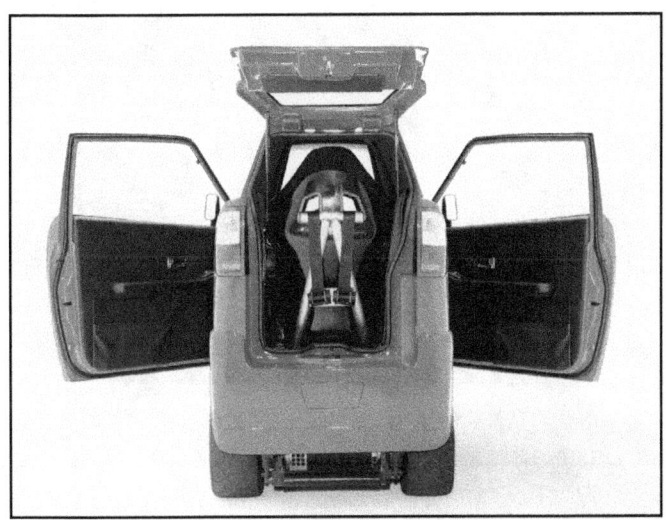

Figure 6-26. The Tango has a range rating of between 40 to 80 miles (64 to 130 kilometers), a top speed of 150 miles per hour (240 kilometers per hour), and can accelerate from 0 to 60 mph (0 to 97 kph) in about four seconds according to the manufacturer. (Courtesy Commuter Cars Corporation)

Figure 6-27. The electric Tesla Roadster revs up the heart rate of any sports-car-loving person just on looks alone. The fact that it is one of the fastest production cars in the world makes the story that much better. The car is based on a modified Lotus Elise and compromises nothing in performance or range. (Courtesy Tesla Motor Company)

Figure 6-28. This shadow view of the Tesla Roadster exposes the transverse-mounted 180-kilowatt, 3-phase, alternating current motor and the battery bank comprising 6,831 lithium-ion laptop computer batteries. The combination gives the Tesla better acceleration than a Lamborghini Gallardo and a range of 250 miles (400 kilometers) before it requires a 3.5-hour recharge. (Courtesy Tesla Motor Company)

Figure 6-29. A red line of 13,500 RPM, normally reserved for motorcycles, is about how fast the Tesla Roadster is. Although the need to drive any car as quickly as the Tesla is questionable, if it helps to rebuild the panache of oil-free battery electric vehicles, then it will have done its job. (Courtesy Tesla Motor Company)

Figure 6-30. The battery-charging socket of the Tesla has been fitted with glowing blue LED lights, possibly to remind the owner just how cool the vehicle is. It will be cooler still if zero-carbon electricity is used to power the vehicle. Zero-carbon emissions and an energy efficiency rating about four times greater than the average North American automobile is a very good start for advanced low-emission automobiles. (Courtesy Tesla Motor Company)

BEV Summary

The prognosis for battery electric vehicles appears to be reasonably good provided they are used in the specialized applications they are best suited for, notably local commuter vehicles, courier services, inner-city transit, or specialized applications.

On the positive side of the equation, BEVs can easily be recharged using low- or zero-carbon electrical sources, ensuring that the carbon profile from power plant to wheels is truly zero. Even where jurisdictions do not have their own clean-technology electrical-power sources, there are innovative ways of ensuring that battery charging remains zero-carbon, which I will discuss in Chapter 8.

Unfortunately, the limited range and extended charging times of electric vehicles cannot compete with the very rapid refueling time of a gasoline-powered vehicle, and BEVs will, in my opinion, remain a limited, yet important player in the market.

Regardless of which battery composition is finally chosen for BEVs, the fact remains that storing electrical energy using chemical means will never be able to compete with the energy density of gasoline. For this reason, multiple configurations of vehicle design will be required in pursuit of the zero-carbon-emission vehicle.

The Plug-In Hybrid Electric Vehicle (PHEV)

If we take a few moments to consider the previous sections on hybrid and battery electrical vehicles, it should come as no surprise that combining the best attributes of the two technologies results in a vast improvement in vehicular energy efficiency.

Internal combustion engines are notoriously inefficient, converting approximately 70% of the fuel consumed into waste heat rather than useful work (Figure 5-8). Furthermore, these engines must remain at a minimum idle speed in order to develop sufficient torque to propel the vehicle forward from a stop, consuming more fuel and doing nothing in return. As the vehicle accelerates, the engine rpm varies, never remaining at its peak efficiency point for more than brief moments. As a result, a clutch mechanism and speed-reducing transmission are required to align the mismatched engine and road speed, further compounding energy losses.

An electric motor, on the other hand, can develop maximum torque even as it begins to accelerate from a complete stop. Most electric motors have "linear" speed-to-power relationships, allowing the motor to operate at very high efficiency levels, up to 96%, over most of their operating range.

On the other side of the balance sheet, the internal combustion engine can operate on either gasoline or diesel fuel, both of which are very energy-dense liquids that provide long operating range and rapid vehicle refueling. Current battery technologies can store only limited amounts of energy and require several hours to recharge.

Overview of PHEV Technology and History

If the electrical side of a hybrid vehicle is made stronger than normal, more like a BEV, and is able to propel the vehicle 60 miles (100 kilometers) without using gasoline, total fuel consumption is reduced by 85% while smog-forming emissions are reduced by 55%.[7] A properly designed full-size SUV of this configuration can exceed a fuel economy rating of 80 miles per gallon equivalent even without the development of advanced chassis materials and reduced aerodynamic drag.[8]

The reason that fuel economy and smog-related emissions can be reduced to such an extent has to do primarily with the distribution of people's driving habits and the very high energy efficiency of an all-electric vehicle.

Vehicle Model	PHEV Toyota Prius	Toyota Prius	U.S. Fleet Average
miles per gallon	68.4	43.6	19.8
kilometers per liter	29	18.4	8.4
CO_2 (emissions-equivalent) (lbs/mile)	0.406	0.542	1.192
(grams/kilometer)	0.115	0.153	0.336
gallons of oil saved per year	431	331	n/a
Liters of oil saved per year	1631	1253	n/a

Table 6-1. Using its fleet of Toyota Prius cars that have been converted to PHEV technology, Google.org has determined that PHEV vehicles greatly improve fuel efficiency and carbon dioxide emissions and have the potential to dramatically reduce fossil-fuel consumption. CO_2 emissions equivalent for the PHEV vehicle is well under the limit of 130 grams per kilometer that the European Union is planning to set by 2012. With PHEV technology, we are already there. (Source: Google.org, www.google.org/recharge/)

We learned in Chapter 1 that the average vehicle is driven approximately 23 miles (37 kilometers) each day and covers longer distances only occasionally. Therefore, if a vehicle could be designed that operated from a battery bank charged with zero-carbon electricity for the majority of the driving distance and used gasoline or other fuel for the remainder, the "equivalent" fuel economy of the vehicle would increase dramatically.

The low- or zero-carbon electricity would be provided through overnight recharging by simply plugging the car in at night or during idle periods. Renewable (or nuclear) electric power sources would be used to provide this emission-free energy. Likewise, if the gasoline were replaced with a true renewable fuel that had a zero-carbon life cycle, the vehicle's effective "gas mileage" might truly reach infinity.[9]

Further, if the vehicle were designed like a standard hybrid, the gasoline and electrical systems would work together—symbiotically—each working to offset the shortcomings of the other while enhancing its attributes.

For example, when the vehicle stopped at a traffic light, the gasoline engine would turn off, saving fuel. When the driver removed his or her foot from the brake, getting ready to accelerate, the battery and electric motor would instantaneously restart the motor. If the battery bank were to become overly discharged, the gasoline engine would charge it during low-demand periods; regenerative braking would also assist with recharging.

In short, the PHEV would operate like a finely tuned Swiss watch, performing a technological ballet with the potential to reduce greenhouse gas emissions, improve vehicular fuel economy, and dramatically lower oil consumption if the technology became widely adopted. Fittingly enough, it was the Europeans who developed the early concept of the PHEV. In 1989 Audi introduced the Audi Duo PHEV, which was based on its 100 Avant quattro[10]. This car was equipped with a 12.6 horsepower (9.4 kW) electric motor and nickel-cadmium battery bank which drove the rear wheels, while a 2.3-liter gasoline-powered engine provided front-wheel drive. Audi has continued with this development work, although it has forgone the plug and focused on traditional hybrid vehicles.

It was a University of California professor by the name of Andy Frank who got the modern development of the PHEV rolling. Starting in 1990, Dr. Frank and his engineering students studied the various technologies required to produce commercially viable PHEVs and developed numerous proof-of-concept vehicles, technical papers, and policy documents. His work has attracted government agencies as well as the auto industry, and in 2001 the US Department of Energy created the National Center of Hybrid Excellence at the University of California, Davis, with Dr. Frank as the director.

A few words should also be said about the timing of General Motors' heavy-handed repossession of the EV1 BEV and its subsequent destruction. The entire story is something of a modern tragicomedy: General Motors was able to stop the California Zero-Emission Vehicle Program, repossess the remaining EV1s, and crush them into a heap of scrap metal, but they were entirely unable to destroy the enthusiasm of the people who demanded the cars in the first place. In many ways, if GM had wanted the EV1 to go away in silence, it should have capitulated and let the cars run their course.

One of the EV1 sales specialists was Chelsea Sexton, who became enamoured with the concept of a clean, nonpolluting form of transportation. Chelsea and others started as enthusiastic sales amateurs and evenutally became very proficient at leasing all of the available EV1 models GM allowed to roll off the assembly line.

After the EV1 program was scrapped by GM, former vehicle leaseholders, and environmentalists, including Chelsea Sexton, staged a mock funeral for the cars, trying to convince the manufacturer not to end the program and to allow lessees to purchase their vehicles. It was all for naught, but their efforts did culminate in the film *Who Killed the Electric Car?*, a hit at the Sundance Film Festival and a cult classic in North America. After the EV1 debacle Ms. Sexton focused her energy on a new nonprofit organization known as Plug In America, which advocates the

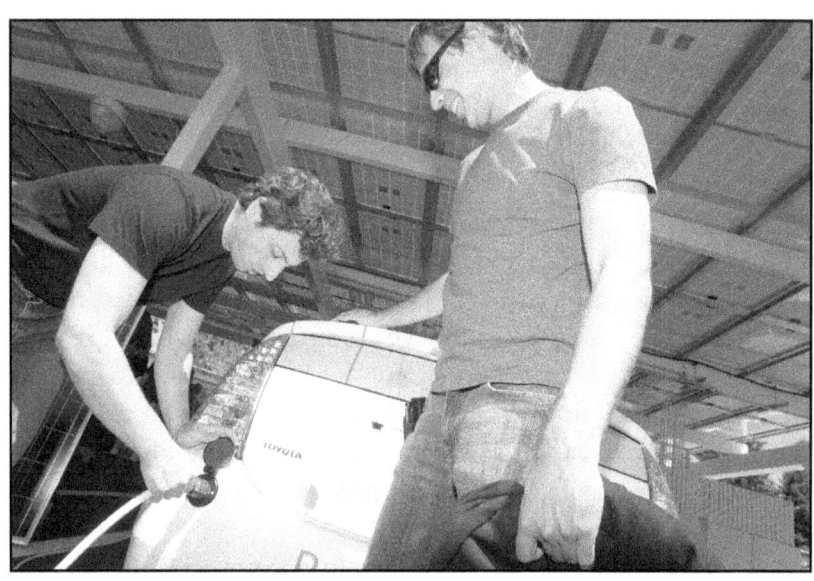

Figure 6-31. Larry Page and Sergey Brin, co-founders of Google Inc., are seen plugging in their Toyota Prius PHEV at the Google campus in California. Google's philanthropic arm has announced its RechargeIT initiative (www.google.org/recharge/) that aims to develop PHEV technology to help reduce US oil dependence and greenhouse gas emissions. Note that the roof covering the parking area is made up of photo-voltaic panels, generating electricity from the sun to aid in the charging of the car. The Google solar sunroof has an electrical rating of 1.6 MW, enough to power approximately 1,000 California homes. (Courtesy: Google)

use of BEV and plug-in hybrid vehicles (www.pluginamerica.com). Plug In America works with numerous organizations and has even recruited former CIA director James Woolsey as one of its members. If General Motors thought for a minute that crushing the EV1 would bring about the end of electrically powered vehicles, it was sadly mistaken.

Figure 6-32. Dr. Larry Brilliant is seen plugging in one of the Google. org fleet of PHEV Toyota Prius vehicles. This model is called a PHEV-30 because it can drive in electric-only mode for up to 30 miles (50 kilometers) on a charge. (Courtesy: Google)

While there is plenty of interest in developing PHEV technology, there are only a very few vehicles on the road at this time, most of them cobbled together by modifying the Toyota Prius hybrid or produced as a concept car, like the General Motors Chevrolet Volt.

A number of activists for clean-car technology, including engineers Ron Gremban, Felix Kramer, and other entrepreneurs, created the California Cars Initiative (www.calcars.org) as a not-for-profit advocacy and technology development group focusing on PHEV development. Their claim to fame was the so-called "Prius+", the first modification of the Toyota Prius hybrid with an extra-large battery bank, on-board charging system, and, of course, a place for a power cord. The vehicles that Kramer and friends produced were capable of exceeding 100 miles per

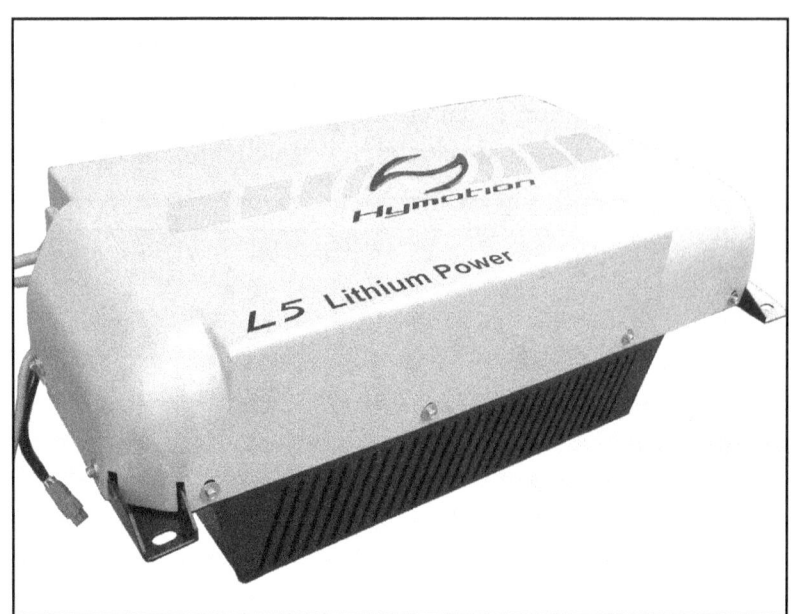

Figure 6-33. The Hymotion company of Toronto, Canada has produced a modular kit that converts a regular Toyota Prius into a PHEV-30. (The company also produces a PHEV kit for the Ford Escape SUV.) Hymotion promises an all-electric range of 30 miles (50 kilometers) with zero gasoline consumption in typical city driving and a combined city/highway fuel consumption of 100 miles per gallon (42 kilometers per liter). (Courtesy: Hymotion Inc.)

gallon (42 kilometers per liter), and by publishing their design concepts on the Web they provided an open-source forum for those interested in creating their own version of PHEV.

The Hymotion company of Toronto, Canada also capitalized on this concept and has created a modular kit that converts a regular Toyota Prius into a PHEV-30 (Figure 6-33). (The company also produces a PHEV kit for the Ford Escape SUV, with other configurations in the works.) Hymotion promises an all-electric range of 30 miles (50 kilometers), with zero gasoline consumption in typical city driving and a combined city/highway fuel consumption of 100 miles per gallon (42 kilometers per liter).

Google's philanthropic arm has announced its US $10 million RechargeIT initiative (www.google.org/recharge/) that aims to develop PHEV technology to help reduce US oil dependence and greenhouse

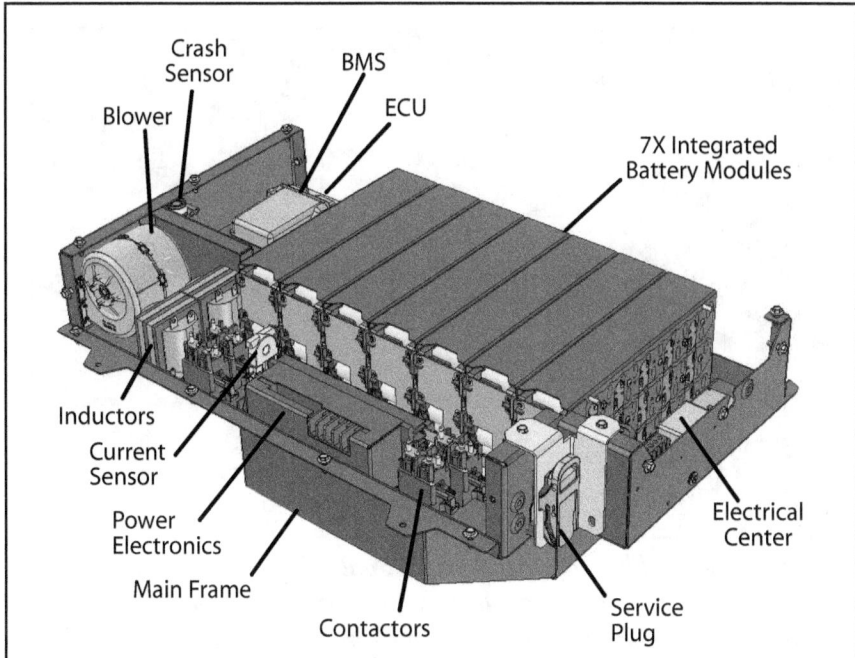

Figure 6-34. This image shows the internal workings of the Hymotion 5kWh plug-in hybrid system. The lithium polymer battery bank takes the majority of the space, with the battery management and power conversion equipment taking the balance. (Courtesy Hymotion Inc.)

gas emissions. Its fleet of Toyota Prius cars converted to PHEVs using Hymotion technology has demonstrated improved fuel efficiency and reduced carbon dioxide emissions and fossil-fuel consumption. CO_2 emissions equivalent for the fleet are well under the limit of 130 grams per kilometer that the European Union is planning to set by 2012. Even though these vehicles are concept or test platforms, the results clearly show that PHEV technology is already here—and works.

General Motors also seems to have had a change of heart regarding the concept PHEV technology known as E-flex. The E-flex System used in the Chevrolet Volt concept vehicle includes a 16-kWh lithium-ion battery pack which powers a 120-kW electric motor used to propel the vehicle. A small 1-liter, turbocharged engine is linked to a 53-kW generator which supplies power to the vehicle and provides charging energy when the battery bank is depleted. The vehicle is a true series PHEV, providing 40 miles (64 kilometers) of range on battery alone.

Figure 6-35. The Google campus not only keeps vehicles cooler by providing shade, but also has the roof of the parking area covered with photovoltaic panels which can generate up to 1.6 MW of electricity, enough power to operate a thousand average California homes as well as recharge its fleet of PHEVs. (Courtesy: Google)

The E-flex title evokes "flexibility," and to this end General Motors has suggested that the engine/generator could be modified to operate on a diverse range of fuels including gasoline, E85 (85% ethanol/15% gasoline), E100 (pure ethanol), and biodiesel, as well as hydrogen fuel cell technology. No firm production date has been set for the Volt, although it has been speculated that it may hit the roads between 2010 and 2012.

Not wanting to be outdone, Toyota has also started tinkering with its own version of PHEV. Toyota indicates that it will upgrade the Prius platform on a limited number of vehicles and perform ongoing tests in Japan and at the Advanced Power and Energy Program at the University of California, Irvine.

The factory-modified Prius will have its battery capacity doubled to 13 amp-hours through the installation of a larger nickel-metal hydride pack which will increase electricity-only range to approximately 8 miles (13 kilometers). Modifying the computer software in the vehicle's management system will also allow the car to operate at speeds of up to

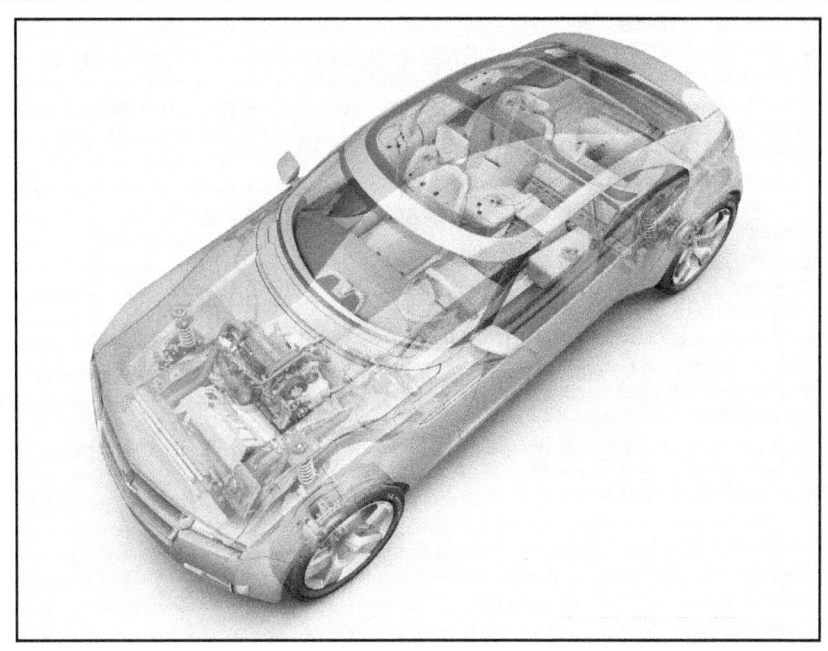

Figure 6-36. The General Motors E-flex System includes a 16-kWh lithium-ion battery pack which powers a 120-kW electric motor used to propel the vehicle. A small 1-liter turbocharged engine is linked to a 53-kW generator which is used to power the vehicle and provide charging energy when the battery bank is depleted. The vehicle is a true series PHEV, providing 40 miles (64 kilometers) of range on battery alone. The concept vehicle shown here is called the Chevrolet Volt, and if rumors are correct you may see these in the showroom by 2012. (Courtesy: General Motors Corporation)

62 miles per hour (100 kilometers per hour) before the gasoline engine turns on. A charging socket on the vehicle will make it possible to recharge the batteries in three to four hours.

Although Toyota has admitted to tinkering with PHEV technology, the company says it expects "standard" hybrid sales to reach the one-million-per-year mark early in the next decade and take the lion's share of advanced technology vehicle sales. The company believes that a considerable amount of work must be done before it delivers PHEVs to the mass market.

PHEV Summary

PHEV technology works, and even limited test runs show that large decreases in fuel consumption and greenhouse gas emissions are not only possible but readily quantifiable and predictable. So what is stopping the wider application of PHEV technology? The answer, not surprisingly, is cheap gasoline and a complete lack of government leadership on carbon valuation.

Vehicles like the Prius+ or Chevrolet Volt are interesting, premium-priced cars and there is no question that there is at least a small market for them today. However, carbon taxation must be punitive in order for people to make the switch and car companies to respond in earnest with viable products. It may be a while before you see these vehicles humming along the streets.

The Propaganda of Hydrogen-Powered Vehicles

You would think, based on TV and magazine ads, that hydrogen-powered vehicles were already plying the nation's roads, with sleek cars noiselessly gliding down the road while water drips benignly from their tailpipes. Cars that run on water? This sounds more like science fiction than serious automotive technology.

The future of vehicular energy might well belong to hydrogen or some other as yet unimagined energy source, although I seriously doubt it. Notwithstanding all of the posturing and hype in the media, the hydrogen infrastructure necessary to power personal vehicles and homes

Figure 6-37. The Hy-wire hydrogen fuel cell car provides a long-term view of the future of automotive technology. The entire propulsion system for this car is housed in an 11-inch-thick (28-centimeter) skateboard-like chassis. All mechanical linkages for the operation of brakes, transmission, and steering are replaced by software and wires. The engine, steering columns, and other components found in a conventional vehicle are eliminated, allowing unprecedented design freedom and the ability to use one mechanical platform and interchange body styles at will. (Courtesy General Motors Corporation)

is a long, long way off and the ravages of climate change and other converging realities will happen much faster than the transition to a hydrogen economy that purports to correct these issues. Even if (and that is a very big "if") hydrogen does enter full-scale production in the next 25 years, it will be too late to keep atmospheric carbon below the catastrophic level.

Figures 6-38 a/b. BMW has not only invested big dollars in hydrogen-powered vehicles but also enlisted the aid of big-name stars like actors Sharon Stone and Arnold Schwarzenegger. Clearly politics and hype are as important as physics and fact in the race to replace oil. (Courtesy BMW)

Obstacles on the Road

What exactly are the problems with hydrogen, and if it is so bad why are companies and governments supporting its development with such vigor?

Although the situations are different, hydrogen and the "dot-com bubble" at the turn of the millennium have very similar plot lines. For those who don't remember, or perhaps have tried to forget, the dot-com bubble was a speculative financial period during which Internet-sector companies saw their values and hence share prices soar without any basis in reality. Fueled by seemingly endless hype, low interest rates, and buckets of venture capital, everyone appeared to throw caution to the wind and jump on the dot-com bandwagon.

Based on the novelty of the Internet in 2000 and its potential to change how businesses operate, people believed that dot-coms, as the new Internet-savvy businesses were called, would be able to sustain almost endless growth and provide their services at a nominal cost, reaping heady profits.

Many established companies got caught up in the infinite growth and "risk everything to ratchet up the share price" mentality. Old "bricks and mortar" companies such as Lucent Technologies, Nortel Networks, JDS Fitel, and Cisco Systems, to name a few, played fast and loose with the dot-com run-up. Nortel Networks is a prime example of what can happen to a company that should have known better but decided to roll the dice and risk it all.

Nortel was at one time Canada's superstar technology company and accounted for almost a third of the value of the Toronto Stock Exchange (TSE) 300 Index, as it was then known, an index that contained 300 companies. Nortel used to have 3.8 billion shares outstanding when it was trading at its height, and with a stratospheric stock price of $124.50 per share it made up 36.5% of the index value; the remaining 299 companies accounted for the balance. With a market capitalization of almost $400 billion, the company could do no wrong as it went on an acquisition spree, buying up almost every startup Internet company it could get its hands on[11].

When the fall came, it was spectacular. October 25, 2000 marked the day that Chicken Little was right and the sky started to fall; Nortel's share price dropped to $71. For the remainder of the year and into 2001 and 2002, sales steadily collapsed while other companies in the same market also ran for cover, further depressing Nortel's sales and share price. Streams of profit warnings were issued by the once-mighty company, and eventually its corporate debt was downgraded to junk status and its share price fell to 69 cents.

Of course the damage was not limited to Nortel but also affected hundreds of other companies and millions of shareholders and financial advisors who thought they could play fast and loose with old-world business rules (such as the requirement to make a profit) and create a new world economy. The Internet may play in the virtual world, but capitalism plays in the real one, and as everyone should now understand, uber-enthusiasm must be tempered by common sense.

But human nature is not quite this logical and people get caught up in the "herd mentality," believing that if everyone else is running for the latest investment flavor of the day they should follow. The run-up in the hydrogen story is really no different.

Ballard Power Systems Inc., the largest manufacturer of PEM fuel cells, is a prime example of the herd following the hype. The rise and fall of investor interest in Ballard and its peers created a "hydrogen bubble" that closely mimicked the dot-com bubble. A news report dated November 10, 2000[12] states:

"The stock has made dramatic price surges when strategic alliances were announced: the stock took off when Daimler-Chrysler, and later Ford announced that they would join Ballard in a technology alliance."

The report goes on to say:

"Tuesday's positive endorsements by Morgan Stanley in New York—reaffirming its 'strong buy' rating for Ballard Power—maintained that news-driven trading pattern. By the end of trading, the stock gained $4.20 to close at $53 even on the TSE [Toronto Stock Exchange]."

The hype about Ballard during this time (2000) was relentless, with the company announcing its objective to "have commercial vehicles available by 2004 in showrooms across North America."[13]

That was then. Reality has taken hold of the overly enthusiastic investors in hydrogen as most stock traders and advisers have gone on to "the next big thing." Ballard is still in the business of burning up cash and trying desperately to find a niche market to create profitability for itself. The more somber tone of current CEO John Sheridan is more realistic:

"Our business strategy is to drive to profitability in these non-automotive nearer term markets, we see automotive then providing a very significant benefit on top of that."[14]

Profitability means finding sales, and for the foreseeable future this will not be in the automotive market. As a result, the company has focused on niche applications in areas where fuel cell technology makes sense today, for example, forklifts and backup power systems. Serious automotive sales, in my opinion, might occur in 25 years' time, if ever.

Ballard, with an all-time high of $200 in 2000, was trading on the Toronto Stock Exchange at $5.14 in May, 2007, a precipitous drop and a new 52-week low for the company. The company has seen its workforce drop from 1,400 to fewer than 1,000 and has sold its German subsidiary, Ballard Power Systems AG. With an operating loss of $17.2 million in 2006, the company is burning more than hydrogen.

Ballard is of course not the only troubled player in the hydrogen market, and the reasons for this are linked to three things: government policy, public attitude, and technology issues.

The problem with government policy is obvious. Low carbon valuation and unrealistically low gasoline and diesel fuel prices make a transition to hydrogen completely impossible, even if the technology were available. Why would we use a more expensive alternative product like hydrogen when governments are telling us there is plenty of cheap oil left for the taking? If the government really wanted to support hydrogen (or any other transportation technology), then properly valuing carbon would be the way forward.

The public, for their part, like to say that they want to protect the environment and not purchase foreign oil (a majority of which comes from Canada through US-owned suppliers), but only if there is no change in cost or perceived value. In other words, as long as hydrogen is just as cheap and available as gasoline and there is a wide selection of low-cost hydrogen vehicles on the market, people will make the switch. There can be no change of entitlement. The bottom line: without a punitive valuation on carbon and given the rest of these conditions people will stick with gasoline.

The technical roadblocks related to hydrogen are much more complex and could easily fill a bookshelf with arguments both for and against the technology. For this review, I will provide a basic overview and refer

Figure 6-39. The public, for their part, like to say that they want to protect the environment, but only if there is no change in cost or perceived value. In other words, as long as hydrogen is just as cheap and available as gasoline, provided there is a wide selection of low-cost hydrogen vehicles on the market, people will make the switch. The Honda FCX hydrogen-powered car shown here is a concept vehicle and will likely remain that way for many years to come. The bottom line, given these conditions, is that people will stick with gasoline. (Courtesy: Honda Motor Company)

the reader to the Resource Guide in Appendix 1 for suggested further reading.

A Few Facts about Hydrogen Technology
Hydrogen as a Fuel
Many people think of hydrogen as a fuel source; it is not. Gasoline, wood, coal, and whale blubber are examples of fuel sources and may be burned to extract the energy (carbon) trapped within. Hydrogen, on the other hand, does not freely exist in nature and must be extracted from other sources. It is known as an "energy carrier" or "medium," because you must use a primary source of energy such as natural gas, coal, or wind to extract it from a hydrocarbon or water source. In today's economy, vir-

tually all of the hydrogen in use is extracted from natural gas (methane), a process which is inherently dirty and provides significantly less energy output than if the natural gas had been burned directly.

You cannot create something from nothing, and hydrogen production is no exception. In order to extract hydrogen from fossil fuels (an inherently dirty approach based on short-term supply at best), a given volume of fuel must be "burned" or reformed, producing a volume of

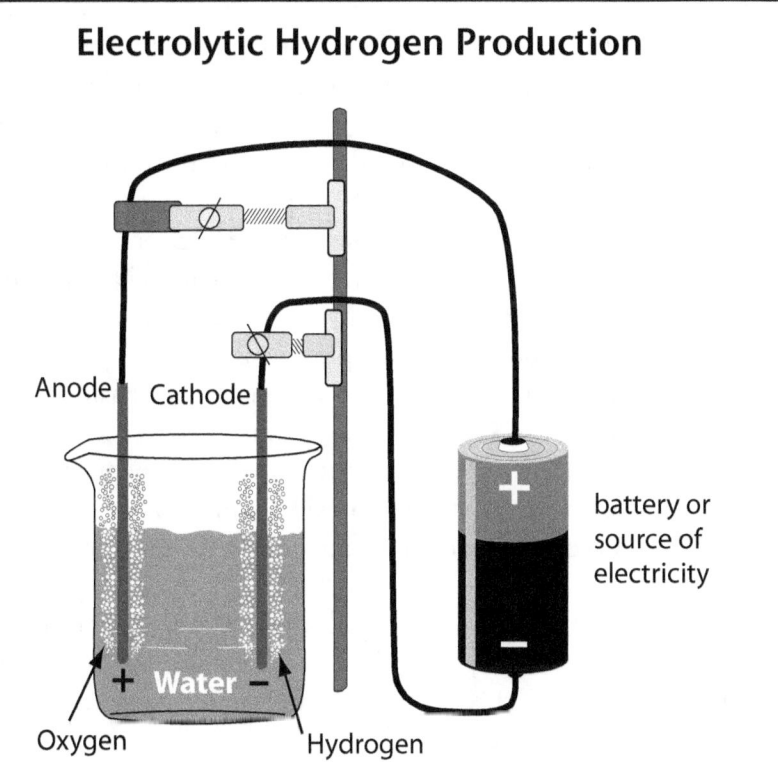

Figure 6-40. Anyone who achieved a passing grade in chemistry will recall the world's most famous chemical symbol: H_2O—plain, old water, which can be converted into hydrogen using the electrolytic system shown here. Passing an electric current through water causes hydrogen gas and oxygen to be liberated at the negative cathode and positive anode respectively. Using alkaline catalysts such as sodium hydroxide can improve water-to-hydrogen conversion efficiency, although systems on the market operate in the range of 66%.

hydrogen with lower energy content than the original fuel source[15]. Although future consumption of the hydrogen will be clean, greenhouse gas emissions resulting from its production will be higher than if the natural gas or other hydrocarbon input had been consumed directly in an internal combustion engine, especially if the vehicle is equipped as a hybrid[16].

Hydrogen can be produced from the reduction of coal or other fossil fuel, but the problem with this approach is that the production process emits vast amounts of greenhouse gases, the very by-product we are trying to avoid. Although carbon-capture or sequestration processes may work in the coming years, they are nowhere near technically or financially feasible at this time. To be fair, if the technology were ever to become functional, a carbon tax would actually aid the technology. In this instance, the fact that carbon could be sequestered at a cost less than the carbon taxation value would make the hydrogen "green", as there would be no carbon emissions. The large deposits of coal available around the world could be a relatively low-cost hydrogen feedstock.

Futurists, on the other hand, are pinning their hopes on water. Anyone who achieved a passing grade in chemistry will recall the world's most famous chemical symbol: H_2O—plain old water. The majority of the earth's surface is covered in water, which also happens to be the largest supply of hydrogen known to man. Ironically, we have been unable to find an economic means of extracting hydrogen from its watery vault.

The most promising means of releasing hydrogen from water is a process known as electrolysis, which on the surface appears to be very simple. Passing an electric current through water causes hydrogen gas and oxygen to be liberated:

$$2 \text{ molecules of } H_2O + \text{electricity} \approx O_2 + 2 \text{ atoms of } H_2$$

It stands to reason that if the supply of electricity used in the conversion, compression, and balance-of-plant processes is "clean" then the production of hydrogen will also be clean, leading to the moniker "green hydrogen," while its natural gas- or coal-derived cousins (created without the benefit of carbon capture and sequestration) are labeled dirty or "brown" hydrogen.

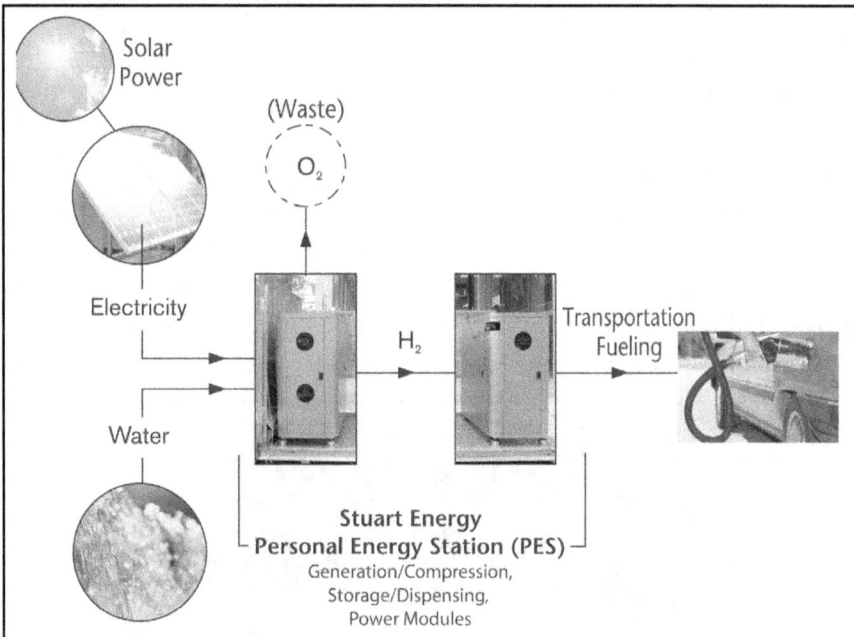

Figure 6-41. Using clean energy from the sun or wind and combining it with water results in clean or "green hydrogen." In theory, hydrogen could be used to fuel our entire economy, heating homes and operating vehicles. Long-term visionaries such as Jeremy Rifkin believe that the nation's automobiles could use their hydrogen fuel to generate electricity for sale to the utility company while sitting parked in the garage. Current-term skeptics such as the author believe that infrastructure costs and low energy-conversion efficiencies along with complex and expensive vehicular technologies make hydrogen impractical for personal transportation. (Courtesy Stuart Energy Systems Corporation)

Using electricity to electrolyze water drastically reduces or even eliminates greenhouse gas emissions provided that the electricity is generated by clean energy sources such as wind turbines or hydroelectric stations. Unfortunately, for every 1 kilowatt-hour of electricity generated by a clean energy source considerably less energy is derived in the form of hydrogen, increasing its retail price significantly. The mass media ads that suggest that hydrogen is cost competitive with gasoline are referring to fossil-fuel-derived hydrogen and are not factoring in the significant negative value of greenhouse gas emissions produced in the conversion process.

To generate enough electricity to produce hydrogen to replace all of the gasoline sold in the United States today would require more electricity than is sold today.[17]

Hydrogen Storage

Hydrogen is the lightest element, having a density of 0.07 grams per cubic centimeter (g/cc), whereas water has a density of 1.0 g/cc and gasoline 0.75 g/cc. Because of its low mass, storing enough hydrogen to provide sufficient vehicular range is very problematic. At room temperatures and pressures, hydrogen takes up approximately 3,000 times more

Figure 6-42. Because of its low mass, storing enough hydrogen to provide sufficient vehicular range is very problematic. At room temperatures and pressures, hydrogen takes up approximately 3,000 times more volume than gasoline for the same energy storage. The General Motors E-flex System shown in this photograph uses high-pressure hydrogen stored in the rear-mounted tanks to feed the front-mounted fuel cell. (Courtesy: General Motors Corporation)

volume than gasoline for the same energy storage. Therefore, to store gaseous hydrogen requires enormous pressures, in the range of 3,000 to 5,000 pounds per square inch (psi) (207 to 345 bar). At these pressures, a hydrogen tank with sufficient safety for road use takes up to eight times more space than the equivalent gasoline storage tank. The compression of the hydrogen gas also consumes approximately 15% of the energy contained in the storage tank[18].

Hydrogen can also be stored as a liquid if it undergoes successive stages of compression and refrigeration. The liquid state has a much higher energy density than hydrogen gas. Unfortunately, liquid hydro-

Figure 6-43. To store gaseous hydrogen requires enormous pressures, in the range of 3,000 to 5,000 pounds per square inch (psi) (207 to 345 bar). At these pressures, a hydrogen tank with sufficient safety for road use takes up to eight times more space than the equivalent gasoline storage tank. The compression of the hydrogen gas also consumes approximately 15% of the energy contained in the storage tank, reducing well-to-wheels efficiency. (Courtesy: General Motors Corporation)

gen must be stored at the ultracold temperature of -423 °F (-253 °C), which is slightly above absolute zero.

The compression and refrigeration used in the liquefaction process is very energy intensive and uses some 40% as much energy as the hydrogen contains.[19] Further, the hydrogen must be stored in complex, superinsulated cryogenic storage tanks that take up a large amount of vehicular space.

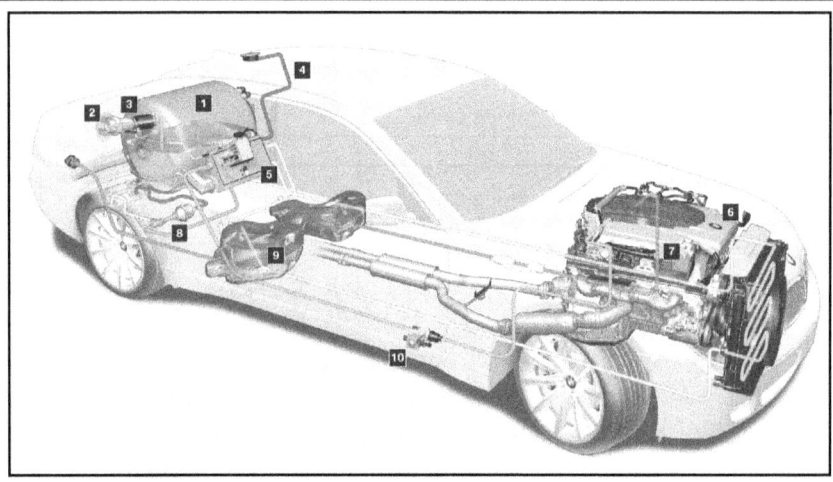

1. LH_2 fuel tank
2. LH_2 tank cover
3. LH_2 tank coupling
4. Safety line to blow valve
5. Auxilary units capsule containing heat exchanger for H_2 and control unit of the tank
6. Bivalent internal combustion engine (H_2/Gasoline)
7. Intake manifold with H_2-Rail
8. Boil-off Management System (BMS)
9. Gasoline tank
10. Pressure control valve

Figure 6-44. The BMW Hydrogen 7, shown here in section view, is equipped with a standard 12-cylinder engine which is fueled from either a 19.5-gallon (74-liter) gasoline tank or a cryogenic hydrogen tank with the capacity to hold 17.6 pounds (8 kilograms) of liquid hydrogen. Since each kilogram (2.2 pounds) of liquid hydrogen contains approximately the same amount of energy as one gallon of gasoline, the vehicle has an effective hydrogen range of only 124 miles (200 kilometers), while the range on gasoline is a more respectable 310 miles (500 kilometers). (Courtesy BMW)

Figure 6-45. Current gasoline fuel tanks are made from very low-cost, lightweight plastics and can be molded into complex shapes to fit the vehicle profile and its design needs. The cryogenic tank and support paraphernalia of the Hydrogen 7 weighs in at 440 pounds (200 kilograms) and would likely have to increase to a large percentage of the vehicles mass in order to provide sufficient driving range. (Courtesy BMW)

Consider the BMW Hydrogen 7 luxury vehicle shown in Figures 6-38 a/b. BMW developed this demonstration car using its well-known flagship model, the 7 Series. The vehicle is equipped with a standard 12-cylinder engine (ridiculous overkill to begin with) which is fueled from either a 19.5 gallon (74 liter) gasoline tank or a cryogenic hydrogen tank with the capacity to hold 17.6 pounds (8 kilograms) of liquid hydrogen. As each kilogram (2.2 pounds) of liquid hydrogen contains approximately the same amount of energy as one gallon of gasoline, the vehicle has an effective hydrogen range of only 124 miles (200 kilometers), while the range on gasoline is a more respectable 310 miles (500 kilometers).

The infrastructure required to provide these ranges is formidable. The section view of the Hydrogen 7 shown in Figure 6-44 demonstrates the complexity of storing liquefied hydrogen. The cryogenic tank is mounted just above the rear wheels of the vehicle, taking up most of the normally cavernous trunk space of the car. In contrast, the gasoline storage tank is mounted just under the rear passenger seating area. To put all

Figure 6-46. The cryogenic tank of the BMW Hydrogen 7 is mounted just above the rear wheels of the vehicle, taking up most of the normally cavernous trunk space of the car. In contrast, the gasoline storage tank is mounted just under the rear passenger seating area. To put all of this in perspective, remember that the cryogenic tank would have to be 2.5 times larger than it currently is in order to provide the same fuel economy as the existing gasoline fuel tank. (Courtesy BMW)

of this in perspective, remember that the cryogenic tank would have to be 2.5 times larger than it currently is in order to provide the same fuel economy as the existing gasoline fuel tank.

Unfortunately, the story does not end here. Liquefied hydrogen "boils off," causing evaporation losses of approximately 4% per day. In other words, for every week that passes, a quarter of the hydrogen is simply lost into the air. If the vehicle were parked for an extended period of time with a partially filled tank (for example, when you park at a train station or are away on vacation), there might not be any fuel left upon your return.

Current gasoline fuel tanks are made from very low-cost, lightweight plastics and can be molded into complex shapes to fit the vehicle profile and its design needs. The cryogenic tank and support paraphernalia of the Hydrogen 7 weighs in at 440 pounds (200 kilograms) and would likely have to increase to a large percentage of the vehicles mass in order to provide sufficient driving range. Interestingly, the driving range of vehicles using compressed or even liquid hydrogen might actually be worse than that of similar battery electric vehicles.

Experiments are now being conducted using a class of materials known as metal hydrides, which allow hydrogen to bond to their surface and in theory could provide a means of storing greater volumes of hydrogen than are possible with compression or liquefaction. Although this process is interesting, it remains at the laboratory level and is not likely to see vehicle-scale commercialization for decades, as the process is no closer to becoming commercially viable than putting a man on Mars.

Hydrogen Distribution

In theory, hydrogen can be produced just about anywhere, allowing the development of distributed production facilities at the local "gas station" or even in your home (distributed generation of hydrogen is also known as "forecourt production") using small-scale water-to-hydrogen reforming stations. In this application low-cost, energy-efficient electrolytic hydrogen production and storage technology would have to be developed. The advantage of such a system is that there would be no major investment in gas pipelines and tanker trucks. To date, the cost-effectiveness of distributed, electrically produced hydrogen and small-scale compression cannot compare with that of central production facilities using natural gas, but this may well reverse depending on carbon taxation policies.

Another likely scenario is the development of centralized coal or other fossil-fuel facilities to generate hydrogen with the carbon dioxide emissions being captured. The CO_2 could be sequestered on-site or potentially sold to enhance oil recovery in mature wells (see Figure 8-10). In this case, some form of national carbon dioxide pipeline infrastructure would be required.

Carrying liquefied hydrogen by means of tanker truck is very common today. Trucking liquefied hydrogen to customers with small to moderate needs is currently cheaper than producing hydrogen on-site

Figure 6-47. In theory, hydrogen can be produced just about anywhere, allowing the development of distributed production facilities at the local "gas station" or even in your home (distributed generation of hydrogen is also known as "forecourt production") using small-scale water-to-hydrogen reforming stations. This image demonstrates BMW's liquid hydrogen fueling station. (Courtesy BMW)

using electrolytic means. Trucking assures the delivery of high-quality product for industrial purposes; concerns about energy efficiency and greenhouse gas emissions are simply not addressed.

The story is worse when it comes to gaseous hydrogen truck transport. Current trucks carry less than 660 pounds (300 kilograms) of hydrogen, enough to fill 60 vehicles. This is entirely wasteful compared to a typical gasoline tanker than can carry some 26 tonnes (equivalent to 10,000 gallons). If gaseous hydrogen were transported exclusively by truck, there would be a ten-times increase in the number of fuel transportation trucks on the road, creating a logistical, cost, and safety nightmare.

Hydrogen can be pipelined, but there are cost complications here as well. Hydrogen is very leak-prone and highly reactive to metals, especially pipeline steel. Accordingly, costs for a network of hydrogen pipelines would be extremely high, and it is unlikely that one will be developed until there is sufficient demand from vehicle use—a classic chicken-and-egg situation.

Fuel Cell Vehicles

BMW has taken the approach of burning hydrogen directly in an internal combustion engine to propel the vehicle. As we saw in Figure 5-8, internal combustion engines have losses amounting to 62.4% of the energy input and a further 18.2% through the drive train system, amounting to a loss of approximately 80% of potential energy. Fuel cell vehicles have a different approach to energy conversion; think of the fuel cell stack as the battery pack in a BEV.

This analogy isn't quite right, but it is close enough to help explain the complex technology. In a fuel cell vehicle (FCV), the fuel cell chemically recombines an on-board fuel supply (hydrogen) with oxygen taken from the atmosphere in a process that is almost exactly a reversal of the electrolytic process used to make hydrogen:

$$2 \text{ atoms of } H_2 + O_2 \rightarrow \text{electricity} + 2 \text{ molecules of } H_2O + \text{Waste Heat}$$

During this recombination process, electricity is formed along with the byproducts of water and heat. The electricity is used to power the vehicle in the same manner as a BEV, and the water drips benignly from the "exhaust pipe." Fuel cells sound like the perfect technology.

Well, not so fast. In his book *The Emperor's New Hydrogen Economy*, author Darryl McMahon describes a 2003 General Motors television commercial where a vehicle powered by a hydrogen fuel cell leaves nothing but a puddle of water on the roadway. Mr. McMahon contemplates this seemingly gentle form of "exhaust pollution," wondering how these cars would perform in the North United States, Europe, and Canada: "Consider the implications for this small puddle where significant numbers of these vehicles are operating in climates where the road surface is below the freezing point for water. The ice that would form on the roads would create road conditions that would cause collisions on a massive scale [necessitating season-long road salting]. The alternative would presumably be to store the condensed water on-board the vehicle until it could be drained at a safe location."

He goes on to point out that while this sounds like a simple enough solution, the storage-container size and added weight would equal nine times the mass of the hydrogen consumed, adding some 140 pounds (64 kilograms) of mass to the vehicle between fill-ups. It might also be necessary to heat the water storage tank (using yet more energy) to prevent it from freezing. And of course the fuel cells themselves would be damaged if they were to freeze, making one wonder how they would operate in the winter in the first place, without consuming vast amounts of energy (hydrogen) to stay warm.

While this is just one of many technical and cost issues relating to bringing fuel cells to the forefront of automotive technology, their proponents like to state that fuel cells can convert fuel to energy twice as efficiently as an internal combustion engine. But the story is being slanted to suit the hydrogen heads. While it is true that FCVs are more efficient than older internal-combustion-powered vehicles, they are not as efficient as either hybrids, BEVs, or PHEVs when you look at the total energy cycle from well to wheels, which is the only measure that counts.

Toyota is a leader in the development of standard internal combustion engine and hybrid vehicles and is now beginning to look at PHEV technology as well. According to research conducted by the company, in a well-to-wheel or life cycle comparison the gasoline-powered Prius is actually more efficient than a typical hydrogen-powered FCV. Toyota states that the Prius has a life cycle efficiency of 29% versus 22% for the FCV, based on hydrogen produced from natural gas[20]. Of course a current Toyota Prius is still fueled with gasoline, and the efficiency picture does not address greenhouse gas emissions.

Summary

Vehicle Type	Fuel Production Efficiency	Vehicle Efficiency	Well-to-Wheel Efficiency
Toyota Prius	79%	37%	29%
Fuel Cell Vehicle	58%	38%	22%
Zero-Carbon Car	80%	53.7%	43%

Table 6-2. The fuel efficiency of a vehicle must be calculated by analyzing the overall efficiency from oil well or power source to tailpipe rather than by considering the vehicle alone. The first two entries of this table, produced by the Toyota Motor Corporation, indicates that the Toyota Prius achieves an overall efficiency of 29%, while a fuel-cell-based automobile powered with natural-gas-derived hydrogen is 22%. If "green hydrogen" derived from wind, hydro, or solar technologies becomes a reality it may be possible for an FCV to reach a well-to-wheel efficiency level of approximately 42%. A PHEV such as the Zero-Carbon Car described in the next chapter can easily reach these levels using technology that is here today.

If the least-cost means of producing, transporting, and retailing hydrogen is not developed until there is sufficient vehicular consumption, then hydrogen costs will remain high since alternative infrastructure means will have to be used in the formative years. Naturally, this creates higher operating costs, which will dissuade all but the early adopters from using hydrogen fuel. The feedback cycle will delay hydrogen's mass application perhaps well beyond the time when personal vehicles will be a viable option at all.

When US President George Bush speaks of hydrogen being the gateway to avoiding imports of foreign oil, he is correct; the US will then be addicted to foreign natural gas instead, a no-win proposition at best. The Freedom CAR program demands cheap, current-size vehicles, which is technically and financially impossible using hydrogen as the fuel source. There is no attempt to put improved mass transit or freight infrastructure on the table, further reinforcing the status quo. And without an effective carbon valuation policy, there is simply no economic means of jump-starting any program that presents an alternative to oil.

The development of a hydrogen-based economy is to my thinking a dead end. Hydrogen is an energy carrier with a negative energy balance and a potentially dubious greenhouse gas emission profile, and people need to recognize that this is simply a story that detracts from proper policy development.

My recommendation to governments is that they stop wasting precious time and research dollars on hydrogen and instead concentrate on developing near-term technologies that have a greater likelihood of being adopted. Advanced PHEVs coupled with a realistic carbon price are the place to start if Western countries really want to stop being addicted to oil and help the environment. PHEVs have already been shown to have a better greenhouse gas emission profile than internal-combustion engines, hybrids or hydrogen-powered cars and could easily reduce vehicular fuel consumption by 80% or more. Better yet, there is no need to wait several decades to make this happen.

Figure 6-48. Billions of dollars are being spent in research and development covering all aspects of the hydrogen economy. Considerable research is required to develop an economic method of generating clean hydrogen, creating a safe storage and delivery infrastructure, and improving the performance and cost of fuel cell technologies. To my way of thinking it is better to bet on PHEVs, which have already been shown to have a better greenhouse gas and efficiency cycle and are far more likely to enter mainstream production sooner. (Courtesy BMW)

Chapter 7
Introducing the Zero-Carbon Car

7.1 Getting the Project Started

The decision to build a plug-in hybrid electric vehicle that uses both zero-carbon electricity and liquid fuels was an attempt to verify the theories developed in other chapters of this book, but it was also the fun part of the project. Critics of ideas that do not conform to the norms of society often claim that academic texts or engineering papers are just not realistic and cannot be realized for one reason or another. This chapter on the practical application of PHEV technology proves naysayers wrong.

I will admit up front that there are limitations in the design of the Zero-Carbon Car. If I had had a larger budget to work with from the outset, the vehicle would have been fabricated differently, but this first version is more than adequate to cruise around the roads and test how well the concept of PHEVs works without having to wait for the commercialization of exotic automotive technologies that are not yet on the drawing board.

The Case for Converting vs. "Scratch Build"

In Chinese philosophy yin and yang are said to portray the unity of opposites, which perfectly describes the feelings I was having at the outset of this project regarding how the vehicle chassis and running gear were to be created. If time and budget had not been a problem, I would have opted for a scratch-built aluminum frame with a custom-fabricated fiberglass body. Everything would have been chosen for low mass yet still designed to meet national safety standards for on-road vehicle operation. Advanced storage batteries and three-phase induction motor drive similar to the that of the Tesla Roadster would also have been a given. The reality turned out to be different. After sitting down with several members of the Electric Vehicle Council of Ottawa at a local coffee shop after one of their monthly meetings last year, it became clear that things were going to happen somewhat differently.

Figure 7.1-1. Rick Lane has been building and converting electric vehicles since he was a teenager, starting with an electric buggy and most recently helping Fred Green with the Volkswagen Jetta conversion pictured here. Rick (left) is shown with author Bill Kemp (center) and vehicle owner Fred Green (right).

There are thousands of electric vehicle enthusiasts around the world and many of these folks join other like-minded people at monthly electric vehicle club meetings to talk about the state of the art in vehicle technology, government regulations, and most of all building electric cars.

One of these eager people was Rick Lane, who has been building and converting electric vehicles since he was a teenager, starting with an electric dune buggy and most recently helping Fred Green with the Volkswagen Jetta conversion shown in Figure 7.1-1.

Rick and I chatted for several hours, along with a number of other club members. We talked about vehicle technologies, drive train design, inverter selection, motor type, and chassis. The conversation was as stimulating as the coffee and sugar-laden donuts, with everyone trying to suggest ways of creating a PHEV that could operate on zero-carbon energy sources. After considering all of the possibilities, the answer was pretty clear. Unless I wanted to spend over a hundred thousand dollars and was willing to wait a couple of years to find parts and fabricate all the custom components, I was limited to converting an existing vehicle.

Although this would save a huge amount of money and time, there were also some drawbacks. The vehicle would be far heavier than necessary, and the size and placement of batteries, electronics, and the power unit would be difficult as I would have to work with the chassis I ultimately selected.

In the end, the decision was made to convert an existing vehicle, and naturally enough this had both good and bad repercussions. On the negative side of the ledger sheet, I would know the shortcomings of the finished PHEV and would find them irritating. True, not having the exact design you want does provide a wonderful segue to "PHEV Version 2.0," perhaps in a future edition of this book. Look forward to it in stores after I solicit my family and friends to bankroll the necessary research and development. The pro side of the sheet was simple: much faster development and fabrication time as well as lower cost. This was the winning alternative.

Series vs. Parallel Hybridization

The Zero-Carbon Car is designed as a plug-in hybrid electric vehicle that uses a combination of electric motor and high-efficiency internal combustion power source to operate the vehicle. As we learned in Chapter 6, there are two fundamental forms of hybridization: series and parallel. These two basic versions include many more variations on a theme, including the system used in the Toyota Prius. The Prius has most of the characteristics of a parallel hybrid in that the electric motor and internal combustion engine can directly power the vehicle. However, due to a special "power splitting" transmission arrangement, power can be sent from the engine or electric motor to the wheels in any desired proportion. With this arrangement the electric motor can also provide all of the tractive force of the vehicle. With an expanded battery capacity added to the stock model, the Prius can be converted into a PHEV, which is exactly how the Hymotion Company has developed its power system.

The complexity of the parallel drive technology was beyond the scope of this project and it was decided to utilize the series hybrid design with the drive wheels powered exclusively by an electric motor. An internal combustion engine would be used to operate a power generating system to both recharge the batteries as required and provide electrical power for long-range driving.

One further advantage of the series hybrid design is that there are thousands of electric-car owners who could use this technology. Some people who convert vehicles from internal combustion to electric happily accept the limited range and slow recharging time either because it suits their lifestyle or because they have an alternate means of transportation when the batteries are dead. While these drivers suggest that range and recharging time are not real issues, most people who own these vehicles would concede that these limitations can be rather trying at times.

The development of the Zero-Carbon Car set out to test the theories of highly efficient personal transportation technology, zero-carbon electricity, and liquid fuels that are considered in this book. These technologies are ready, but the policy environment necessary to make them commercially viable is not.

One side benefit is that people who have either converted or purchased BEVs can use the information contained in this chapter and on the accompanying website to upgrade their vehicles to PHEVs. I challenge any and all to dissect the design my team and I have developed and see if you can improve upon it as you adapt these ideas to your own vehicles. The engine management and control system design and software files are located on the support website at www.thezerocarboncar.com. It is an open-source site, free for all to use, and I encourage you to produce your own PHEV and send your results to the website, where they can be showcased for all to see. Changing government mindset so that it stops waiting for hydrogen and starts developing PHEVs now is one area where the grassroots could use their collective skills and energy to make change happen.

Overview of the Design

In order to develop a PHEV that was economical to build and had a design that could be retrofitted to the thousands of existing BEVs scattered around the world, I decided to use a direct-current, series-wound motor for the traction system (Figure 7.1-3). The vehicle is powered by a 120-volt battery bank using ten series-connected 12-volt lead-acid batteries.

An on-board battery charger allows the battery bank to be recharged from the electrical supply grid when the vehicle is stationary.

The vehicle's original twelve-volt battery provides power for standard electrical equipment such as lights, horn, fuel pump, and stereo. This auxiliary battery is charged from the main 120-volt traction battery by a device known as a DC to DC converter which "steps" the high voltage down to a charging level of 13.5 volts.

This basic arrangement is familiar to most people who have experimented with BEVs and is similar to the design used in Fred Green's modified Volkswagen Jetta discussed in Chapter 6.

In order to make the vehicle a series hybrid, an electrical power generator and system control and monitor are required, as outlined in Figure 7.1-2.

In this system a three-cylinder Kubota D722 67-cubic-inch (1,100 cubic-centimeter) engine is used to provide the mechanical power to drive a specially designed permanent-magnet generator (Figure 7.1-7). The nominal output of the generator is 120 volts direct current with a capacity of 10 kW (10,000 watts) at 83 amperes. This configuration allows the generator to charge the traction battery bank as well as power the vehicle. If required, the generator can also provide full-cycle charging if the traction battery becomes depleted and cannot be recharged by the AC mains supply.

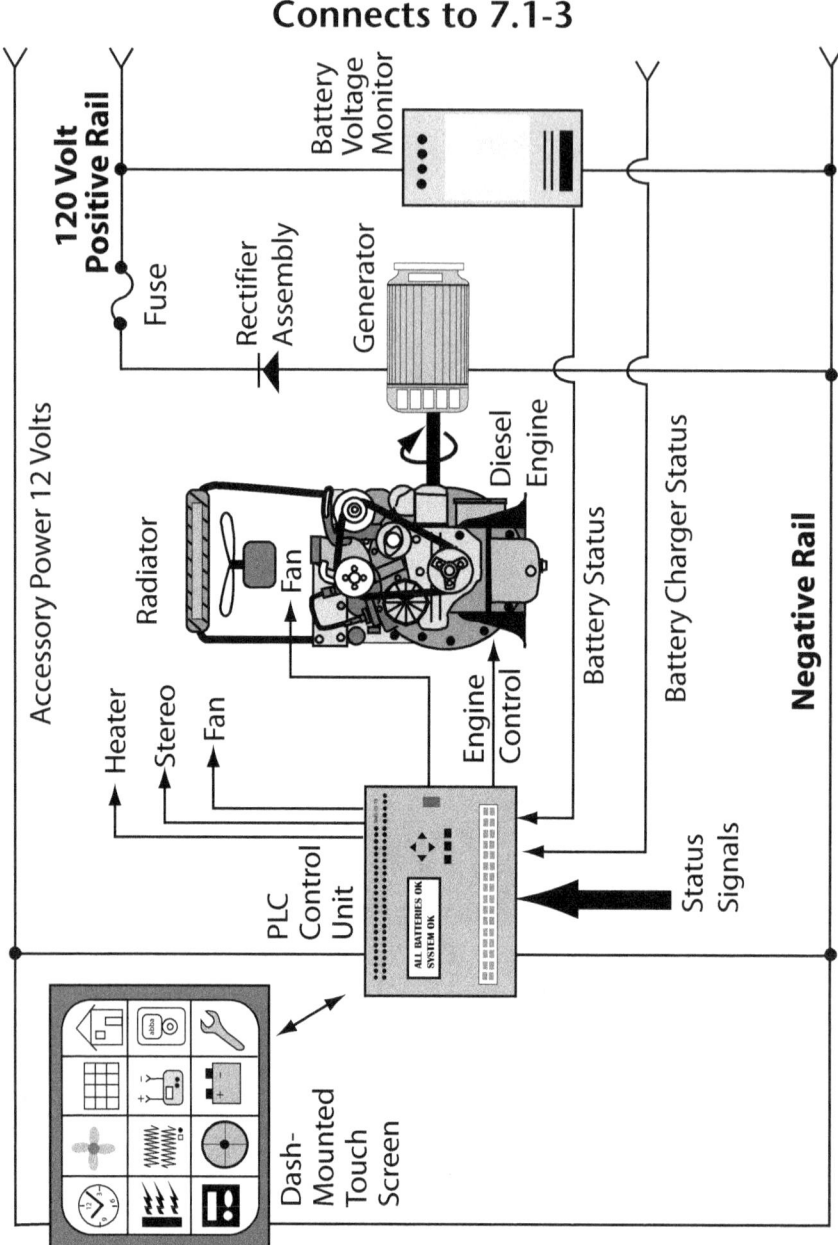

Figure 7.1-2 In order to make the vehicle a series hybrid, an electrical power generator and system control and monitor are required, as noted in this schematic fragment. In this configuration a diesel engine is used to provide the mechanical power to drive a specially designed permanent-magnet generator. This configuration allows the generator to directly charge the traction battery bank as well as power the vehicle.

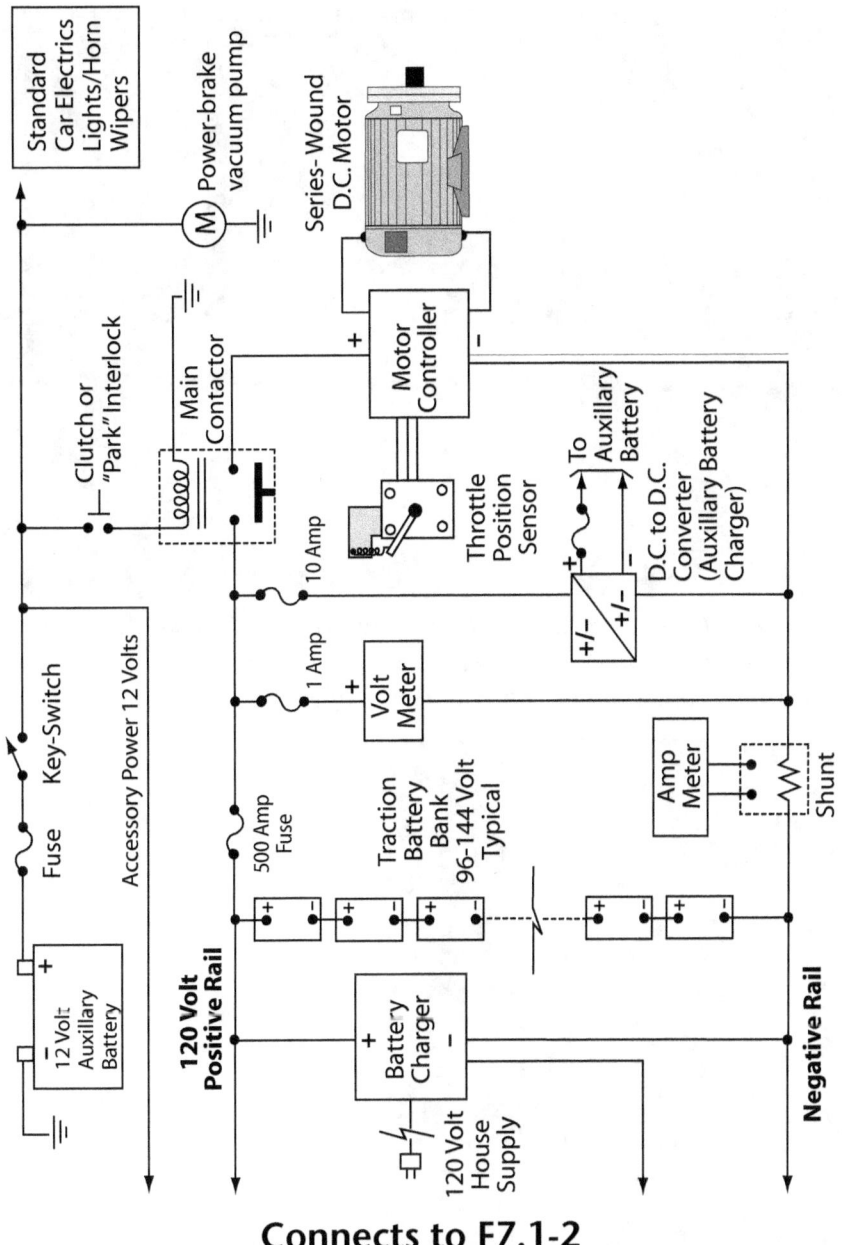

Connects to F7.1-2

Figure 7.1-3. In order to develop a PHEV that was economical to build and had a design that could be retrofitted to the thousands of existing BEVs scattered around the world, I decided to use a direct-current, series-wound motor for the traction system shown here. This basic arrangement is familiar to most people who have experimented with BEVs and is similar to the design used in Fred Green's modified Volkswagen Jetta discussed in Chapter 6.

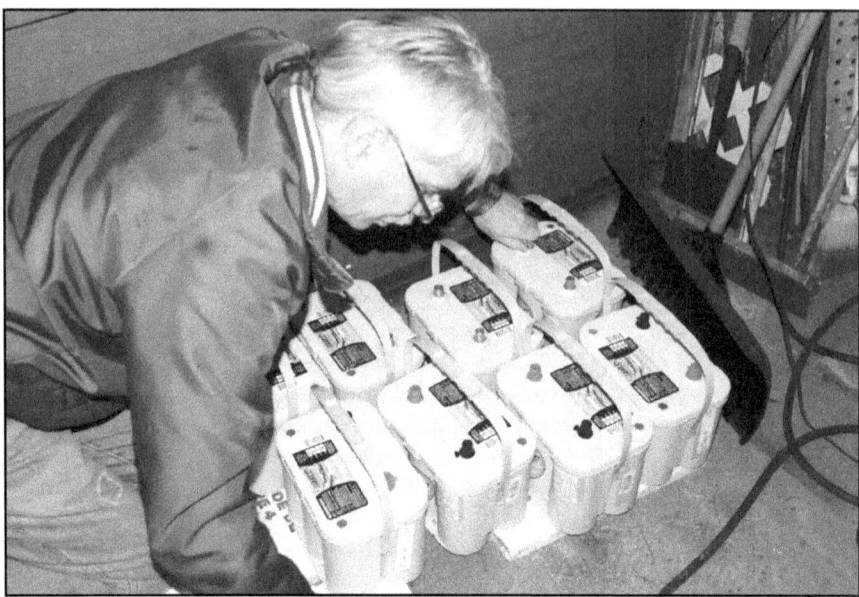

Figure 7.1-4. The Zero-Carbon Car is powered by a 120-volt battery bank using ten series-connected 12-volt lead-acid batteries. Rick Lane is seen testing the condition of the batteries shortly after their arrival.

Figure 7.1-5. An on-board battery charger allows the battery bank to be recharged from the electrical supply grid when the vehicle is stationary.

Figure 7.1-6. The vehicle's original 12-volt battery provides power for standard electrical equipment such as lights, horn, fuel pump, and stereo. This auxiliary battery is charged from the main 120-volt traction battery by a device known as a DC to DC converter which "steps" the high voltage down to a charging level of 13.5 volts.

Figure 7.1-7. In this system a three-cylinder Kubota D722 67-cubic-inch (1,100 cubic-centimeter) engine is used to provide the mechanical power to drive a specially designed permanent-magnet generator. The nominal output of the generator is 120 volts DC with a capacity of 10 kW (10,000 watts).

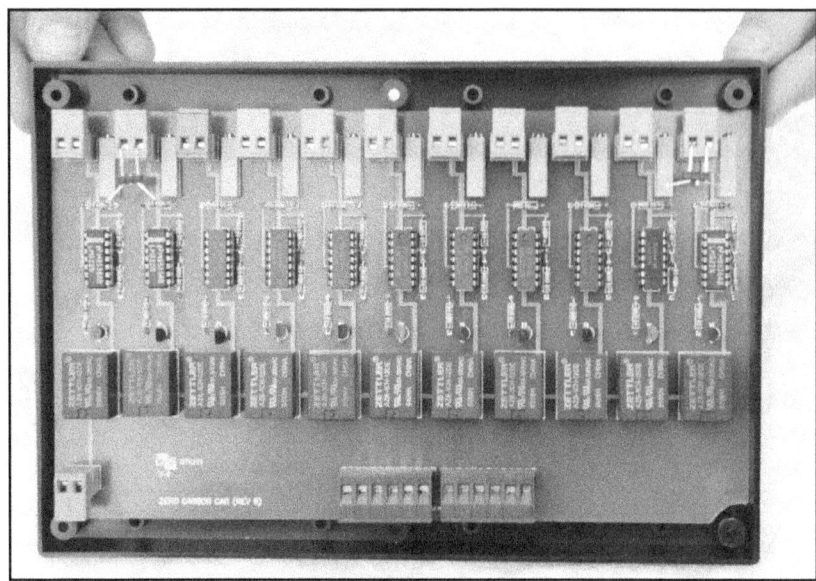

Figure 7.1-8. A custom-fabricated battery voltage monitoring device senses the battery voltage from each of the ten batteries and provides status signals to a logic controller.

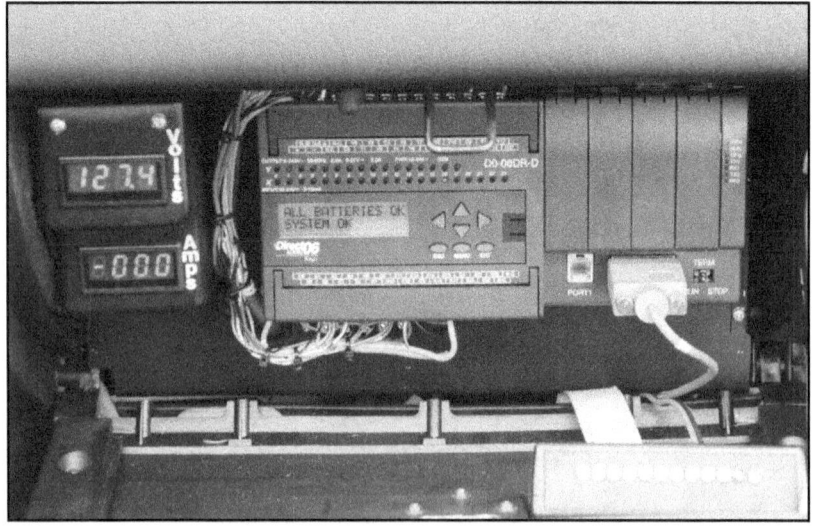

Figure 7.1-9. A device known as a programmable logic controller (PLC) receives status signals from the battery monitoring device and battery charger and also provides output control signals to start and stop the generator and operate the electric fuel pump, radiator fan, rear window defogger, ventilation fan, in-car heater, stereo, and other devices on-board the vehicle.

A custom-fabricated battery voltage monitoring device senses the battery voltage from each of the ten batteries and provides status signals to a logic controller.

A device known as a programmable logic controller (PLC) receives status signals from the battery monitoring device and battery charger and also provides output control signals to start and stop the generator and operate the electric fuel pump, radiator fan, rear window defogger, ventilation fan, in-car heater, stereo, and other devices on-board the vehicle.

A dash-mounted touch screen is connected to the PLC unit and provides a bit of "sizzle" to the vehicle with graphic touch control of many desired functions. For example, pressing the image of the stereo turns on the iPod music player. The touch screen also provides feedback on vehicle status, and a diagnostic screen shows the condition of each of the vehicle's ten traction batteries.

Software is written based on a series of "logic tables" that were developed early on in the design stage. By developing a set of rules and converting them to "machine-readable" format, it is possible to allow the PLC to monitor all of the vehicle's functions and respond to these in the correct manner. For example if a traction battery is failing, the PLC can signal the touch screen to alert the user to a problem. Likewise, if the battery bank is nearly depleted and the driver wishes to continue using the vehicle, the PLC can automatically start the engine/generator set (hereinafter referred to as "genset") and begin providing electrical energy to power the vehicle and recharge the battery bank.

Later in this chapter we will review each component of the system in detail and discuss its installation and operation.

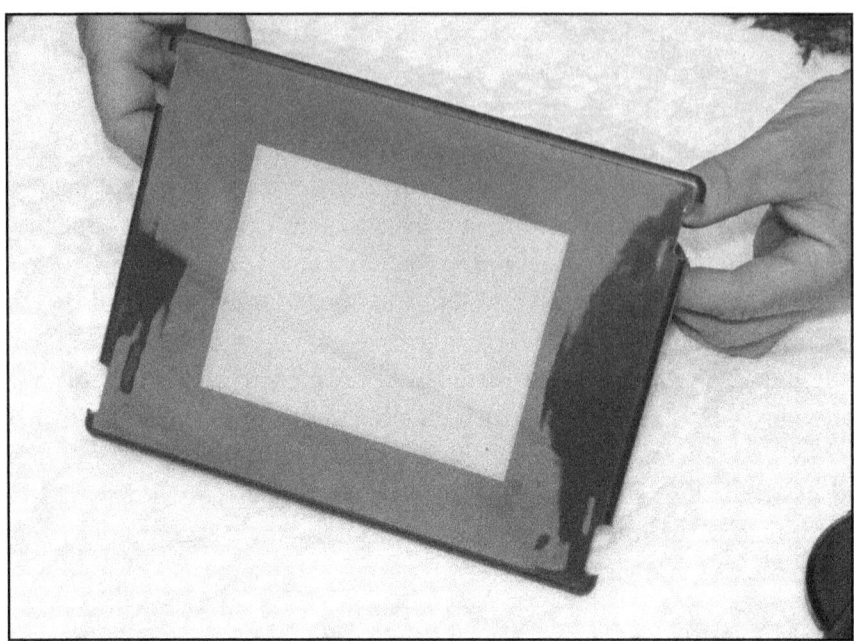

Figures 7.1-10a/b. A dash-mounted touch screen is connected to the PLC unit and provides graphic touch control of vehicle functions. The touch screen also provides feedback on vehicle status and a diagnostic screen shows the condition of each of the vehicle's ten traction batteries.

The Donor Vehicle

A very low-mileage 2000 Mazda Miata was selected to become the shell of the Zero-Carbon Car. There are numerous automobiles on the road that make acceptable conversions to an electric vehicle; the Mazda Miata is not one of them. Don't get me wrong. The Miata is a nice little sports car, but with the operative word being "little" it was a challenge finding space to jam in all the bits and pieces we needed to work with. I would suggest using a larger car if you decide to undertake a project like this.

The first step in the conversion process is to accurately weigh the vehicle and measure the distance from the road surface to the underside of each of the four wheel wells. This is an important step because as weight is removed and added to the vehicle during the conversion process it is critical to maintain the manufacturer's pre-engineered ride height levels

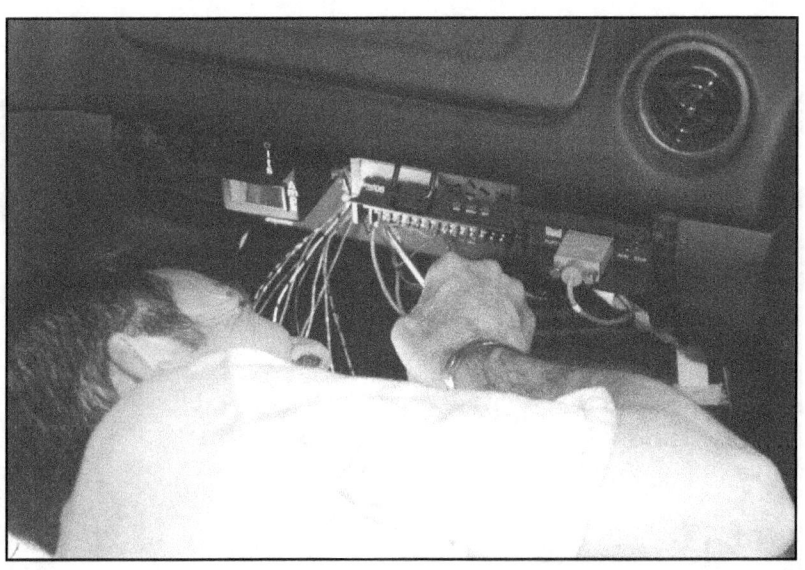

Figure 7.1-11. There are numerous automobiles on the road that make acceptable conversions to an electric vehicle; the Mazda Miata is not one of them. Don't get me wrong. The Miata is a nice little sports car, but with the operative word being "little" it was a challenge finding space to jam in all the bits and pieces we had to work with. The author is shown squeezing himself under the dash to wire the PLC unit.

to ensure safe road handling. Increasing suspension capacity to accommodate excess loading can be accomplished by using higher-capacity springs or adjustable air shock absorbers.

Out with the Old

The second step in developing the Zero-Carbon Car was the removal of the internal combustion engine parts and accessories. A surprising amount of material and weight was removed from the car during the conversion process, which is a good thing given that the vehicle was loaded back up with a heavy battery bank, an electric motor, and assorted materials. During this removal process, a total of 560 pounds (254 kilograms) of engine, muffler, pipes, and wires were taken from the vehicle.

In this conversion process, the stock weight of the empty car was 2,340 pounds (1,061 kilograms), having a perfect 50/50 distribution

Figure 7.1-12. A surprising amount of material and weight was removed from the car during the conversion process. The donor vehicle's engine was sold to help offset the cost of conversion.

between front and back axles. Removing the internal combustion engine and related materials dropped the vehicle mass to 1,780 pounds (807 kilograms).

After the upgrade to PHEV, the vehicle weighed in at 2,820 pounds (1,279 kilograms), 480 pounds (218 kilograms) more than the factory-built vehicle. Weight distribution was very close to the ideal 50/50 weight split. The additional weight necessitated an upgrade to slightly heavier springs and shims, placing the vehicle at the proper wheel-well-to-road clearance.

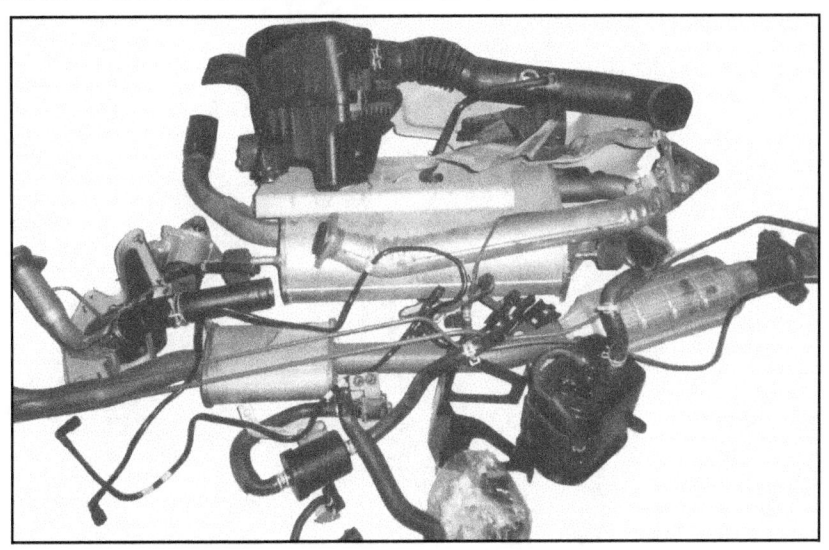

Figure 7.1-13. A total of 560 pounds (254 kilograms) of engine, muffler, pipes, heater plumbing, and wires were removed from the vehicle.

7.2 Battery Technologies: An Overview

There are thousands of battery styles and technologies available, and the more you look the more selection and complexity you will find. Car, truck, boat, golf cart, and telephone batteries are a few examples that come to mind. Then, of course, there are the "exotic" kinds such as NiCad, nickel metal hydride, and lithium ion, to name a few. Given the importance of the battery, whether it is used in a BEV, hybrid, or PHEV, it is important to determine what technologies are available and can be used in production vehicles. Even hydrogen-fuel-cell-powered vehicles use batteries to help smooth out rapid changes in energy demand, allowing the fuel cell to operate at a constant power output.

However, before we delve too deeply into battery technology, I would like to say a few words about electrical energy to assist those who are not familiar with some of the technical terms used in describing electrical circuits.

The Story of Electrons

If you can remember back to your high school days in science class, you may recall that an atom consists of a number of electrons swirling around a nucleus. When an atom has either an excess or a lack of electrons in comparison to its "normal" state, it is negatively or positively charged, respectively.

In the same way that the north and south poles of two magnets are attracted, two oppositely charged atoms are also attracted. When a negatively charged atom collides with a positively charged one, the excess electrons in the negatively charged atom flow into the positively charged atom. This phenomenon, when it occurs in far larger quantities of atoms, is called the flow of electricity.

The force that causes electricity to flow is commonly known as *voltage* (or V for short). The actual flow of the electrons is referred to as the *current*. So where does this force come from? What makes the electrons flow in the first place? The trigger that brings about the flow of electrons can come from several energy sources. Typical sources are chemical batteries, photovoltaic cells, wind turbines, electric generators, and the up-and-coming fuel cell. Each source uses a different means to

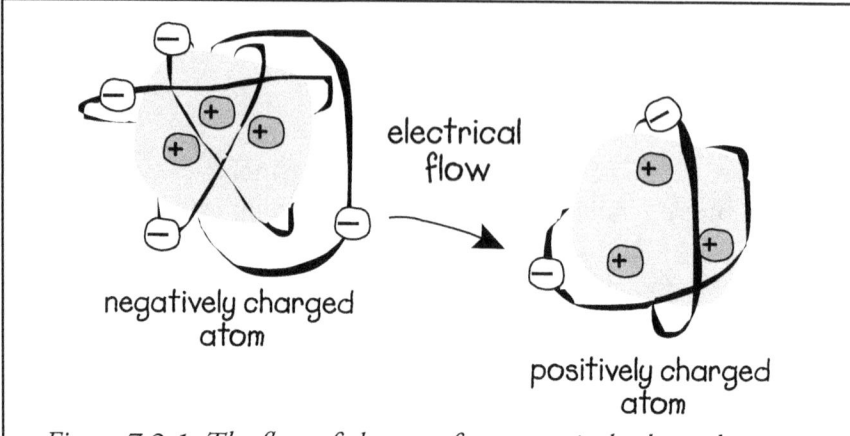

Figure 7.2-1. The flow of electrons from negatively charged atoms to positively charged atoms is known as the flow of electricity.

trigger the flow of electrons. Waterfalls, coal, oil, and nuclear energy are commonly used to generate commercial electricity. Fossil or nuclear fuels are used to boil water, which creates the steam that drives a turbine and generator; falling water drives a turbine and generator directly. The spinning generator shaft induces magnetic fields into the generator windings, forcing electrons to flow.

Let's use the flow of water as a visual aid to understanding the flow of electricity that is otherwise invisible.

Let's presume that a greater amount of water moving past you per second equates to a higher flow and that a river with a high flow of water

Figure 7.2-2. The flow of electricity is very similar to the flow of water in a pipe or stream. A waterfall has a higher flow (current) and greater pressure (voltage) than a creek; similarly, a higher number of electrons moving from atom to atom increases electrical current.

has a large current. With electricity, a large number of electrons flowing from one atom to the next is similar to a large flow of water. Therefore, the greater the number of electrons flowing past a given point per second, the greater the electrical current.

The flow of water is typically measured in gallons per minute or liters per second; electrical current is measured in amperes (or "A" for short). If the measured current of electrons in a material is said to be 2 A, we know that a certain (very large) number of electrons have passed a given point per second. If the current were increased to 4 A, there would be twice the number of electrons flowing through the material.

Gravity is a factor in considering water pressure. A meandering creek has very little water pressure, owing to the small hill or height that the water has to fall. A high waterfall has a much greater distance to fall and therefore has increased water pressure. For example, if a water-filled balloon were to fall on you from a height of 2 feet you would find this quite refreshing on a hot day. A second balloon falling from a height of 100 feet would exert a much higher pressure and probably knock you out! This increase in water pressure is similar to the electrical pressure or voltage that forces the electrons to flow through a material.

Conductors and Insulators

Electricity flows through conductors in the same manner as water flows through pipes and fittings. When electrons are flowing freely through a material, we know that the material is offering little resistance to that flow. These materials are known as conductors. Typical conductors are copper and aluminum, which are the substances used to make electrical wires of varying diameters. Just as a fire hose carries more water than a garden hose, a large electrical wire carries more electrons than a thin one. A larger wire can accept a greater flow of current (i.e. handle higher amperage).

Before electricity can be put to use, we must create an electrical circuit. We do this by connecting a source of electrical voltage to a conductor and causing electrons to flow through a load and back to the source of voltage.

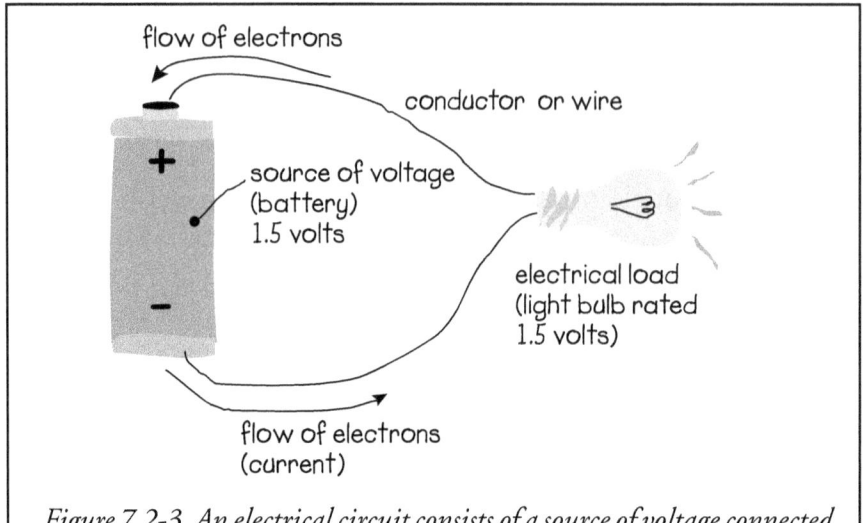

Figure 7.2-3. An electrical circuit consists of a source of voltage connected to an electrical load using conductors to carry the current flow.

The drawing in Figure 7.2-3 shows how a simple flashlight works. Electrons stored in the battery (we will cover that one later) are forced by the voltage (pressure) to flow into the conductor from the battery's negative contact (-) through the light bulb and back to the battery's positive contact (+). Current flowing in this manner is called *direct current* (or DC for short). The light will stay lit until the battery dies (runs out of electrons) or until we turn off the switch.

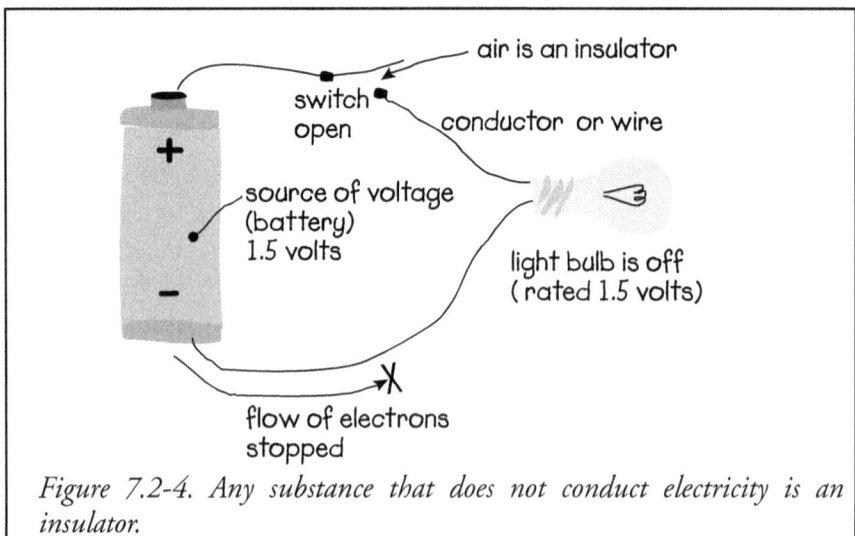

Figure 7.2-4. Any substance that does not conduct electricity is an insulator.

Oh yes, a switch would be a good idea. This handy little device allows us to turn off our flashlight to prevent using all of the electrons in the battery. From our description of a circuit, we can assume that if the electrical conductor path is broken, the light will go out. How do we break the path? The flow of electrons may be interrupted using a nonconductive substance wired in *series* with the conductor. Any substance that does not conduct electricity is known as an insulator.

Typical insulators include rubber, plastics, air, ceramics, and glass.

Batteries, Cells, and Voltage

You may have noticed that your brand of flashlight has two or even three cells. Placing cells in a stack or in *series* causes the voltage to increase. For example, the flashlight shown in Figure 7.2-4 contains a cell rated at 1.5 V. Placing two cells in series, as shown in Figure 7.2-5, increases the voltage to 3 V (1.5 V + 1.5 V = 3 V). The light bulb in Figure 7.2-5 is glowing very brightly and will quickly burn out because it is rated for 1.5 V. In this example, a higher voltage (pressure) is causing more current to flow through the circuit than the bulb is able to withstand. Likewise, if the cell voltage were lower than the rating of the bulb, insufficient current would flow and the bulb would be dim. This is what happens when your flashlight batteries are nearly dead and the light is becoming dim: the batteries are running out of electrons. In a similar manner, the

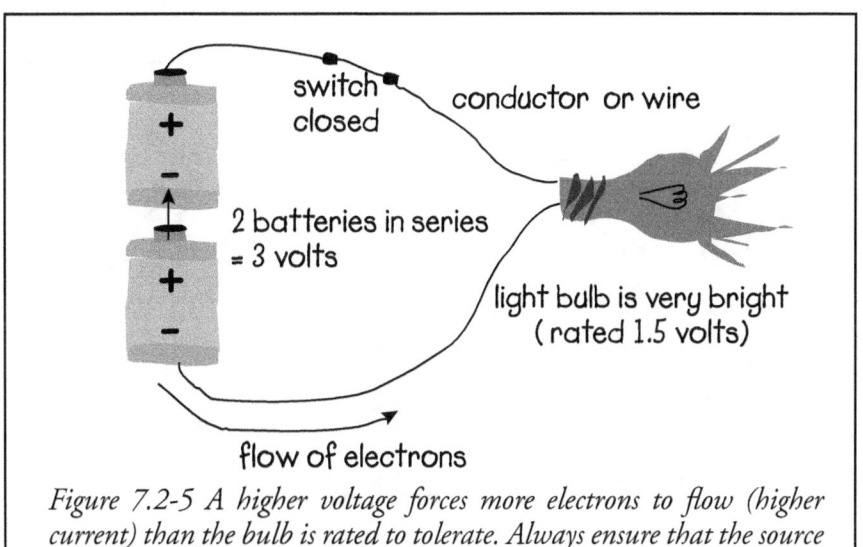

Figure 7.2-5 A higher voltage forces more electrons to flow (higher current) than the bulb is rated to tolerate. Always ensure that the source and load voltages are rated equally.

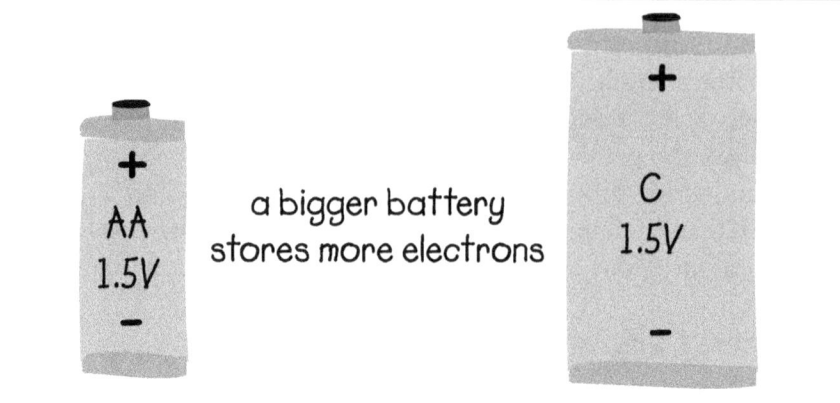

Figure 7.2-6. Cells of different sizes have different amounts of electrical energy storage. Just as a two-gallon jug contains twice as much water as a one-gallon jug, so a larger battery stores more energy and works for longer periods of time than a smaller one.

Zero-Carbon Car uses ten 12-volt batteries wired in series, providing a nominal 120-volt battery bank (Figure 7.2-8).

Next time you are waiting in the grocery store lineup, put down that copy of *National Enquirer* and take a look at the battery display. The selection will include AA, C, and D sizes of cell, which all have a rating of 1.5 V. Can you guess the difference between them?

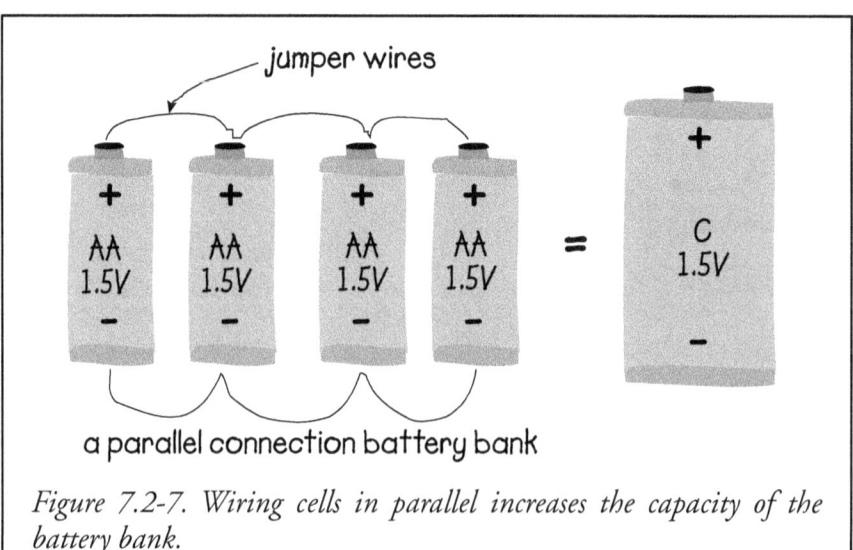

Figure 7.2-7. Wiring cells in parallel increases the capacity of the battery bank.

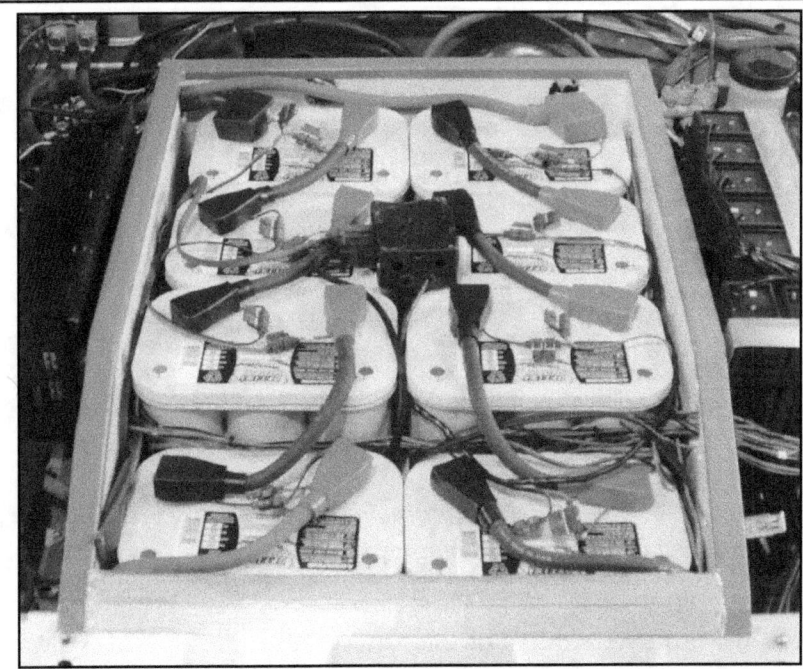

Figure 7.2-8. A BEV or PHEV requires far more electricity than a simple flashlight does. The battery bank of the Zero-Carbon Car, which weighs 510 pounds (231 kilograms), is shown here. The battery bank has a rated capacity of 120 volts, which is obtained by series connecting ten 12-volt batteries together.

A larger cell holds more electrons (via chemical capacity) than a smaller one. With more electrons, a larger battery cell can power an electrical load longer than a smaller one. The circuit shown in Figure 7.2-7 shows a set of jumper wires connecting the cell terminals in *parallel*. This parallel arrangement creates a battery bank of 4 AA-size cells, which has the same number of electrons and the same capacity as the C-size cell. Any grouping of cells, whether connected in series, in parallel, or both, is called a battery bank or battery.

Alternating Current

The modern home is supplied with electricity in the form of 120/240V *alternating current* (VAC). In a DC circuit, as defined earlier, current flows from the negative terminal of the battery through the load and

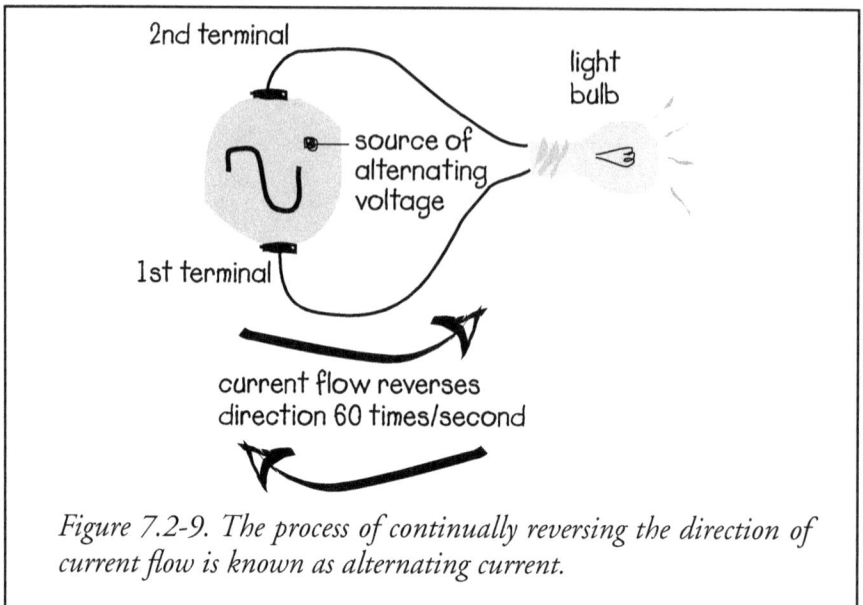

Figure 7.2-9. The process of continually reversing the direction of current flow is known as alternating current.

back to the positive terminal of the battery. Since the current is always flowing one way, in a direct route, this is called direct current.

In an AC circuit, the current flow starts at the first terminal and flows through the load to the second terminal, similar to the flow in a battery circuit. However, a fraction of a second later the current stops flowing and then reverses direction, flowing from the second terminal through the load and back to the first terminal, as shown in Figure 7.2-9.

Generating electricity in the modern, grid-connected world is accomplished by using various mechanical turbines to turn an electrical generator. In the early years of the electrical system, there was considerable debate as to whether the generator output should be transmitted as AC or DC. For a time, both AC and DC were generated and transmitted throughout a city. However, as time passed, it became clear that for safety, transmission, and other practical reasons AC was more desirable. As the old saying goes, "The rest is history."

Advanced electrically powered vehicles have also started to adopt alternating current for traction power. These vehicles are equipped with direct current batteries, but through the use of a device called an inverter it is possible to generate an AC power waveform with a variable frequency. An AC electric motor rotates at a speed (rpm) that is synchronized with the AC power waveform; adjusting the frequency therefore allows the motor to increase or decrease rotational (and vehicle) speed as desired.

Power, Energy, and Efficiency

Conserving energy and doing more with less not only is good for the planet but also helps keep the size and cost of the electric vehicle's power system within reasonable limits. We discussed earlier how a bigger battery could run a light bulb longer than a smaller battery. This is fairly obvious. What might not be so obvious is that if we replace the "ordinary" light bulb with one that is more efficient we might not need the larger battery in the first place. The same logic can be applied to an electric vehicle, using energy efficiency and advanced engineering practices to lower the amount of energy required to operate the vehicle in the manner of the "Hypercar" concept vehicle discussed in Chapter 5.

Let's review how any typical electrical circuit operates: A source of electrons under pressure (i.e. voltage) flows through a conductor to an electrical load and back to the source. For household circuits, the voltage (pressure) is usually fixed at 120 VAC or 240 VAC. For battery-powered vehicle circuits, the voltage is usually fixed at some level between 96 VDC and 350 VDC for more advanced designs, depending on the amount of electron flow (i.e. current) that is required to make the vehicle operate. Using our light bulb analogy, let's say that an ordinary bulb requires 12 V of pressure and 1 A of current flow to make it light. Now suppose that we can find a light bulb that uses 12 V and only 0.5 A of current

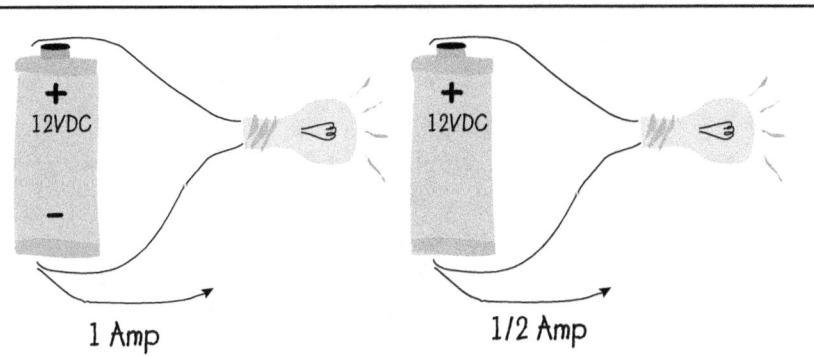

Figure 7.2-10. The lamp on the right uses 50% of the electrical energy of the one on the left and both have equal brightness. The lamp on the right is said to be twice as efficient. The same logic can be applied to an electric vehicle, using energy efficiency and advanced engineering practices in the manner of the "Hypercar" concept vehicle to lower the amount of energy required to operate the vehicle.

flow. Assuming that both lights are the same brightness, we infer that the second light is twice as efficient as the first. Stated another way, we would need only half the battery-bank size (at lower cost and weight) to run the second lamp for the same period of time, or the same size of battery bank would run the second lamp for twice as long.

The relationship between the pressure (voltage) required to push the electrons to flow in a circuit and the number of electrons actually flowing to make the load operate (current or amps) is the power (commonly expressed as *watts*, or W) consumed by the load. Using the above example, let's compare the power of the two circuits:

Ordinary Lamp:
$12\ V \times 1\ A = 12\ W$
More Efficient Lamp:
$12\ V \times 0.5\ A = 6\ W$

The second lamp uses half the number of electrons to operate. The power of a circuit is simply the voltage multiplied by the current in amps, giving us the instantaneous flow of electrons in the circuit measured in watts.

Remember that the more efficient bulb operates twice as long on a battery as the other bulb. How does time factor into this equation? If a battery has a known number of electrons stored in it and we use them up at a given rate, the battery will become empty over time. The use of electrons (power) over a period of time is known as *energy*.

Energy is *power* multiplied by the *time* the load is turned on:

Ordinary Lamp:
$12\ V \times 1\ A \times 1\ hour = 12\ Watt\ hours;$ or
$12\ W \times 1\ hour = 12\ Watt\ hours$
More Efficient Lamp:
$12\ V \times 0.5\ A \times 1\ hour = 6\ Watt\ hours;$ or
$6\ W \times 1\ hour = 6\ Watt\ hours$

Your electrical utility charges you for energy, not power. You run around turning off unused lights to cut down on the amount of time the lights are left on. If we think about it long enough, we can also use these calculations to figure out how much energy is stored in a battery bank. For example, if a 12 V battery bank can run a 10 A load for 30 hours, how much energy is stored in the battery?

Battery Bank Energy (in Watt Hours):
12 V x 10 A x 30 hours = 3,600 Watt hours or 3.6 kilo-Watt hours of energy

An interesting thing about batteries is that their voltage tends to be a bit "elastic." The voltage in a battery tends to dip and rise as a function of its state of charge. For example, a typical lead-acid battery bank will register 11 V under load and when the batteries are nearly dead. When they are under full charge, the voltage may reach nearly16 V. Because of this swing in the voltage, many batteries are not sized in *Watt hours* of energy, but in *Amp hours*.

The math is similar; just drop the voltage from the energy calculation:
Battery Bank Energy (in Amp Hours):
10 A x 30 hours = 300 Amp hours of energy

To convert Amp hours of energy to Watt hours, simply multiply Amp hours by the battery voltage. Likewise, to convert Watt hours of battery bank capacity to Amp hours, divide Watt hours by the battery voltage.

I will cover this section again as we apply it to the discussion of battery bank size for an electrically powered vehicle.

How Batteries Work

A battery comprises two or more plates which form an electric cell. Electric cells are placed in an electrolyte material, which converts chemical to electrical energy. Cells provide a nominally constant direct current voltage from a positive and a negative terminal. It is the chemical reaction that occurs in the cell that generates the electrical current. Primary cells will provide electrical energy to a circuit one time until the chemical reaction has completed, at which time the cell has expired. A secondary or rechargeable cell is one where the chemical reaction can be reversed by the application of electrical energy in a process known as recharging.

The most common deep-cycle electrical storage battery, like a car battery, uses a lead-acid composition. A single battery cell uses a plastic case to hold a grouping of lead plates of slightly different composition. The plates are suspended in the case, which is filled with a weak solution of sulfuric acid, called *electrolyte*. The electrolyte may also be manufactured in a gelled form, which prevents spillage. (Batteries of

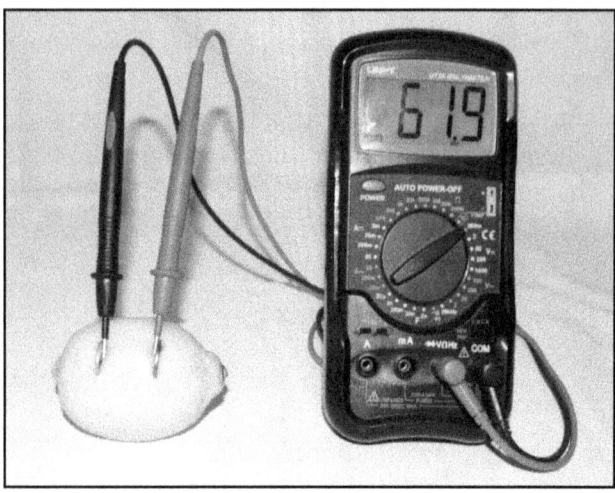

Figure 7.2-11. A battery comprises two or more plates which form an electric cell. Electric cells are placed in an electrolyte material which converts chemical to electrical energy. When two dissimilar metals are pressed into the acidic pulp of a lemon, a crude battery is formed and electrons will flow in the circuit as shown in this picture.

this type are often sold as "maintenance free.") The lead plates are then connected to positive and negative terminals in exactly the same manner as the AA and C cells described in Figure 7.2-7. A single lead-acid cell with one negative and one positive plate has a nominal rating of 2 V; a 12- volt battery therefore comprises six cells wired in series. Cells using different materials and electrolyte compositions have different nominal voltages.

Connecting an electrical load to the battery causes sulfur molecules from the electrolyte to bond with the lead plates, releasing electrons. The electrons then flow from the negative terminal through the conductors to the load and back to the positive terminal. This action continues until all of the sulfur molecules are bonded to the lead plate. When this occurs, it is said that the cell is discharged or dead.

As the cell is discharged of electrical energy, the acid continues to weaken. Using a device called a hydrometer (see Figures 7.2-14 a/b), we can directly measure the strength or specific gravity of the battery electrolyte. A fully charged battery may have a specific gravity of 1.290 as seen in the detail view of the hydrometer in Figure 7.2-14b (or 1.290 times the density of pure water). As the battery discharges, the specific

Figure 7.2-12. These battery models from Surrette Battery Company (www.surrette.com) are typical long-life, deep-cycle batteries for use in forklift trucks and industrial applications. The two batteries in the foreground have the same voltage rating, as they each have three cells wired in series (6 VDC). The model on the right has twice the capacity of the model on the left. (Courtesy: Surrette Battery Company Limited)

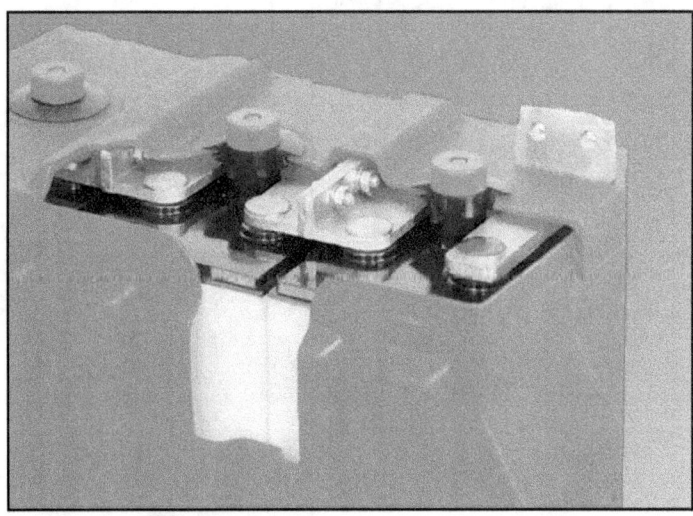

Figure 7.2-13. All lead-acid batteries comprise lead plates suspended in a weak solution of sulfuric acid. The size of the plate and acid capacity directly affect the amount of electricity that can be stored (energy). Each cell of the battery can be interconnected with others, increasing capacity and voltage. (Courtesy: Surrette Battery Company)

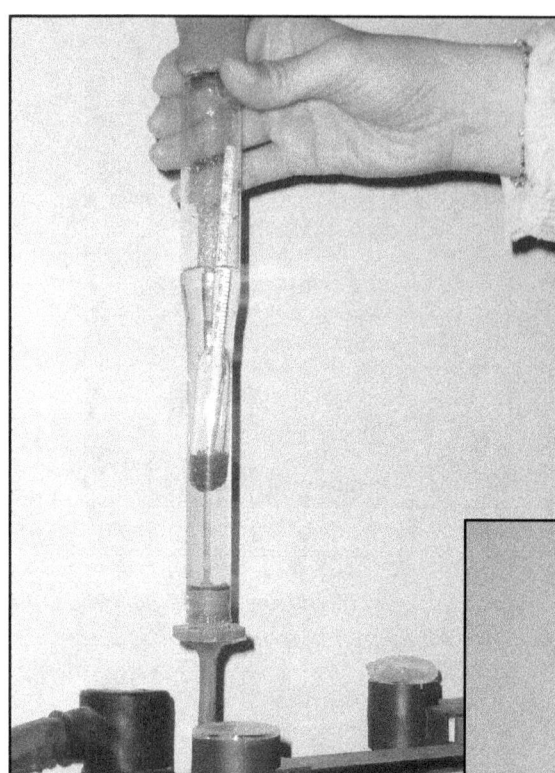

Figure 7.2-14a. The hydrometer measures the density or specific gravity of fluids such as the electrolyte in this cell. The higher the reading, the more energy is stored in the cell, indicating a higher state of charge.

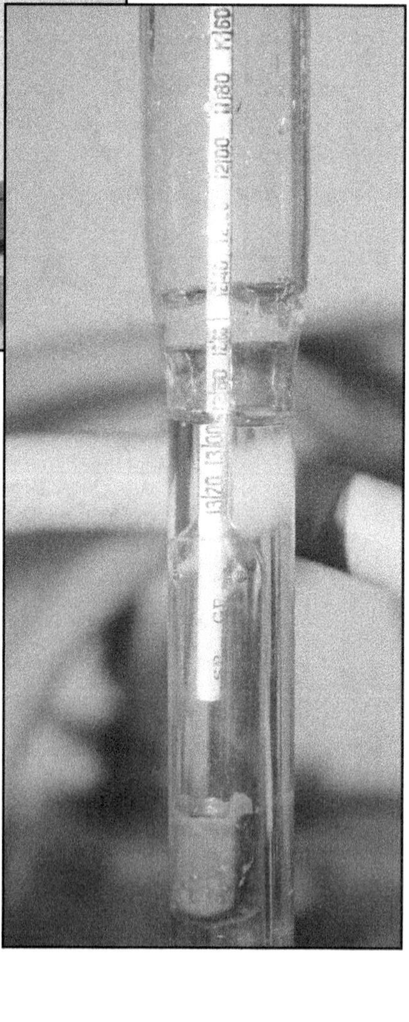

Figure 7.2-14b. This detail view of the hydrometer indicates a reading of 1.290.

gravity continues to drop until the flow of electrons becomes insufficient to operate our loads.

When a regular AA or C cell is discharged, the process is irreversible, meaning the cell cannot be recharged. Discharging a deep-cycle battery bank is reversible, allowing us to put electrons back into the battery, thus recharging it. Forcing electrons into the battery causes a reversal of the chemical discharge process described above. When the voltage or pressure from a battery charger is higher than that of the battery, electrons are forced to flow from the charger into the cell plate. Electrons combine with the sulfur compounds stored on the plate, in turn forcing these compounds back into the electrolyte. This action raises the specific gravity of the sulfuric acid and recharges the battery for future use. Although there are many types of battery chemistry, the charge/discharge concept is similar for all types.

Depth of Discharge

A deep-cycle battery got its name because it is able to withstand severe cycling or draining of the battery. A car battery can only withstand a couple of "Oops, I left my lights on. Can you give me a boost?" mistakes before it is destroyed, whereas a deep-cycle battery may be subjected to much higher levels of cycling.

Depth of Discharge %	Specific Gravity @ 75° F (25° C)	Cell Voltage
0	1.265	2.100
10	1.250	2.090
20	1.235	2.075
30	1.220	2.060
40	1.205	2.045
50	1.190	2.030
60	1.175	2.015
70	1.160	2.000
80	1.145	1.985
90	1.140	1.825
100	1.130	1.750

Table 7.2-1 Using the hydrometer shown in Figures 7.2-14 a/b, it is possible to accurately determine the state of charge, specific gravity, and "no load" voltage of each cell in a lead-acid battery bank.

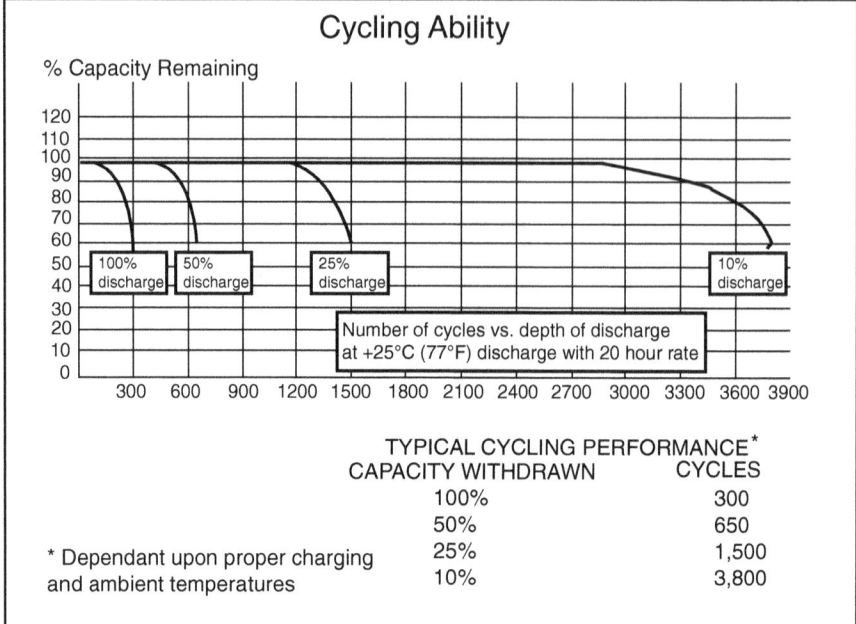

Figure 7.2-15. This graph relates the life expectancy in charge/discharge/charge cyles to the depth-of-discharge level of a typical lead-acid battery.

Figure 7.2-15 graphs the relationship between the life of a battery in charge/discharge/charge cycles and the amount of energy that is taken from the cell. For example, a battery that is repeatedly discharged completely (100% depth of discharge) will only last 300 cycles. On the other hand, if the same battery is cycled to only 25% depth of discharge, the battery will last 1,500 cycles. It stands to reason that a bigger battery bank will provide longer life, albeit at a higher cost, size, and weight for the added capacity. There is a fixed relationship between depth of discharge and battery life for all battery technologies. This relationship is crucial when considering the concept of vehicle-to-grid electricity transfer which is discussed in Chapter 8.

Operating Temperature

A battery is typically rated at a standard temperature of 75°F (25°C). As the temperature drops, the capacity of the battery drops as a result of the lower "activity" of the molecules making up the electrolyte. The graph in Figure 7.2-16 shows the relationship between temperature and available battery capacity. For example, a battery rated at 1,000 Amp-hours (Ah) at room temperature would have its capacity reduced to 70% at -4°F (-20°C), resulting in a maximum capacity of 700 Ah. If the battery-powered vehicle is expected to operate where winter temperatures may reach this level, the reduction in capacity must be taken into account.

Freezing is another concern at low operating temperatures. A fully charged battery has an electrolyte-specific gravity of approximately 1.25 or higher. At this level of electrolyte acid strength, a battery will not freeze. As the battery becomes progressively discharged, the specific gravity gradually falls until the electrolyte resembles water at a reading of 1.00.

The batteries in the Zero-Carbon Car are protected from freezing in several ways:

1. The battery bank is connected to a battery charger that can be quickly connected to a source of power, ensuring that electrolyte is fully regenerated.
2. The genset can be programmed to automatically switch on and recharge the batteries if they fall below a given charge threshold.
3. The container that holds the batteries is lined with a closed-cell insulation, ensuring that any self-heating of the batteries during

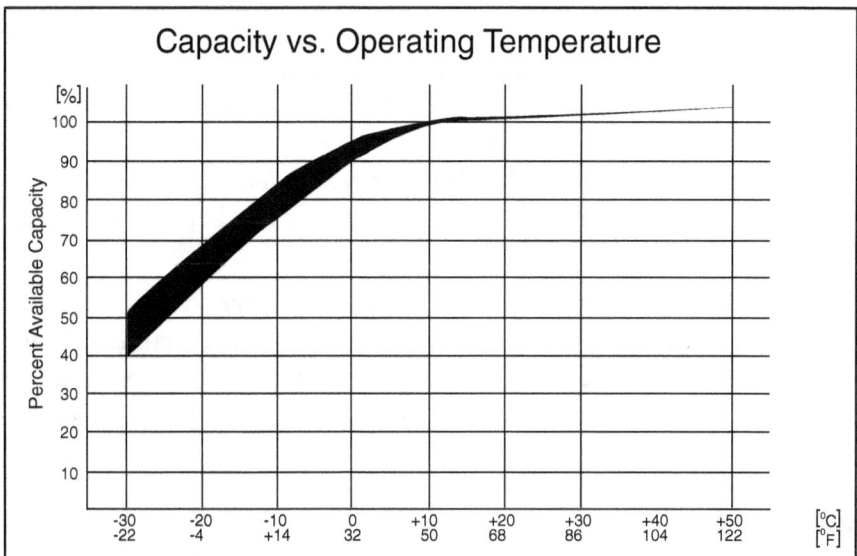

Figure 7.2-16. This graph shows the resulting reduction in battery capacity as the ambient temperature is lowered. A thermometer should be used to correct the specific gravity reading of very cold or overly hot electrolyte.

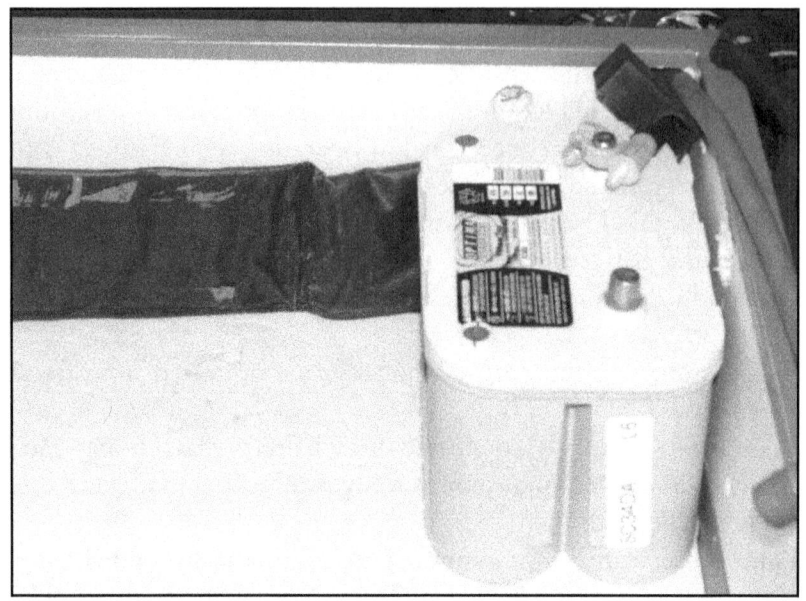

Figure 7.2-17. The batteries in the Zero-Carbon Car are protected from freezing in several ways, including a supporting chassis that is lined with closed-cell foam insulation and a battery heating blanket that is thermostatically controlled. The heating blanket is the dark pad at the rear of the chassis.

charge/discharge cycles can be used to keep the bank warm during extended idle periods.

4. A thermostat-controlled heating blanket is added to the battery box to ensure that battery temperature does not fall below the freezing point (Figure 7.2-17).

For cold-weather applications consider a nickel-metal cadmium or NiCad battery. Although more expensive than standard lead-acid batteries, the NiCad cell is relatively unaffected by extremes in temperature, particularly where deep discharge cycles and cold weather threaten to destroy lead-acid varieties.

Battery Sizing

A single cell does not have enough voltage or capacity to perform useful work. Single cells may be wired in series to increase voltage and/or in parallel to increase capacity. Batteries are just big storage buckets. Pour in some electrons and the "buckets" will fill with electricity. If cells are wired in parallel, the capacity is increased. If the cells are wired in series, the voltage or pressure rises.

Batteries may be purchased in a range of voltages and capacities. Individual cells have a nominal rating of 2 V for lead-acid and 1.2 V for NiCad brands. The lead-acid battery shown in the background of Figure 7.2-12 has two cells, each of which can be identified by the servicing cap on the top. Using interconnecting plates inside the battery case, the cells are wired in series, providing 2 + 2 volts = 4 V total capacity. The batteries in the foreground have three cells, providing an output of 6 V. Using the same approach, a car battery contains six cells, providing a nominal 12 V rating.

It stands to reason that the greater the energy consumed, the larger the storage facility required to hold all the electricity. While the capacity of buckets is given in gallons or liters, the capacity of batteries is rated using watt-hours or Amp-hours, units of electrical energy. We discussed earlier that a battery's voltage tends to be a bit elastic, with the voltage rising and falling as a function of the battery's state of charge (see Table 7.2-1). Additionally, when the battery is charged from a source of higher voltage such as the battery charger or genset, the voltage increases further. It is not uncommon for a battery bank with a nominal 24 V rating to have a voltage reading of 28 V when it is charging.

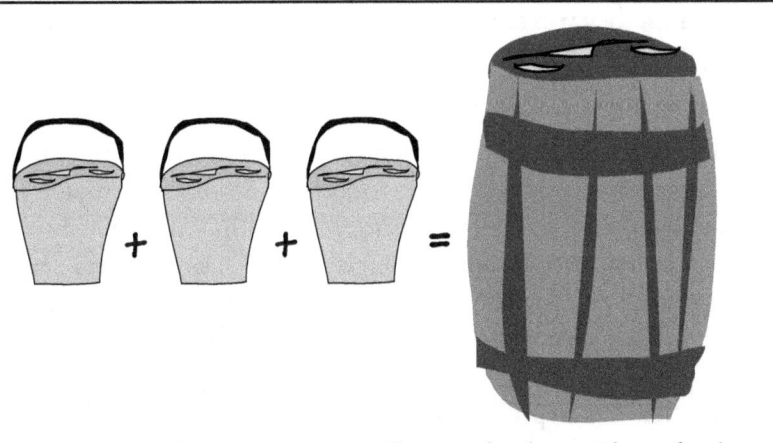

Figure 7.2-18. Batteries are similar to buckets. Three buckets paralleled together have the same capacity as a large barrel.

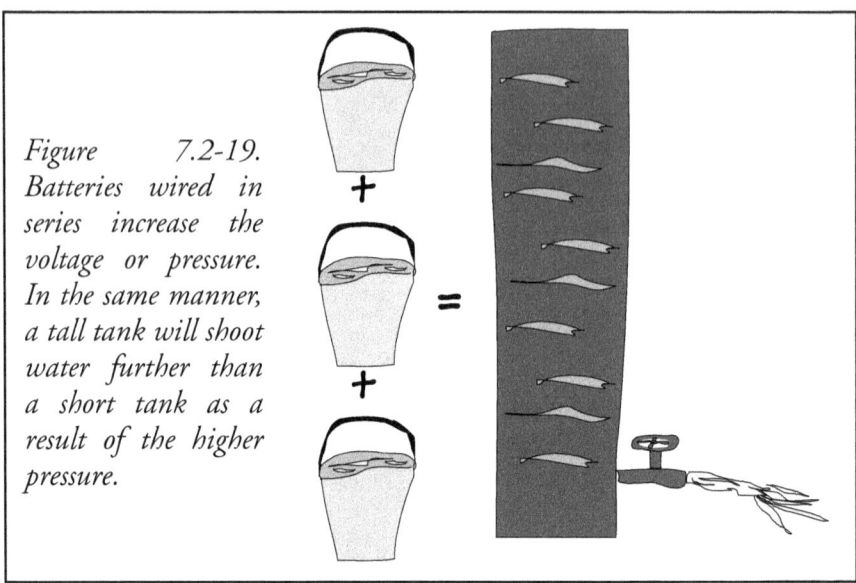

Figure 7.2-19. Batteries wired in series increase the voltage or pressure. In the same manner, a tall tank will shoot water further than a short tank as a result of the higher pressure.

This fluctuation of battery voltage makes it more difficult to calculate energy or capacity ratings. You will recall that energy is the voltage (pressure) multiplied by the current (flow) of electrons in a circuit multiplied by the amount of time the current is flowing. The problem with this calculation when it is applied to a battery is deciding which voltage to use.

Energy (watt-hours) = Voltage x Current x Time

Battery manufacturers are a smart bunch. To eliminate any confusion with ratings, they will often drop the voltage from the energy calculation, leaving us with a rating calculated by multiplying current x time.

Battery Capacity (Amp-hours) = Current x Time

The assumption is that there is no point in trying to shoot a moving target. If we really want to look at our energy storage in more familiar watt-hour terms, we can simply multiply the Amp-hour rating by the nominal battery-bank voltage. In any case, you will see battery manufacturers supplying capacity measurements in both units.

Determining battery capacity requirement for a given electric vehicle is simply a matter of relating the energy consumption of the vehicle to the capacity and depth of discharge of the battery. The Zero-Carbon Car consumes approximately 387 watt-hours (Wh) per mile (240 Wh per kilometer). Very efficient vehicles such as the Toyota Prius consume approximately 125 Wh per mile (77 Wh per kilometer), making it approximately three times more energy efficient than the Zero-Carbon Car. The calculations below will put to rest any discussion that energy efficiency is unimportant.

The Zero-Carbon Car has a gross battery capacity of approximately 6,000 Wh capacity (120 Volts x 50 Amp-hours). Based on the energy consumption per unit distance, and using an 80% depth of discharge, we find that the effective range of the Zero-Carbon Car is:

(6,000 Wh capacity x 80% depth of discharge) ÷ 387 Wh per mile = 12.4 miles

(6,000 Wh capacity x 80% depth of discharge) ÷ 240 Wh per km = 20 km

A Toyota Prius equipped with the same battery pack yields an effective range of:

(6,000 Wh capacity x 80% depth of discharge) ÷ 125 Wh per mile = 38.4 miles

(6,000 Wh capacity x 80% depth of discharge) ÷ 77 Wh per km = 62 km

As expected, the 300% improvement in energy efficiency of the Toyota Prius allows the vehicle to travel three times as far as the Zero-Carbon Car using the same capacity battery bank. Likewise, per-mile operating costs are lower as a result of lower electricity consumption.

Undersizing the battery can lead to overdischarging and shortening the battery's life. If you have to estimate and select sizes, it is always in your best interests to round up. Battery capacity is like money in your bank account: it never hurts to have extra on hand.

One further point about range relates to the real range of an all-electric vehicle. People claim that a given vehicle will travel a certain distance on a charge. Usually this number is inflated to begin with and does not take into account the effects of depth of discharge, which can severely reduce battery life. A typical BEV contains several thousand dollars' worth of batteries, and there is no point in causing them to fail prematurely by overdriving (overdischarging) them. When the cost of replacement is factored back into the "cost per mile driven," abusing batteries will drive this figure skyward.

Hydrogen Gas Production

Connecting a battery to a source of voltage will push electrons into the cells, recharging them to a full state of charge. As a liquid-electrolyte battery approaches the fully charged state it will be unable to absorb additional electrons, causing the electrolyte to start bubbling and emit hydrogen gas. "Maintenance-free" gelled-cell batteries are designed to minimize the possibility of hydrogen buildup and require accurate charging voltages.

The battery bank should be accessible to allow for servicing and periodic inspection. However, hydrogen safety is of primary importance. Batteries are stored in some form of battery box on-board the vehicle. Installing a brushless motor ventilation fan on the intake side of the battery box will pressurize the container, allowing a vent tube on the other side of the chassis to vent hydrogen outside the vehicle. This fan can also be thermostatically controlled to operate when the battery compartment becomes too hot or whenever the battery charging unit is active.

An alternate method of controlling hydrogen gassing is to install a set of reformer caps (Hydrocap Corporation). These caps contain a catalyst that converts the hydrogen and oxygen by-products of battery charging into water that simply drips back into the cell. The net effect is eliminated hydrogen gassing and lower battery water usage. Note, however, that hydrogen reformer caps control but do not completely eliminate hydrogen gassing, making some supplementary ventilation necessary.

Safe Installation of Batteries

Large deep-cycle batteries are heavy and awkward. Typical "golf cart" batteries weigh 70 pounds (32 kilograms) each. It is important that any frames and mounting hardware be of sufficient strength to hold them securely. During installation, make sure you have adequate manpower to lift and place the batteries without undue straining. Tipping a battery and spilling liquid electrolyte (acid) all over you is very dangerous. Follow these rules to ensure a proper and safe installation:

- Batteries must be installed in a well-insulated and sealed battery container. There is an enormous amount of energy stored in the batteries. Children (or curious adults) should not be allowed near them.

- Remove all jewelry, watches, and metal or conductive articles. A spanner wrench or socket set accidentally placed across two battery terminals will immediately weld in place and turn red-hot. This can cause battery damage, possible explosion, and severe burns.
- Hand tools should be wrapped in electrical tape or sufficiently insulated to prevent contact with live battery terminals.
- Wear eye-splash protection and rubber gloves when working with batteries; splashed electrolyte can cause blindness. Also, wear old clothes or coveralls since electrolyte just loves to eat your $100 jeans.
- Keep a 5 lb (2.3 kg) can of baking soda on hand for any small electrolyte spills. Immediately dusting the spilled electrolyte with baking soda causes aggressive fizzing, neutralizing the acid and turning it into water. Continue adding soda until the fizzing stops.

Battery Installation in the Zero-Carbon Car

The following collage of images shows the battery installation in the Zero-Carbon Car:

Figure 7.2-20. The insulated battery box used in the Zero-Carbon Car was designed to fit on top of the electric drive motor (discussed in Chapter 7.3). The box was fabricated using welded steel sheets painted to resist rusting and corrosion and lined with one inch (25 millimeters) of blue STYROFOAM brand insulation. Battery cables are placed through holes drilled in the box and are surrounded with a protective grommet to prevent sharp metal edges from cutting through them.

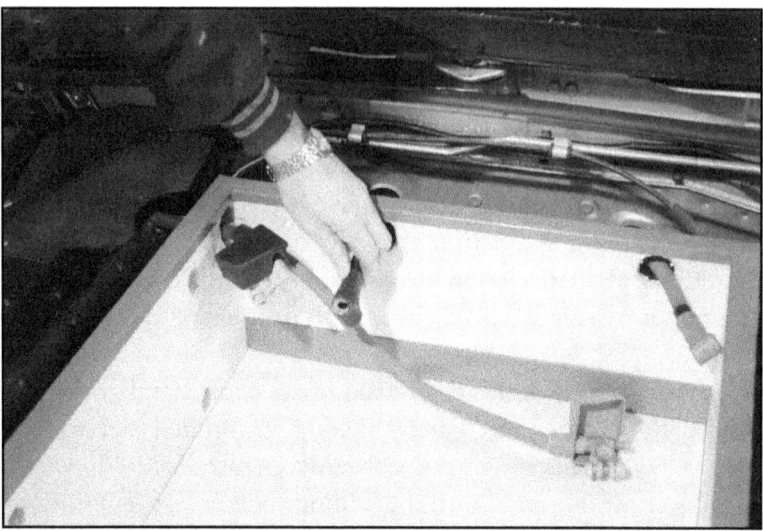

Figure 7.2-21. Rick Lane is shown holding the negative lead of the wire supplied by the biodiesel-powered generator. Both the negative and positive cables run to the rear of the vehicle and must therefore be protected by flexible, anti-abrasion sleeving. The wire pair in the top left corner of the photograph supplies power to the motor controller and traction motor. Together, these wires form the "positive and negative rails" shown in Figures 7.1-2 and 7.1-3.

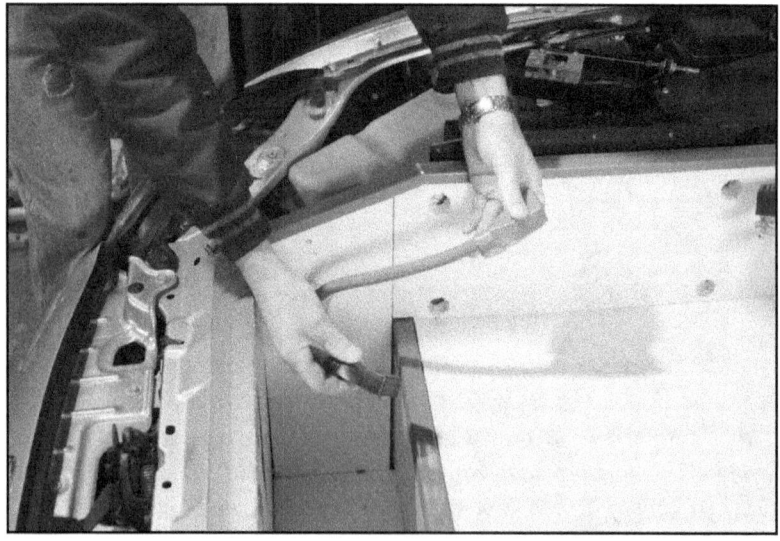

Figure 7.2-22. In order to protect the battery bank from a catastrophic overload or short-circuit, a separate short cable is run outside of the battery bank to a 500-ampere protective fuse.

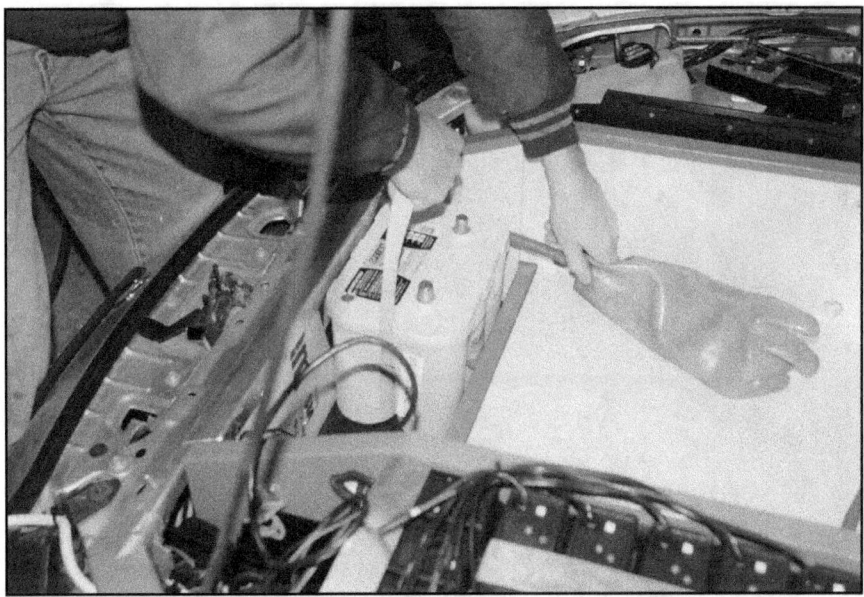

Figure 7.2-23. Once the battery posts have been cleaned with a wire brush or sandpaper, the batteries are carefully placed into the battery compartment. The layout is designed to ensure that battery cables fit and are as short as possible.

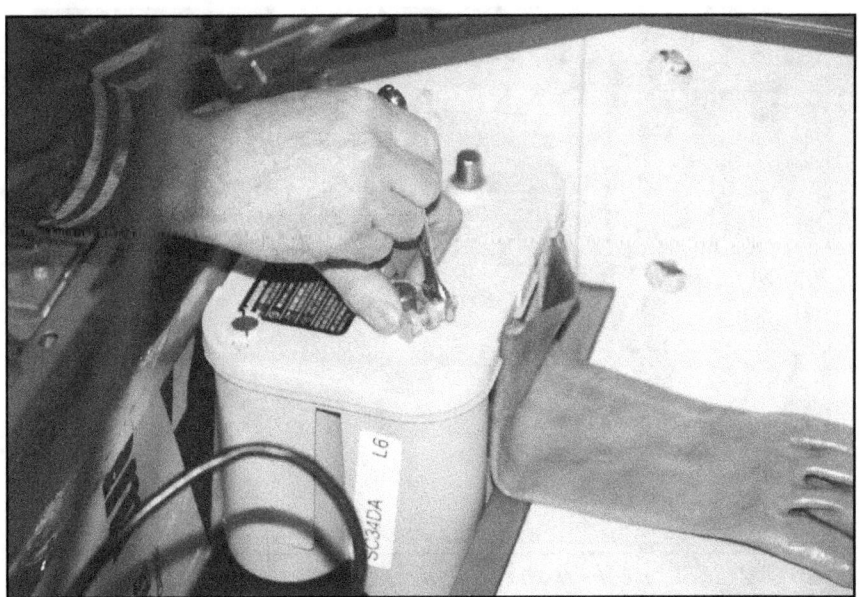

Figure 7.2-24. The battery clamps are placed on the battery posts and tightened to the torque value recommended by the manufacturer.

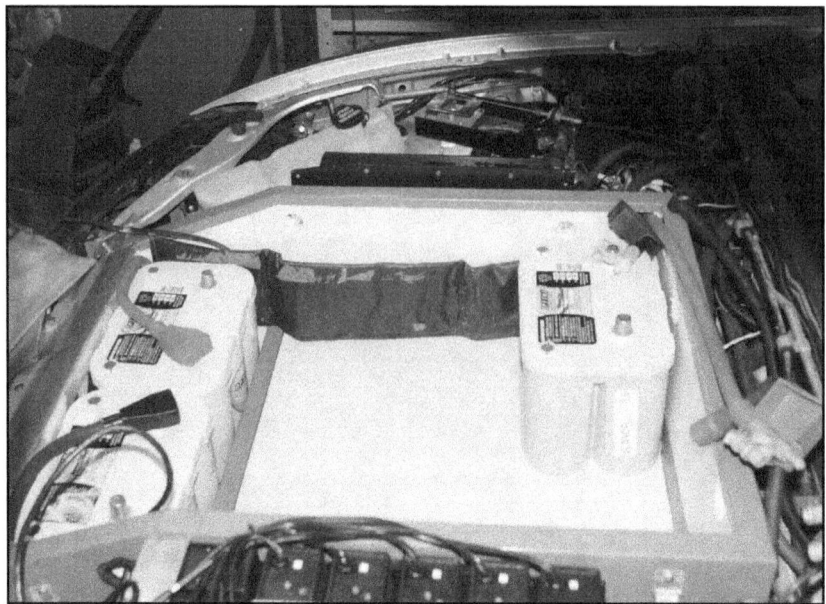

Figure 7.2-25. The placement of the batteries continues until all of them are in place. A 120-volt battery-warming pad is placed on either side of the battery bank. The pad is connected to a thermostat and wired to the same supply circuit that provides power to the battery charger.

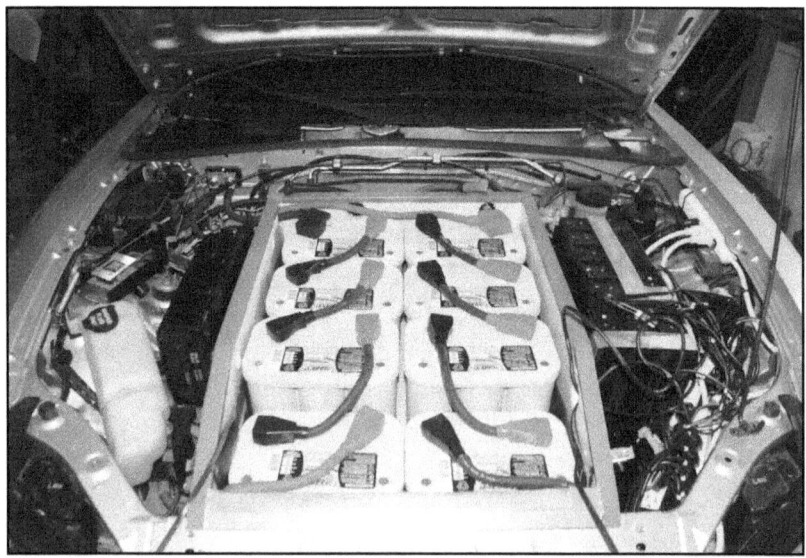

Figure 7.2-26. Once the battery wiring is completed, insulating "boots" are placed over the battery connectors and posts. The wiring should look neat and professional.

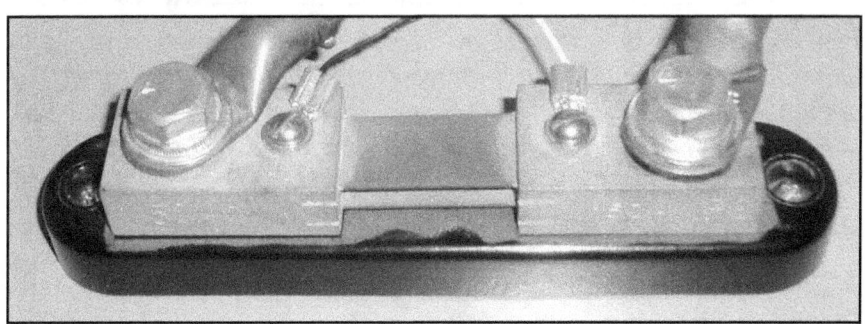

Figure 7.2-27. The negative lead that supplies the negative rail of the main controller is fed through a device known as a resistive shunt. This device provides a milli-volt reference voltage that is proportional to the current being drawn by the traction motor. The shunt can be used to feed metering or automatic control devices. In the case of the Zero-Carbon Car the shunt is used only for display metering.

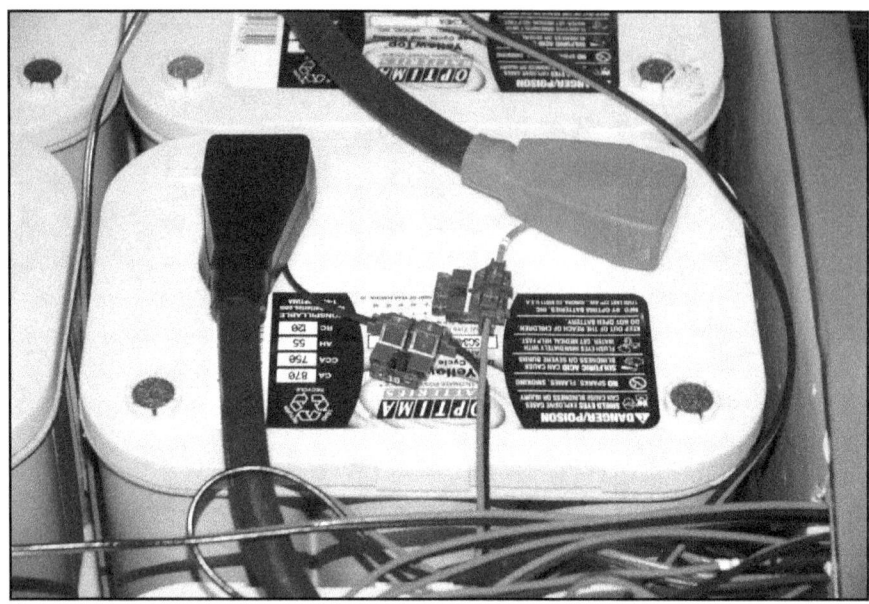

Figure 7.2-28. Each battery is provided with a smaller-gauge wire pair which is connected to the battery charger for that battery. Note that each leg of the battery charger supply wire is provided with a ten-ampere safety fuse. Wires are also tagged with a wire number and logged against the electrical schematic.

Figure 7.2-29. Each of the ten traction batteries is provided with its own battery charger. These models are manufactured by the Soneil Company and are fully automatic units that provide rapid yet accurate battery charging and status monitoring. A color light provides at-a-glance status (orange = charging; green = charge complete and "floating"; flashing orange = charger fault or battery voltage too high).

Figure 7.2-30. The completed battery bank with the battery charger cable harness is shown in this view. The black rectangular object in the center of the battery bank is the thermostat which activates the battery heater blankets when the temperature drops below 50° F (10° C). The ten Soneil battery chargers are mounted on a shelf to the right of the battery bank. Each of the ten grid supply connections is terminated in a power bar arrangement with the charging supply cable temporarily placed through a seam in the vehicle's hood.

Figure 7.2-31. The battery box cover, also insulated, is fitted into position. Four hold-down clamps secure the lid to the battery box, ensuring a weathertight seal.

Figure 7.2-32. Rick is shown here measuring the battery voltage to ensure that the batteries are at their correct voltage before connecting the battery charger to the supply mains.

Figure 7.2-33. Once the battery box cover has been tested and checked for clearance with the hood and battery voltage is determined to be within specification, the car is plugged in to check battery charger performance.

The Fine Art of Battery Cable Manufacturing

Making battery cables that are capable of handling the power levels required to drive an electric vehicle is something of an art. Given that very high currents will be flowing through these cables, even a small amount of electrical resistance will cause a loss of power and range, cable heating, possible failure, and potential fire.

The importance of making these cables correctly cannot be understated, so let's take a few moments to consider the right way to do the job:

Figure 7.2-S1. Rick Lane starts by applying a healthy globule of plumbing flux to the inside of a battery cable connector, using a piece of lead-free solder to do the wiping.

Figure 7.2-S2. Rick then applies heat using a soldering torch. The flux will melt and eventually reach the temperature at which solder melts. Once this temperature is reached, the cable connector area is filled with solder.

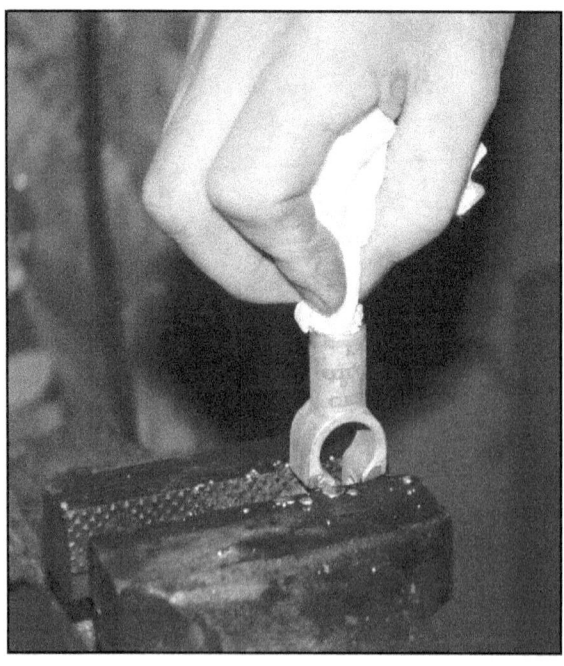

Figure 7.2-S3. A small piece of paper towel is used to quickly draw the last remains of flux and other foreign materials that are floating on top of the liquid solder.

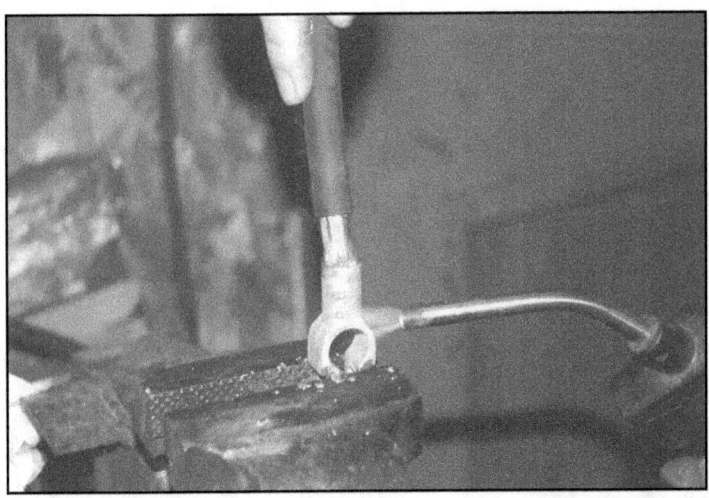

Figure 7.2-S4. The cable connector is reheated and the battery cable wires are fed into the connector "cup." Note that superfine "welding wire" is used to make the battery cables. Although welding wire is slightly more expensive than standard battery cable wire, it is much more flexible, which is a blessing if you're trying to route cables in tight, confined spaces.

Figure 7.2-S5. As the wire begins to heat it will start to absorb the solder from the cup. The wire is pressed firmly into the cup, ensuring that it goes in right to the edge of the insulation.

Figure 7.2-S6. Spilled flux or solder is wiped away from the connector and cable insulation.

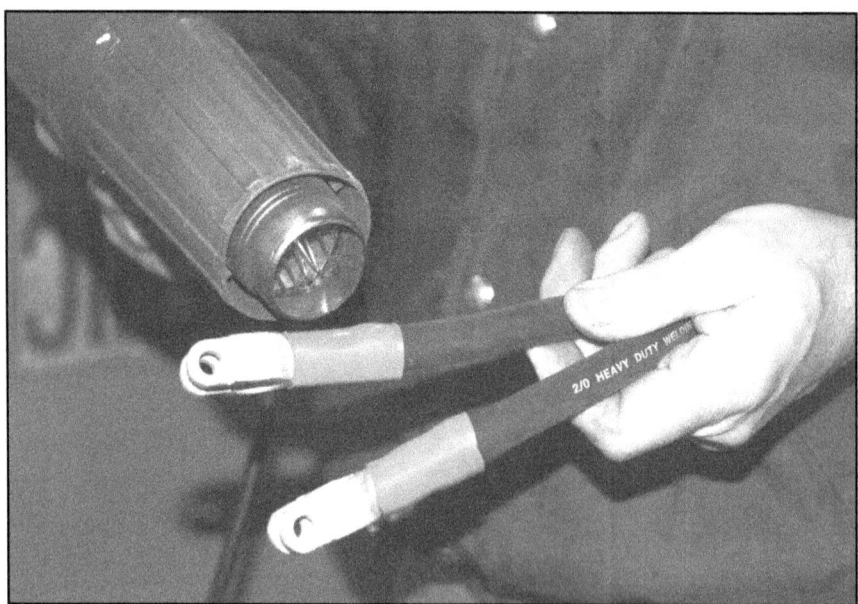

Figure 7.2-S7. Color-coded heat-shrinkable tubing is applied to the cables to complete the job.

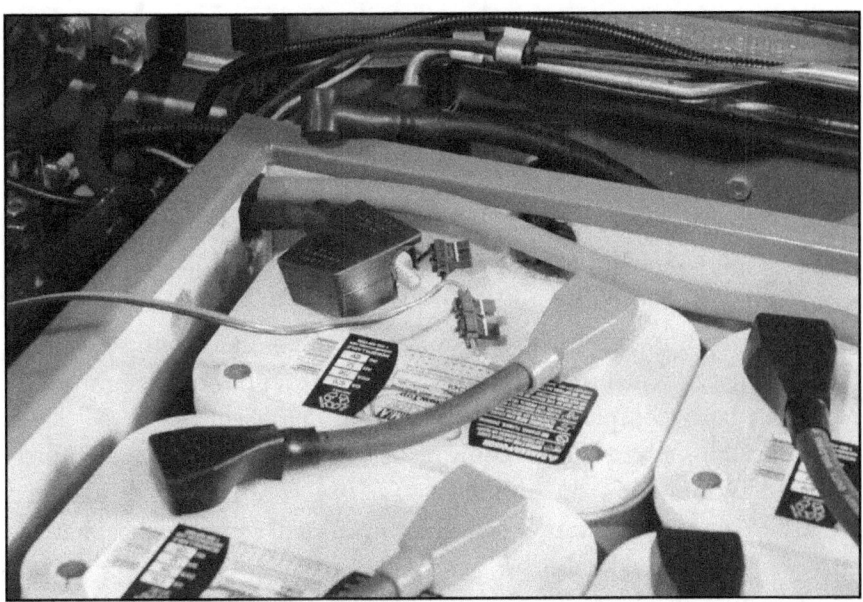

Figure 7.2-S8. Properly manufactured cables have a very professional appearance and provide years of trouble-free operation.

Battery Charging

All electrically powered vehicles must have some form of battery charging system, whether it is supplied by an on-board, low-power charger, a high-powered internal combustion engine genset, or a large stationary charger located at the home or office. Notwithstanding the type of charger or battery technology used, if the charging technique is not properly implemented, battery failure or reduced vehicular performance will result.

Charging a battery involves more than the simple application of a fixed voltage to replenish the electrical energy that has been removed. It requires a fairly complex control system and methodology.

Although purchasing a commercial charger such as the models supplied by Soneil is the simplest way to ensure a properly charged battery in the least amount of charging time, I will cover a typical battery charge regime to make this process clear.

Referring to Figure 7.2-34, you will see a typical battery charging time versus voltage plot. Increasing voltage is plotted on the Y axis, while time is shown on the X axis. Note that the nominal battery voltage is shown on the plot, indicating that charging always requires a higher-voltage source of electrical power to ensure that current flows into the battery bank.

The charge cycle begins with the "bulk charge mode" which allows the battery charger to supply full charging power to the battery. As the battery begins to absorb this energy, the electrolyte becomes regenerated and the battery terminal voltage begins to rise.

When the charging voltage reaches a given level (for example, 2.4 volts per cell for a lead-acid battery), the control circuitry within the charger "holds" the charging voltage at the "absorption mode" fixed level for a given period of time, typically two hours.

At the conclusion of the absorption mode, the charger enters one or two remaining modes, with the "float mode" being the one used on a day-to-day basis.

"Float mode" supplies a small charging current to the battery to overcome any self-discharging of the battery. It can be maintained indefinitely and will ensure that the battery bank is ready for operation as required.

Due to the effects of a process known as self-discharge, a battery that is left disconnected will become fully discharged over a period of time.

If left in this state, the discharged cells may become defective over time. Also, as each cell has a different self-discharge rate, battery capacity will vary from cell to cell over a period of time. For example, if the battery bank is depleted to a 50% depth of discharge, some of the cells may have 55% capacity remaining while another may have only 45%. If the batteries are recharged using the standard "bulk/absorption/float" cycle, the cell with the highest depletion rate may only be charged to 90% capacity. If this continues for an extended period of time the cell will be left in a constant state of discharge and will fail.

To counteract this problem a periodic equalization charge is completed, in which all of the cells are deliberately overcharged, ensuring that the depleted cell has enough time and available energy to become fully charged.

> **CAUTION!**
> **Sealed or "maintenance–free" batteries do not require equalization charging. Placing them in such a condition can result in fire or cause the battery to explode.**

To visualize the equalization process, think of batteries containing electrons as buckets containing water, with one bucket equivalent to one battery cell.

Over the course of a few dozen charge cycles, energy is taken out of and replaced into the battery. This is the same as if I asked you to take four cups (one liter) of water out of each "cell" or bucket, representing a day's electrical load consumption. Assume that our charging cycle produced enough energy to replace three cups (750 ml) of "energy" into the cells, leaving us with a deficiency of one cup (250 ml). This process is repeated with different amounts of water being added or removed over the course of a month's driving until such time as we believe the batteries are back to a full state of charge. In our example, the buckets are also full.

In reality, the water in each of the buckets would not be at exactly the same level. A certain amount of spillage and uneven amounts of water removal will leave the buckets with varying amounts of water in them. This is the exact scenario that plays out in your battery bank.

Over time there is a gradual change in the state of charge in the cells, which can be determined by comparing the specific gravity of one cell with that of adjoining cells.

If the difference is allowed to continue for an extended period, the cell with the lower state of charge can fail prematurely and the amount of available energy stored in the battery bank can be reduced.

To correct this situation, a periodic (typically once per month) controlled overcharge is conducted. The charge controller is either manually or automatically set to equalization mode and is maintained for approximately two hours, after which battery voltage is reduced and the charge controller automatically enters float mode.

If we go back to our bucket example, the effect of equalization is similar to using a garden hose to add water to the buckets and deliberately overfilling them. When you stop adding water (equalization completed) the buckets are topped up right to the rim.

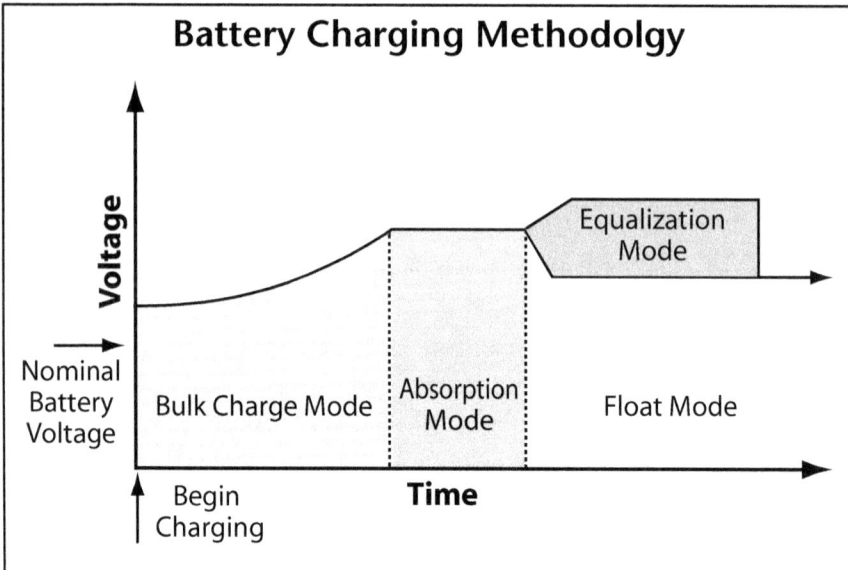

Figure 7.2-34. Charging a battery involves more than the simple application of a fixed voltage to replenish the electrical energy that has been removed. It requires a fairly complex control system and methodology.

Where does the "extra" electricity go during the equalization mode? During this charging stage, the excess energy applied to the batteries causes aggressive bubbling of the electrolyte, producing large amounts of hydrogen and oxygen gas as well as heat and water vapor. It is necessary to monitor electrolyte levels and battery temperature during this charging stage to ensure that battery parameters are within manufacturer ratings.

If your batteries are equipped with hydrogen reformer caps, ensure that they are removed during equalization mode.

Battery Technology Selection

There are numerous battery technologies that have been in commercial use for a number of years, and many more technologies and novel packaging means are entering the market on a regular basis. However, for home builders of BEV or even PHEV vehicles, the lead-acid battery will remain the technology of choice for a few years to come.

The lead-acid battery is a tried and true workhorse that offers low cost, easy availability, and ease of recycling. Unfortunately, it has the lowest *specific energy* (excessive weight) and *energy density* of any battery technology currently on the market. Battery cycle life (charge/discharge cycles) remain too low to be of practical value for long-term production vehicle prospects.

Higher performance nickel-cadmium (NiCad) batteries are manufactured by the Saft Company and offer numerous improvements over lead-acid technology. According to information supplied by the company as of April 2007, their NiCad STM Modules power the largest fleet of electrically powered vehicles in the world. The batteries offer a lifetime of over 65,000 miles (105,000 kilometers), are suited for a wide operating temperature range, can be rapidly recharged, and are fully recyclable.[1] As with lead-acid batteries, low specific energy also makes the NiCad battery unsuitable for mass production electric vehicles.

Nickel metal hydride (NiMH) technology further improves upon NiCad technology, offering higher energy density (more energy packed into a smaller volume and mass), albeit at much greater cost. Battery cycle life should easily meet the demands of the buying public. Tests conducted by the Electric Power Research Institute (EPRI) conclude that:

"5-year old Toyota RAV 4 EVs, in real world driving, have traveled over 100,000 miles (161,000 kilometers) on the original NiMH battery with no appreciable degradation in battery performance or vehicle range. The vehicles are projected to last for 130,000 to 150,000 miles (209,000 to 241,000 kilometers)."[2]

Because PHEVs are unlikely to be fully discharged before the battery management system activates the recharging genset, it can be expected that batteries used in this type of vehicle will show higher cycle life than those used in pure BEVs.

Lithium ion batteries and their derivatives offer the highest performance of any current technology, although the long-term reliability and cost of these batteries will have to be evaluated in the coming years.

Battery Technology	Specific Energy Wh/kilogram	Energy Density Wh/Liter	Specific Power Watt-Kgs	Relative Cost
Lead Acid	30	75	250	1
NiCad	55	88	122	3
NiMH	63	150	200	4
Li-ion	90	150	300	20

Table 7.2-2. A comparison of commercially available battery technologies according to company-supplied datasheets and websites.

Summary

Battery-electric vehicles are not suited for the majority of people because of their limited range and slow recharging time. However, this is not to say that BEVs cannot make a significant contribution to the transportation equation when they are used in the right applications.

Research conducted on advanced batteries using NiMH has proven that they can be used with great success even in full battery electric vehicles. When used in plug-in hybrid vehicle applications where discharge stresses and recharging times are greatly reduced, there is no question that these batteries can and should be used immediately.

Advanced batteries including Li-ion, zinc-air, and sodium metal chloride (Zebra) are certainly showing promise for electrically powered vehicles, and many more technologies will be brought to production provided that the correct regulatory environment forces these ideas out of the laboratory and into the streets.

7.3 DC Motor and Controller

Electric motors and their associated controllers are the equivalent of the internal combustion engine and its fuel management system, respectively. An electric motor converts electrical energy stored in the battery bank into rotational energy (torque) required to drive the vehicle. This occurs through the electromagnetic attraction and repulsion of the motor's rotating components via a switching mechanism that controls the flow of electricity within the electrical windings.

Electric Motor Overview

The simplest electric motor is the direct current, permanent magnet brushed design that uses two or more brushes to transfer electrical power

Figure 7.3-1. The series-wound design that is used in the Zero-Carbon Car is a variation on the permanent magnet motor. After the internal combustion engine is removed, the cavernous area allows plenty of room for the new electric traction motor and battery bank. People are amazed that such a small motor is able to replace the much larger gasoline engine yet still maintain adequate vehicle performance. Note the power steering pump located on the "accessory" end of the drive motor. The light-colored cylinder to the left of the motor is a 12-volt vacuum pump which is used to operate the power brake booster unit.

to a rotating commutator. Once power is applied to the brushes it flows into the rotor windings, creating an electromagnetic field that repels the field of the permanent magnets fixed to the stationary outer diameter of the motor chassis. The commutator gear is so arranged that the rotor will continuously attract and repel successive poles of the stationary outer magnets, thus inducing torque on the rotor shaft and causing it to rotate.

The series-wound design that is used in the Zero-Carbon Car is a variation on the permanent magnet motor. The ZCC design replaces the fixed permanent magnets with "stator" coils that become magnetic when electrical power is applied to it. Simultaneously applying power to the stator coils and rotor brushes creates the same magnetic effect described for the permanent magnet motor, inducing torque and causing the rotor to spin.

Figure 7.3-2. This side view of the motor shows the machined aluminum adapter plate that allows the DC motor to be bolted to the transmission bell housing. The adapter plate is designed to permit the clutch assembly to operate in the same manner as in the gasoline engine. Because of the motor's very high torque at low speed and its wide speed range, the car requires only two gears rather than the typical five of the gasoline engine.

Figure 7.3-3. The unpainted battery box is L-shaped to allow it to sit on top of the DC traction motor with batteries stacked two layers deep in the space in front of it. The box is fabricated from welded steel sheet which is insulated and provided with a locking access cover.

One advantage of the series-wound motor is that the strength of the stator windings can be adjusted by varying the applied voltage. Motor power and torque can then be optimized to match the connected load.

The Advanced DC Motors Inc. nine-inch model FB1-4001A has rated horsepower of 28 and a peak rating of 85. This may seem anemic until you realize that the electric motor produces maximum torque at zero speed; a gasoline engine would be of course be stalled. Later in this chapter I will take the car to a race-tuning specialty shop and place it on a dynamometer for further evaluation.

Although direct current motors are powerful and flexible in their design, the use of brushes acting on the rotating commutator gear creates a wear point that eventually causes the motor to fail.

Alternating current motors have begun to replace direct current motors in many applications, including advanced vehicle technologies used in BEV, PHEV, and hydrogen fuel cell designs. Although there are many classes of AC motor, the most commonly used in traction applications is the asynchronous or induction motor invented by Nikola Tesla in 1883.

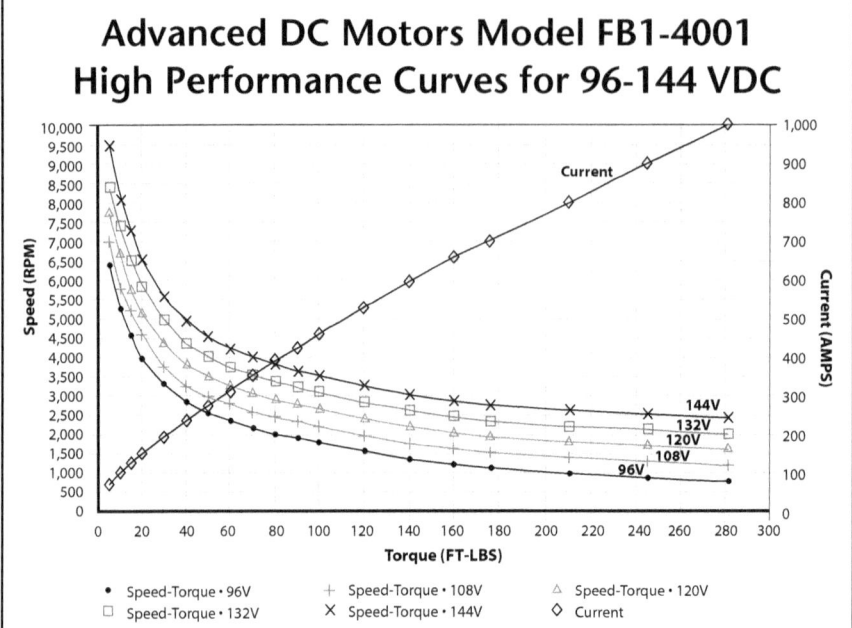

Figure 7.3-4. The Advanced DC Motors Inc. model nine-inch FB1-4001A has rated horsepower of 28 and a peak rating of 85. This may seem anemic until you realize that the electric motor produces maximum torque at zero speed; a gasoline engine would of course be stalled. (Courtesy: KTA Services, www.kta-ev.com)

In his design, the motor contained a fixed stator winding but replaced the brush gear, commutator, and rotor windings with an ingenious "induction rotor." The induction rotor received magnetic energy from the stator windings, "inducing" it into the rotor core where it in turn created its own magnetic field and torque.

The magnetic field of the rotor would remain stationary except for the fact that the stator windings are connected to a source of alternating current, which causes the magnetic polarity of the stator and by induction the rotor winding to oscillate, thereby inducing rotational energy into the motor shaft.

Controller Overview

Historically motors were simply turned on and off by the use of a switch. While this simple control principle might be acceptable for items such as fans, grinders, or other basic applications, it would be wholly unacceptable for traction applications. In the early days of electric vehicle technology, motor speed controls were developed whereby an operator would move a switch mechanism and activate "load resistors" of sequential ratings in series with the motor wiring. The application of a resistance in series with the motor would absorb some of the applied energy, making the motor weaker than if it were directly connected to the source of electricity. Using this crude technology, vehicles such as Rick Lane's Milburn (Figure 6-11) could be smoothly accelerated rather than harshly applying full power to the stopped vehicle with a power switch.

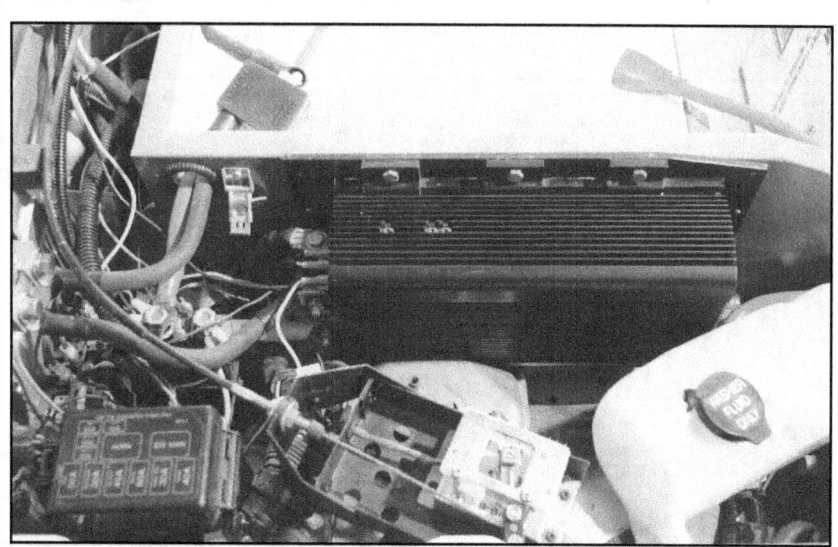

Figure 7.3-5. The Zero-Carbon Car uses a series motor controller manufactured by the Curtis Instruments Company (black finned box, center). Electronic controllers use fast-acting power switching devices known as field-effect transistors or insulated-gate bipolar transistors which can be rapidly turned on or off to regulate power flow through the motor windings. The Curtis controller has a rating of 120 volts at 500 Amps (60 kW). The speed control potentiometer can be seen in the center foreground, just to the left of the windshield washer fluid reservoir.

DC Electronic Controllers

As electronic motor controllers became part of the technological lexicon, they were quickly applied to battery electric vehicles, golf carts, forklift trucks, and the like. The Zero-Carbon Car uses a series motor controller manufactured by the Curtis Instruments Company. Electronic controllers use fast-acting power switching devices known as field-effect transistors or insulated-gate bipolar transistors which can be rapidly turned on or off to regulate power flow through the motor windings. Using the technique of pulse-width modulation, the transistor circuits can be cycled on and off several hundred times a second, thereby reducing current flow through the motor and lowering torque and vehicle speed and acceleration.

Figure 7.3-6. This detail view shows the speed control potentiometer unit which provides speed control signals to the Curtis controller. The device consists of a variable resistor which is connected to a spring-loaded actuator arm. The spring pulls the control arm to the zero-speed position while a control cable is fixed to the opposite side of the lever, which is in turn connected to the accelerator pedal. Pressing on the accelerator pedal moves the control arm, varying the resistance of the potentiometer. The change in resistance causes the controller to vary the current flow through the traction motor, allowing the vehicle power and speed in proportion to the pedal movement.

AC Electronic Controllers

It was discovered that by varying the frequency of the alternating current applied to an AC induction motor its speed would vary in direct proportion. Capitalizing on this finding, many advanced electric vehicle designs now use a high-voltage battery bank of typically 340 volts capacity connected to a device known as an inverter, which converts the direct current into a variable frequency supply voltage. This alternating current supply is then used to operate the induction-type traction motor in the vehicle. Varying the frequency of the applied voltage provides very smooth vehicle acceleration and virtually silent vehicle operation. As a further benefit to the vehicle operator, induction motors have few moving parts and are very simple, energy efficient, and rugged.

Figure 7.3-7. When an electric motor is driven by an external source of energy, an electric current is present at the winding terminals, making the motor a generator. This effect is well known and is often used in industry to produce electrical energy. The small hydro power generating station shown here uses a water-powered turbine to turn the shaft of an AC induction motor, generating electrical energy. This concept can be applied to electric vehicles, allowing braking energy to be captured and stored in the battery bank and thus improving overall energy efficiency.

Regenerative Braking

When an electric motor is driven by an external source of energy, an electric current is present at the winding terminals, making the motor a generator. This effect is well known and is often used in industry to produce electrical energy. The small hydro power generating station shown in Figure 7.3-7 uses a water-powered turbine to turn the shaft of an AC induction motor. In this application, an electric motor is the generator, producing 500 kW of electricity for the local electrical grid.

Slowing a vehicle requires dissipating its forward energy through the application of brakes, with the energy being wasted as heat. In an electric vehicle, some of the energy required to slow the vehicle can be captured using a process known as regenerative braking. When the accelerator pedal is released, the traction motor is no longer driving the vehicle. Rather, the vehicle's forward motion is now driving the motor, exactly the condition required to make it a generator. Electrical energy can thus be captured and stored in the battery bank, further helping to improve vehicular efficiency.

7.4 The Liquid Fuel Power Plant

One half of a series plug-in hybrid vehicle power system comprises the electric motor, controller, and battery bank; the other half is the electrical generating source. The vehicle designer has numerous power plant options to choose from, including gasoline spark-ignition, diesel, or even a hydrogen fuel cell unit. If the vehicle is to operate in a true well-to-wheels zero-carbon cycle, the hardware can stay the same but the fuel must change to sustainable and renewable sources.

Although hydrogen produced from water or cellulosic ethanol could have been used in the Zero-Carbon Car, the cost of the electrical generating system for hydrogen was prohibitive, the hydrogen unavailable, and the cellulosic ethanol difficult to obtain and use at 100% concentrations. The diesel engine offered a number of advantages, including high thermal efficiency and small size for a given level of electrical power generation. A further advantage was the ability to use zero-carbon biodiesel as the fuel source, which is covered in detail in Chapter 9.

The Fischer Panda Generating Unit

The choice of which diesel generator set to use was actually very easy. A quick look on the Web for "marine generators" took me directly to Fischer Panda Generators Inc. in Florida (http://www.fischerpanda.de/eng). One major advantage of the Fischer Panda generating system is its ability to generate direct current which can charge the vehicle's battery bank or drive the traction motor directly.

I assumed that the company could upgrade one of their standard 48-volt systems to the higher 120-volt potential that was required by the design of the Zero-Carbon Car. Having a direct current output would provide two immediate benefits for the system design. First, the generator windings and controls would be optimized to output direct current, avoiding the energy losses that would be incurred if a standard alternating current generator had to be modified. Secondly, the speed of the engine would not have to remain fixed as is necessary with alternating current generators.

You will recall my discussion in chapter 7.3 about the speed of an induction motor being proportional to the frequency of the applied voltage source. This phenomenon requires that all generators that have

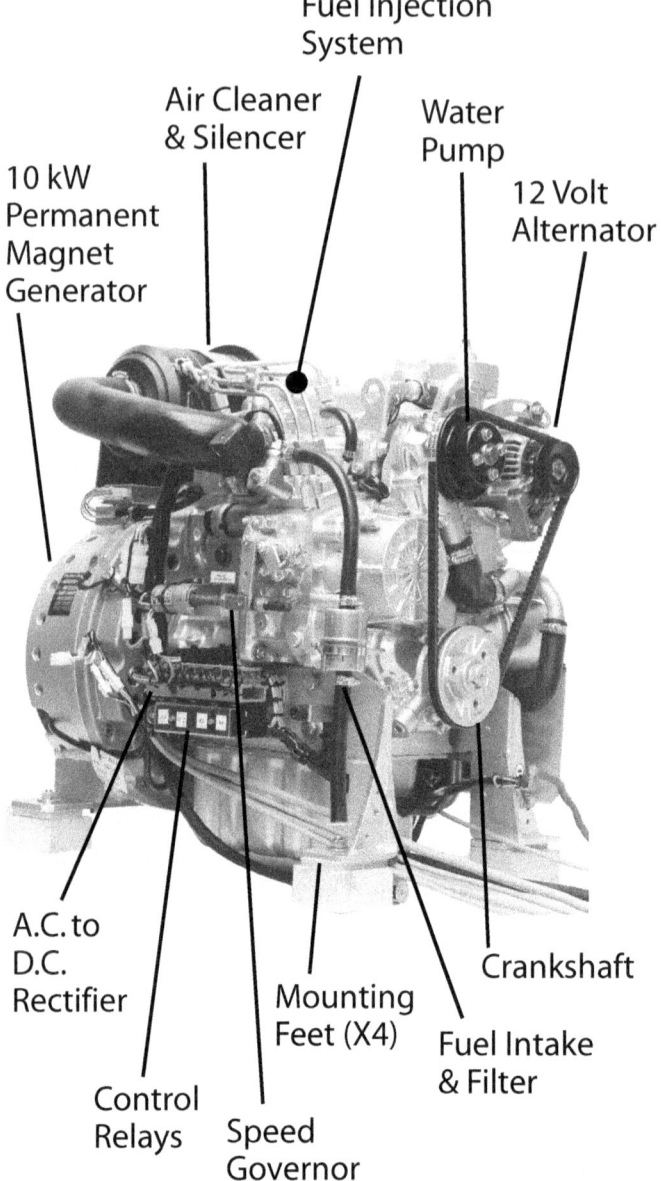

Figure 7.4-1. The choice of which diesel generator set to use was actually very easy. A quick look on the Web for "marine generators" took me directly to Fischer Panda Generators Inc. in Florida. One major advantage of their system is the ability to generate direct current which can be used to charge the vehicle's battery bank or drive the traction motor directly. This image shows the generator unit and identifies its major components.

the same frequency as the North American electrical grid must operate their generators at a fixed speed, typically 1,800 or 3,600 rpm, in order to supply 60 Hertz power. The engine must operate at this fixed speed any time there is an applied load, regardless of whether the load is a single light bulb or an entire worksite full of high-power drills and saws. Clearly the former application is very energy inefficient.

Using a direct current generator solves these problems immediately. As the direct current electricity does not have a frequency component, the engine driving the generator is free to vary in speed as a function of electrical load. This allows the control system to "hunt" for the correct balance between energy efficiency and electrical output required to charge the batteries and drive the vehicle's traction motor.

I contacted Paulo Oliveria, chief engineer at Fischer Panda, to discuss my requirements and determine whether one of their marine generators could be converted for this application. "Our line of AGT-DC generators should do a perfect job for this application, although you will have to design some form of cooling system for the unit," explained Paulo. "Our marine units are normally cooled by raw seawater, which is not going to be possible in your application. As for the high battery-charging voltage, we can easily produce a new [generator] winding configuration for this application."

While Paulo set about working on the electrical and thermal design of the package, Gary Baker created a three-dimensional model of the trunk space of the Miata, which we would provide to Fischer Panda. It is hard to tell from the picture (Figure 7.4-2), but Gary is staring in disbelief at the trunk space as I explain to him that I want to fit a 10-kilowatt generator in there. Anyone who is familiar with gasoline generators knows that a unit of that capacity is extremely large owing to the low efficiency of the gasoline engine. One of the reasons for picking the Fischer Panda system is that the genset is approximately 30% smaller than an equivalent gasoline model and has been optimized for installation in the cramped quarters of sailing vessels.

Figure 7.4-2. It is hard to tell from the picture, but Gary is staring in disbelief at the trunk space as I explain to him that I want to fit a 10-kilowatt generator in there. Anyone who is familiar with gasoline generators knows that a unit of that capacity is extremely large owing to the low efficiency of the gasoline engine.

Fitting the Generating Unit into the Zero-Carbon Car

Within a few weeks, Paulo supplied a set of technical drawings which indicated that the generator would fit within the vehicle envelope Gary had provided. The order was placed and the factory in Germany began to build the unit.

Rick Lane then set about modifying the trunk of the vehicle and developed the frame rails to support the generator. In the first version of the design, rubber washers were used as vibration dampers (Figure 7.4-4), but they caused far too much shuddering through the vehicle body, so commercial isolation dampers were substituted.

Photographic Collage of the Installation

The following photographic collage shows the installation of the generator system in the vehicle:

Figure 7.4-3. Rick Lane then set about modifying the trunk of the vehicle and developed the frame rails to support the generator.

Figure 7.4-4. . In the first version of the design, rubber washers were used as vibration dampers, but they caused too much shuddering through the vehicle body, so commercial isolation dampers were substituted.

Figure 7.4-5. We were like kids at Christmas when the generator unit arrived from Fischer Panda. Rick Lane grabs his trusty tape measure, convinced that there is no way on earth the unit will fit into the vehicle.

Figure 7.4-6. Weighing 238 pounds (108 kilograms), the generator unit is just a bit too heavy to lift by hand. A hydraulic motor lifting crane does just the trick.

Figure 7.4-7. The trunk lid of the Miata is removed from its hinges and folded back on the removable winter roof, ensuring wide access for placing the genset into the trunk space.

Figure 7.4-8. The genset is raised into position and slowly lowered into the trunk...

Figure 7.4-9. slowly...

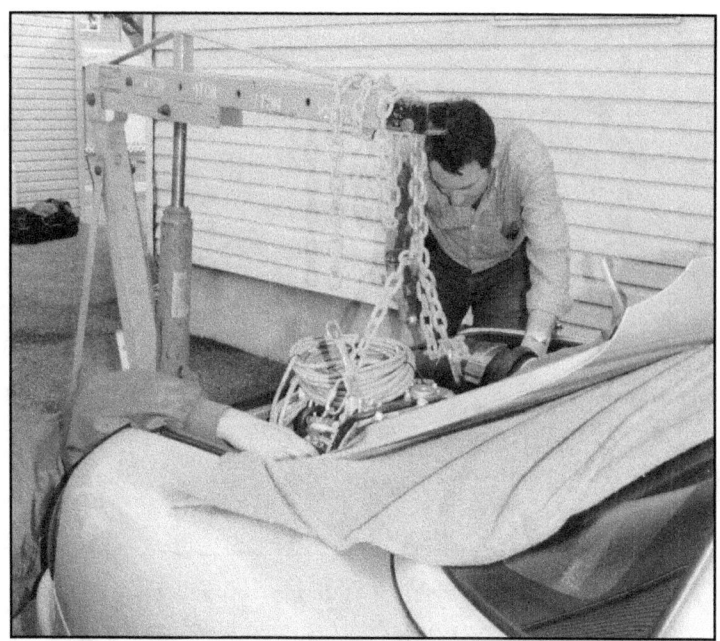

Figure 7.4-10. carefully...until it comes to rest on the motor mounts.

Figure 7.4-11. I was having a bit of trouble believing the unit actually fit, so I had to crawl under the car to make sure the genset had not simply fallen straight through. Well, to be honest I had to tighten the motor mount hold-down bolts.

Figure 7.4-12. The generating unit is shown mounted in the back of the vehicle with plenty of room to spare. It almost looks as though the vehicle came from the factory that way.

Figure 7.4-13. Fischer Panda provides a remote start/stop and status display package as part of its standard unit. As there was plenty of space in the trunk area, it was decided to mount the unit in the left wheel well. It would act as a handy control and reference point for testing the engine and monitoring operating run time on the integrated hour meter.

Figure 7.4-14. The final configuration shows the generating unit with remote radiator cooling hoses and expansion bottle (bottom right).

Figure 7.4-15. This side view of the trunk area shows the generator crankshaft pulley end with water pump and alternator. The cooling system expansion bottle is lower left. Center left is the first version of muffler used in the vehicle. Although this location was very compact and ideal, the exhaust sound level was too great and so a remote muffler was mounted under the vehicle in a more traditional manner.

Figure 7.4-16. Two galvanized iron pipes were run underneath the car from the trunk area to just before the front fender. These pipes carry generator-cooling water/antifreeze mix from the engine to the remote radiator.

Figure 7.4-17. The front bumper and trim work was removed from the car and custom mounts were developed to hold the radiator assembly. This radiator model is used in a Volkswagen Jetta and was further equipped with two electric fans that pulled the air through the radiator assembly. The fans were initially activated by a thermostatic switch mounted in the plumbing lines, but we experienced overheating as a result of the rather restrictive air path. The problem was solved by allowing the PLC control to activate the fans as soon as the generator unit is started.

Figure 7.4-18. This close-up view details the radiator assembly, mounting rail, and water cooling lines that run to the generator unit.

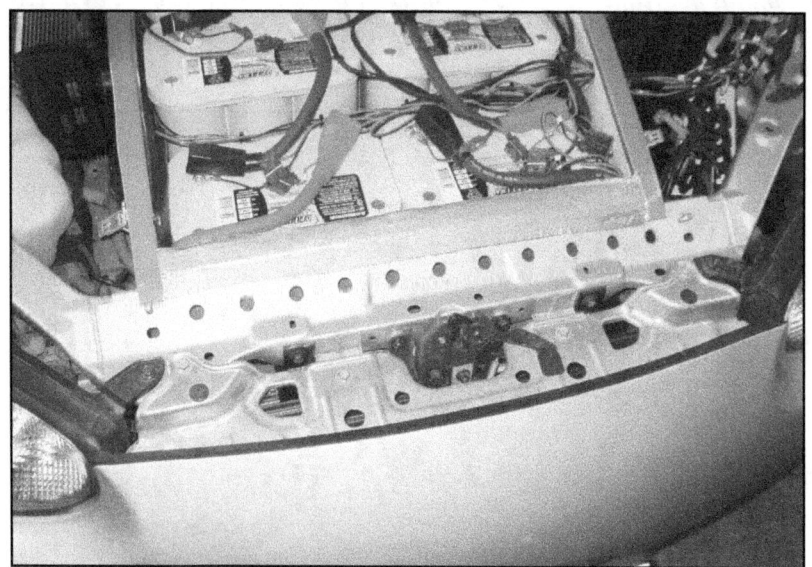

Figure 7.4-19. It was difficult to get enough air flow past the radiator because of the confined space, so Rick drilled a series of ventilation holes in the chassis.

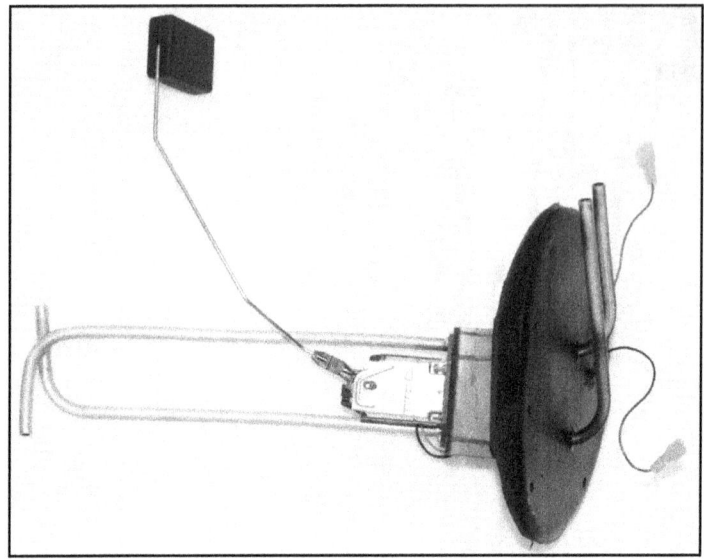

Figure 7.4-20. The Mazda Miata is a gasoline-powered car, so a few modifications were required to store and pump biodiesel to the generator unit. In a gasoline-powered vehicle, fuel is taken from the tank in the quantity required and then consumed. A diesel system draws fuel from the tank, pressurizes it, and sprays the required amount into the engine. Any fuel that is not required by the engine is returned to the fuel tank by a return line. This image shows the fuel pickup and return pipes as well as the fuel level sensing float (top).

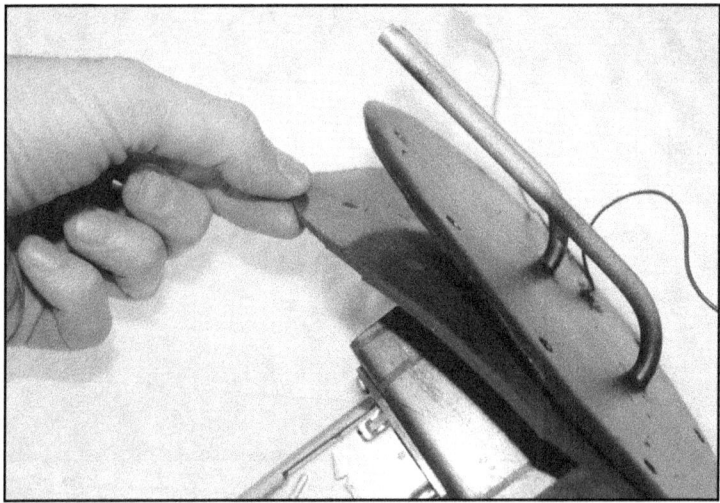

Figure 7.4-21. Biodiesel can react with natural rubber lines and gaskets, so all of these items were made from an artificial cousin known as Viton®.

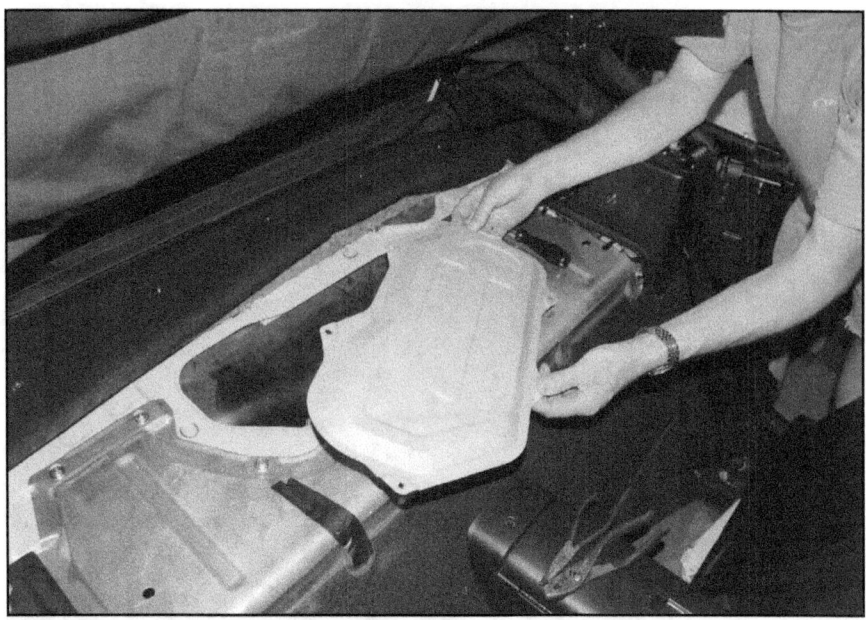

Figure 7.4-22. Rick is shown removing the cover from the fuel tank, which is located just behind the passenger seats.

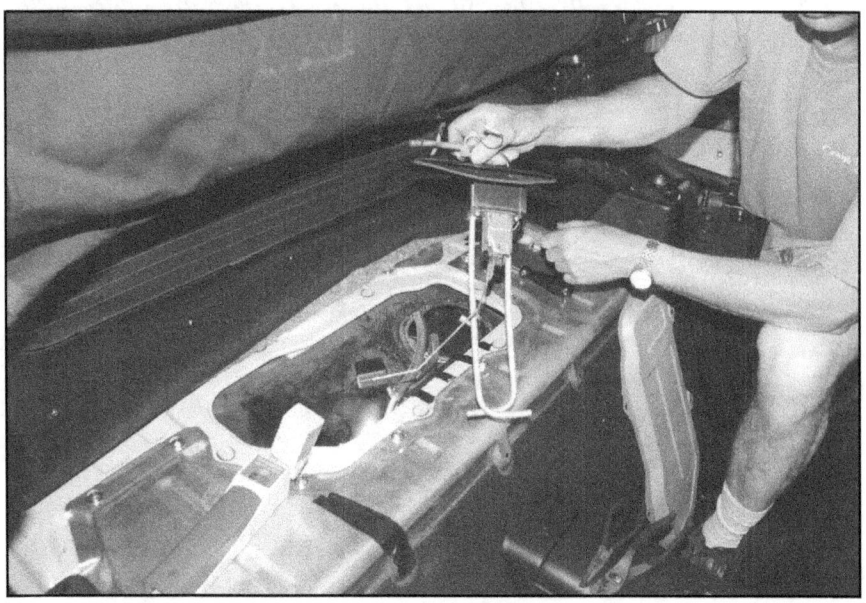

Figure 7.4-23. The modified fuel sensor and supply pipe unit is lowered into the tank and connected to the fuel hoses.

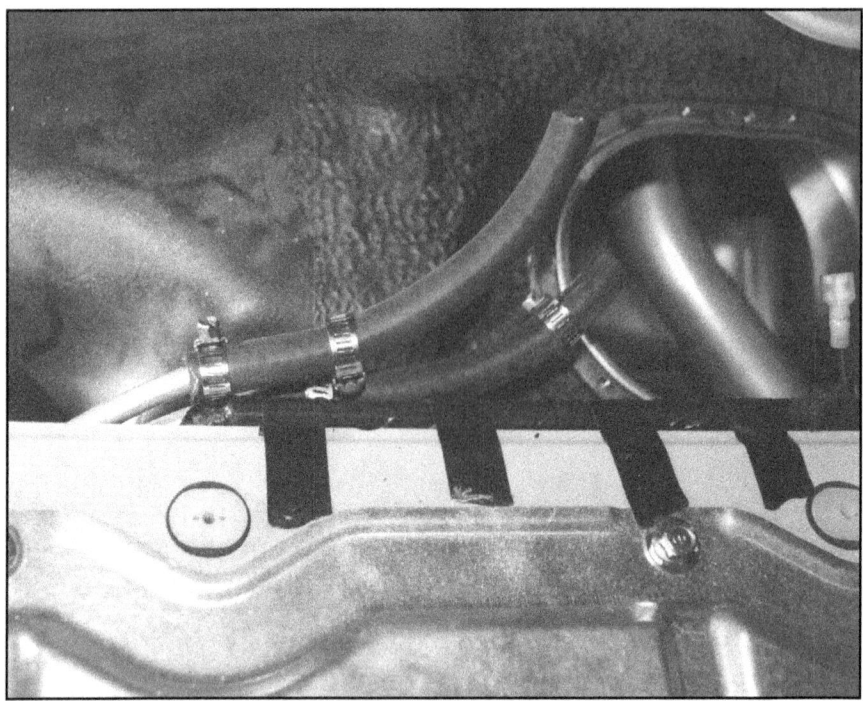

Figure 7.4-24. The Viton® fuel supply and return hoses are routed to the generator set. An after-market 12-volt fuel pump was installed in the supply line and designed to activate any time the genset is activated. An inertial safety switch is connected in series with the fuel pump. In the event of an accident, a sudden change in vehicle velocity will cause the inertial switch to open, disabling the fuel pump.

7.5 Engine Management

The vehicle has been equipped with its electric traction system, generator, and battery charger, but the individual parts will not work as an integrated system until some form of intelligent management system is installed. There are many ways of developing a system like this, but the least expensive and simplest to use is the programmable logic controller. In this section, I will provide an overview of the device and provide some software examples and a basic tutorial on the technology. Those who wish to learn more about these low-cost and adaptable control units will find purchasing information and educational materials in the Appendices at the end of this book. Anyone wishing to delve further into the PLC/PHEV relationship can also visit the Zero-Carbon Car website to download the complete software set used in the car.

The Programmable Logic Controller (PLC)

A Programmable Logic Controller (PLC) is a computer-operated control unit that was originally designed to replace logic relays, counters, timers, and other mechanical control devices used in industrial electric systems. PLCs are used in countless applications ranging from the control of industrial boilers to soft drink bottling plants and machine shop automation; they are designed to control almost all applications that require anything more complicated than the flipping of a switch.

For example, if a given application requires that a blower operate for 15 seconds after a push button is pressed, this can easily be done with a switch and control timer. But if the process requires ten blowers and special interlocks to prevent odd-numbered units from turning on if even-numbered units have already been activated, this becomes a design and wiring nightmare.

Likewise, a bottling plant requires that processes occur in a given order. Filling a soft-drink container involves moving the bottle into position, verifying that it is in the right place, initiating the filling process, stopping it when the bottle sensor indicates the correct fill level, moving to the capping location, and so on.

The PLC is designed to simplify these sorts of complex logic arrangements by replacing the wiring, relays, and counters that were used in the past with very flexible software instructions.

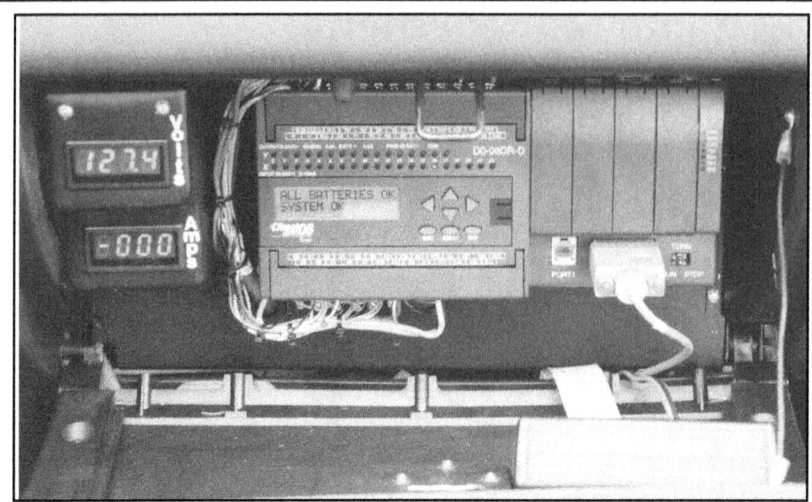

Figure 7.5-1. A Programmable Logic Controller (PLC) is a computer-operated control unit that was originally designed to replace logic relays, counters, timers, and other mechanical control devices used in industrial electric systems. The unit I selected for the Zero-Carbon Car is available from AutomationDirect for the sum of $199.

The PLC industry recognized early on that plant engineers worked with wiring circuits and electrical schematic diagrams and might not have any programming skills, so a special software language was developed to make the transition to PLCs easier. This language, known as "ladder logic," is written on a standard computer using simple wiring diagram icons that can be dragged and placed in the sequence in which the designer wants the logic to operate.

If an end-system design changes, it is no longer necessary to modify wiring harnesses and reset timers since the software program can be quickly and easily updated at any time.

One further advantage of PLC technology is its low cost. The unit I have selected for the Zero-Carbon Car is available from AutomationDirect[1] for the sum of $199. Even at such a low price there is an amazing amount of capability built into these little units.

PLC Structure

The PLC consists of a chassis, input and output connectors, and a power supply that supports the internal electronic subsystems and interfaces the unit to the outside world. The main electronic components internal to the PLC are a central processing unit (CPU), volatile and nonvolatile memory used to store the user program and data variables, and the appropriate interface circuitry to the input and output subsystem. The input and output subsystem can have numerous configurations, although "on/off" voltage inputs and relay outputs are the simplest to understand and therefore the most common arrangement.

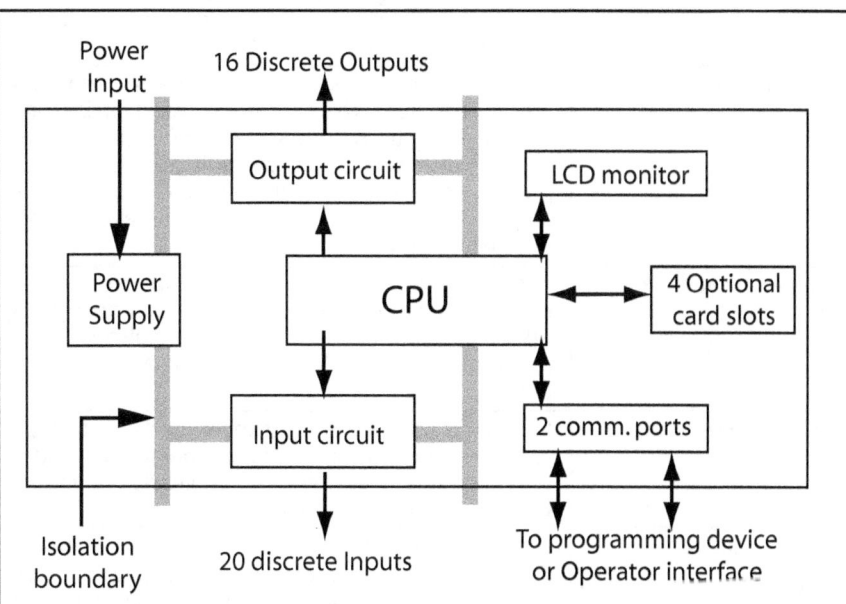

Figure 7.5-2. The PLC consists of a chassis and power supply as well as main electronic components consisting of a central processing unit (CPU), volatile and nonvolatile memory, and the appropriate circuitry to interface to the outside world.

Each of these subsystem components has a specific function, as outlined below:

Subsystem Component	Device Function
Power Supply	Converts electrical power from AC mains or battery supply to the levels required to operate the PLC electronics.
Input Circuit	Accepts voltage and current signals from the outside world and converts them to a level compatible with the CPU. Signals may be either continuously variable (analog) or on/off (digital). Examples of devices that output analog signals include temperature or fluid level sensors. Digital signals are output by items such as on/off switches or devices that detect when a container has been filled.
Output Circuit	Converts data signals from the CPU to an appropriate level to send to the outside world. Relays are typical of digital output signals (those having two states) which can be used for turning on or off lights, pumps, etc. Analog outputs include variable voltage signals for operating devices such as speed or light dimmer controls.
Communication Ports	Allow the PLC to "talk" to other devices such as remote monitoring units, computers, modems, or other PLCs. Communication ports are also used for programming the ladder logic commands into the device.
Expansion I/O Slots	Most PLCs have additional expansion slots which allow the user to custom-configure the device by adding special-purpose interface modules depending on the end-system application.
LCD Monitor	Many PLCs have the option of adding an on-board liquid crystal display (LCD) to monitor device status. These low-cost displays remain with the unit at all times, allowing operators to view status or update software without having to install a computer.
CPU	The central processing unit (CPU) is the brain of the PLC, storing the user program in a nonvolatile memory, reading inputs, executing program directions, and updating outputs accordingly.

Table 7.5-1. The PLC is made up of a number of discrete blocks which together are able to sense input status, execute program directions, and modify outputs. This table illustrates the basic components of the PLC used in the Zero-Carbon Car.

PLC Input and Output Configuration

In order for the PLC to "sense" what is happening in the outside world, the input subsystem has to convert input signals to a format that the CPU can interpret. Likewise, the output subsystem must convert commands generated by the CPU to output signals that control outside devices.

To accomplish this, PLCs are available with a wide array of different I/O modules, but the most common type comprise digital inputs and relay outputs which are shown in Figure 7.5-3. In this view, the inputs are on the left side and the outputs on the right, with the CPU and user program in the box in the center.

The inputs comprise a bank of four separate connections noted as X0 through X3. One side of each input is wired to a figurative "light bulb" (the circle and coil), with the other side of the bulb connected to a common rail and returned to connection C0. Although this configuration is highly simplified, it does represent fairly accurately how a typical digital input operates. For example, suppose a user wants to signal to

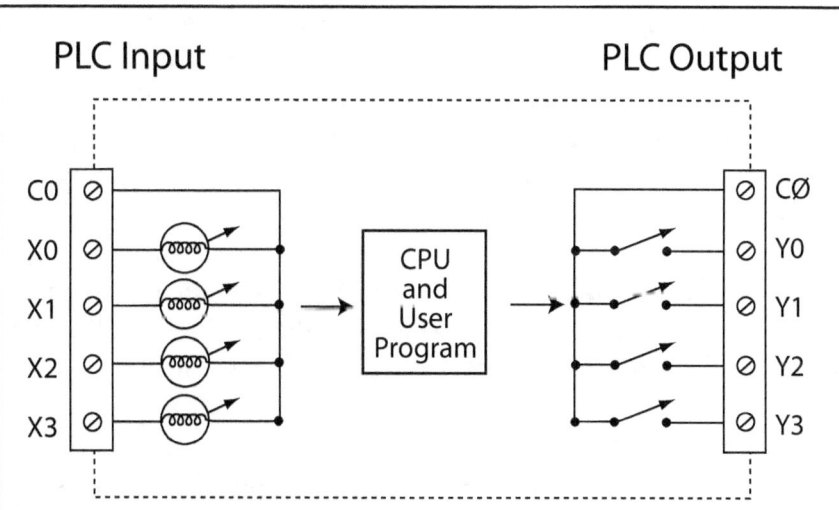

Figure 7.5-3. In order for the PLC to "sense" what is happening in the outside world, the input subsystem has to convert input signals to a format that the CPU can interpret. Likewise, the output subsystem must convert commands generated by the CPU to output signals that control outside devices.

the CPU that input X0 is turned on. If a voltage is applied between the common point C0 and the input X0, voltage flows through the light bulb, causing it to illuminate. Now suppose that the CPU is equipped with a light-sensitive receiver. A signaling method can be developed so that any time the light bulb associated with input X0 is on, the CPU "knows" that the external device connected to that input is also turned on.

This simplified concept describes a device known as an optical isolator, which can convert high-voltage signals from large devices such as motors or compressors while safely providing a status signal to the CPU.

Other approaches are used for temperature sensors, speed controls, or analog inputs.

Likewise, the CPU must provide signals to the outside world and it does so through a variety of methods the most common of which is the relay. Figure 7.5-3 shows four separate relays connected to the output terminals labeled Y0 through Y3. As with the input structure, one side of the relays is connected to a common rail and fed to connection C0. Any time the CPU wishes to change an output device status, it can operate a relay by activating a miniature magnetic coil integral to the relay, which in turn activates the relay's "switch". For example, if a motor is connected to relay Y2 and common connection C0, the CPU can activate the motor by applying a voltage to the coil corresponding to the relay connected to Y2. When the relay closes, voltage flows through the relay into the motor, and it starts.

Of course, very large electrical loads cannot be connected to the PLC relays directly, as they are quite small and of limited electrical capacity. The installing electrician would use the PLC relays to operate larger slave or pilot relays located close to the electrical load.

Program Flow

The PLC does not know what the inputs or outputs are connected to; nor does it know what to do with them. In order to make sense of the relationship between inputs and outputs, a program must be written that instructs the PLC on what to do. When the program is loaded and running, the PLC follows what is known as a "program scan" (Figure 7.5-4) to perform the desired functions.

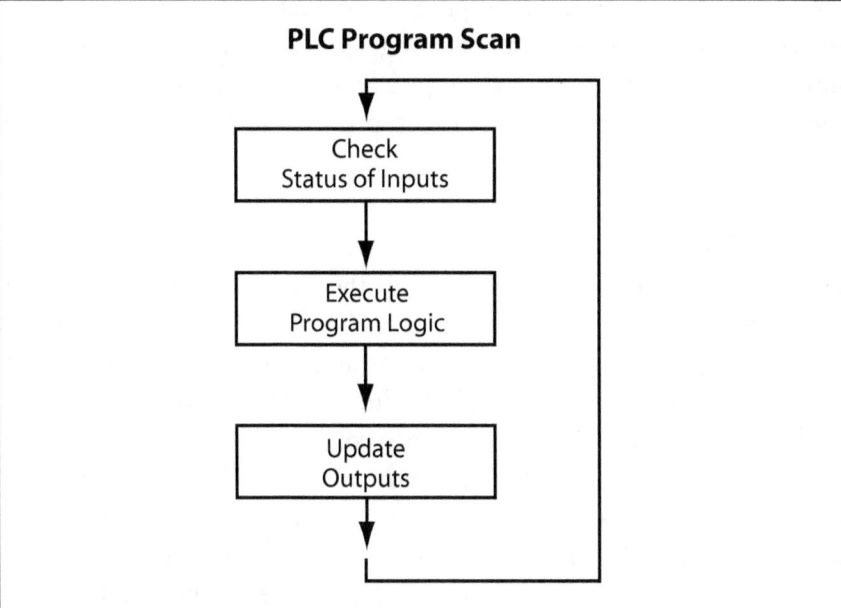

Figure 7.5-4. The PLC must perform a "program scan" in order to perform the desired functions. The program scan checks the status of inputs connected to the PLC, executes the user program logic, and updates the outputs accordingly.

The PLC program scan starts by checking the current status of each of the inputs, which tells the program which devices are turned on or off. If the inputs are analog, the current readings (for example water tank temperature) are recorded in the CPU memory.

The program scan is then incremented to the next stage, where the input status data are compared with the program logic. In this step, the programmer has designed the program using a series of logic functions and timers that provide the logical outputs required.

The next stage updates the output status based on the outcome of the logic program in the previous step.

The program scan then jumps back to the first step of checking the input status, and the loop is repeated endlessly and at very high speed, usually a few hundred to a thousand times per second. Although this is entirely adequate for most applications, it is possible that the PLC may miss events if they occur while the program is executing other steps of the program scan. For example, if a very fast transient input occurred while the program was updating the output status, this signal would be missed. To prevent this from occurring, special high-speed input modules, timers, and counters can be purchased for more demanding applications. It is very important to keep program scanning rates in mind when developing logic programs.

PLC Tutorial

This section is a basic tutorial on both the hardware and software development for the AutomationDirect DL06 PLC. Although this model actually has 24 inputs and 18 outputs arranged in banks of four, Figure 7.5-5 shows the PLC with a total of 4 inputs and 4 outputs.

I have connected three momentary contact push buttons to inputs X0, X1, and X2. The common connection is wired to the negative terminal of a 12-volt battery while one side of each switch is wired to the positive terminal. In this configuration, when the switch is not being pressed it is open and no current can flow in the circuit. If a button is pressed the contact closes, allowing current to flow in the circuit to the appropriate input of the PLC. As discussed above, when the input is ac-

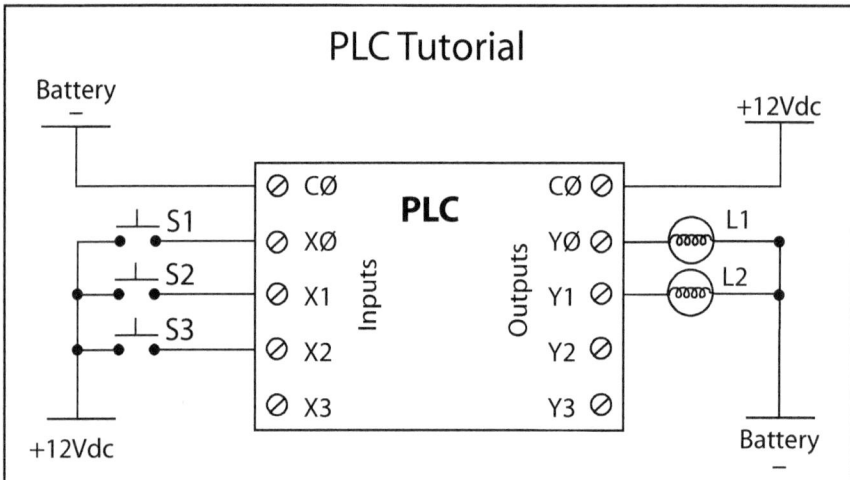

Figure 7.5-5. Although the AutomationDirect DL06 model actually has 24 inputs and 18 outputs arranged in banks of four, the PLC is shown here with a total of 4 inputs and 4 outputs. In the test system, three momentary contact push button switches have been connected to the inputs and two light bulbs have been connected to the outputs.

tivated the voltage flowing in the circuit illuminates the optical isolator "light bulb," signaling the CPU that a given button has been pressed.

Outputs Y0 and Y1 have been connected to one side of a small 12-volt light bulb, with the other side of the bulb connected to the positive terminal of the same battery used on the input circuits. The negative terminal of the battery is connected to the common connection point for the outputs, C0.

When the CPU wishes to activate either light bulb, the relay corresponding to the bulb is closed, allowing current to flow.

To demonstrate how the PLC operates, let's assume we have an application where the following control sequence is desired:

Input Action	Output Function
Press and Release Switch #1	Turn on light bulb L1 for ten seconds
Press Switch #2	Turn on light bulb L2 constantly
Press Switch #3	Turn off light bulb L2

To perform this simple function with switches and light bulbs would be tricky, requiring some form of adjustable timer for the operation of light bulb L1 and a latching relay and pulse control design to operate light bulb 2. As the logic and number of inputs and outputs increased, the complexity would be overwhelming. Using a PLC, this operation can be reduced to a few lines of programming as shown in Figure 7.5-6.

The PLC ladder logic diagram may seem complex at first glance, but once you get the hang of it the programming structure becomes straightforward. As the name implies, the program is structured like a ladder. On the left side of the page is one leg of the ladder and, depending on the manufacturer, a second leg may be shown on the right side as well.

Between each ladder leg is a numbered rung, with inputs shown near the left leg and output or logic functions shown on the right side. To help visualize how this layout works, think of the left leg of the ladder as being connected to the positive side of our example circuit battery, while the right leg is connected to the negative terminal.

The first rung of the ladder shows input X0 in an open state, mimicking the push-button switch connected to it. Like the actual hardware

PLC Tutorial

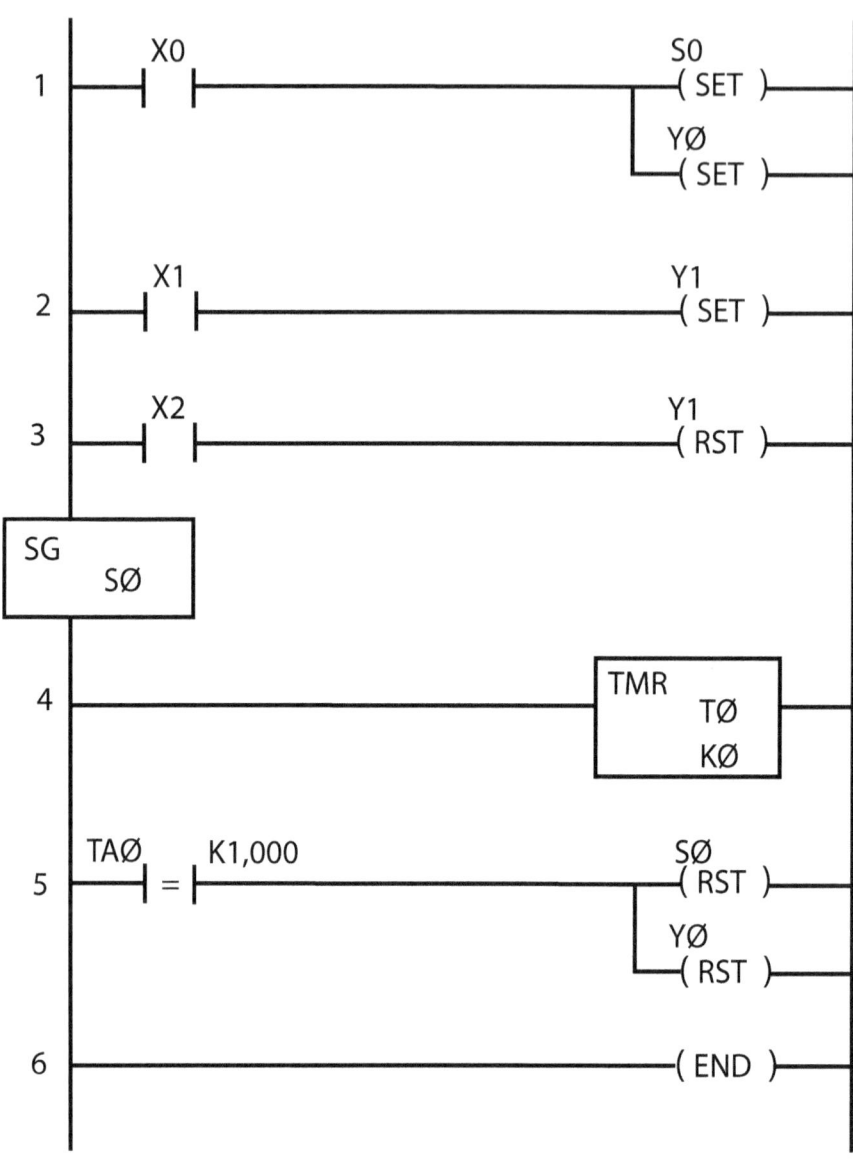

Figure 7.5-6. This PLC ladder logic diagram may seem complex at first glance, but once you get the hang of it the programming structure becomes straightforward.

circuit, no electrical current can flow across input X0 until the switch is pressed. The PLC software interprets that the switch may be either open or closed, and when it is open the CPU will do nothing since "current" in a rung cannot flow to activate the desired logic or output function.

As soon as the button is pressed, current flows in the actual input circuit, signaling the CPU that the input is active. The CPU will not react to this action until the program scan loops to the start of the sequence (rung #1) and checks the status of the inputs.

Assuming that the switch is being pressed, "current" will flow across rung #1 through the now active input X0 and will reach the output and logic control on the right side of the ladder. In this example, there are two paths that occur simultaneously: the first logic action is to "set" (turn on) a function called S0 active; the second function is to set output Y0, turning on light bulb #1.

The program scan then increments to rung number 2. Assuming that input X1 (switch #2) is in an open state, current cannot flow to the logic on the right side of the ladder and therefore no action is taken. If, on the other hand, the button is pressed, input X1 becomes active, current flows along rung #2, and output Y1 is set, turning on light #2. The program scan will advance to rung #3.

This rung is identical to rung #2, except that if the button corresponding to input X2 is pressed, the control function resets (turn off) output Y1, turning off light #2.

When the program scan reaches the box labeled "SG" a check of an internal register is conducted to see if status flag "S0" is either set or reset. Referring back to rung #1, you will see that whenever it is pressed the logic sets S0; therefore the program scan will continue with the functions below this box, namely rungs #4 and #5. Likewise, if S0 is reset, the program scan ignores functions that follow this status flag.

Rung #4 has no input, just a logic statement that indicates that timer (TMR) T0 will be cleared (set to a constant value of zero by the instruction denoted "K0" in the box). The PLC has a number of timers each of which increments at a rate of 0.01 seconds (10 milliseconds). These timers can be monitored by the program scan to perform various tasks.

In rung #5, the function shows what appears to be an input but is actually a timer control stating that when timer T0 equals 1000 counts

(1000 x 0.01 seconds = 10 seconds) current is allowed to flow across the rung. The logic is then implemented, resetting stage S0 and output Y0 and turning off light #1.

The program scan then reaches rung #6 where the end statement indicates to the CPU that it should now loop back to rung #1 and repeat the process. Depending on the version of PLC used, the scan rate for a typical unit will be in the order of a thousand or more loops per second.

If we go back and assess the program logic for a moment, we can start to understand it in greater detail. It should be clear that rungs #1, #4, and #5 deal with switch #1 and light bulb #1, while rungs #2 and #3 deal with light bulb #2.

When designing a program, it is better to concentrate on one function at a time and then integrate the logic, in effect writing two sets of logic, testing them individually, and then combining the program as we have done in the example shown in Figure 7.5-6.

User Touch Screen

Although most PLCs can be configured with low-cost LCD displays, there is nothing like a little sizzle to wrap up the project and give it a polished, professional look. The touch screen shown in Figure 7.5-7 is an excellent way to make this happen. Virtually all PLC manufacturers can supply touch screen units that are compatible with their controllers and connect together using a simple computer-style interface cable.

The advantage of the touch screen is that status information can be shown using graphics and images rather than status lights or text-only information. There may be no need for this level of detail when solving simple control functions, but a touch screen makes more complex operations such as the control and monitoring of an automated plastic injection molding machine, soft drink filling plant, or PHEV much simpler and more user-friendly.

As with the PLC itself, software must be written to make the touch screen operate, but most manufacturers provide their own Supervisory Control And Data Acquisition (SCADA) program with the units, allowing you to "drag and drop" provided images including gauges, buttons, dials, pumps, tanks, and other functions that are required in a given application.

Figure 7.5-7. Although most PLCs can be configured with low-cost LCD displays, there is nothing like a little sizzle to wrap up the project and give it a polished, professional look. The touch screen mounted in the dash of the Zero-Carbon Car is an excellent way to make this happen. Virtually all PLC manufacturers can supply touch screen units that are compatible with their controllers and connect together using a simple computer-style interface cable.

In the case of the Zero-Carbon Car, we opted to develop our own images and buttons, which will be discussed in the next section.

Summary

There are many PLC manufacturers, each with its own lineup of specialty devices and options that can be configured to meet just about any application. The AutomationDirect model DL06 is a low-cost yet powerful unit that is easy to use and provides more than adequate capacity for the operation of the Zero-Carbon Car. Once the basic input/output structure is determined, the user can modify the program, adding more features and options without having to change wiring harnesses or incur additional hardware costs.

The steps required to install and program a PLC take a little bit of time to understand, but having either an actual PLC to experiment with or a software simulator that mimics an actual unit will help you navigate the learning curve quickly. There are many excellent resources for PLC training, one of which is listed in the Appendices at the end of this book.

7.6 Integrating the Functions

The final step in getting the Zero-Carbon Car on the road is integrating the separate components and making the system operate as one seamless machine. This section reviews final installation and discusses the theory of operation of the plug-in series hybrid electric vehicle.

Touch Screen Overview

Figure 7.6-1. Gary Baker owns a design and machine shop that specializes in high-precision work. Gary spent quite of bit of time working on the 3-D modeling of dashboard and engine components, making sure everything fit perfectly. One of his jobs was to modify the Miata dashboard, removing the heater controls and stereo and replacing them with an anodized aluminum bezel that would support the LCD touch screen.

Figure 7.6-2. The touch screen is shown mounted in the dash of the vehicle. Once the PLC is up and running, the home screen appears on the display.

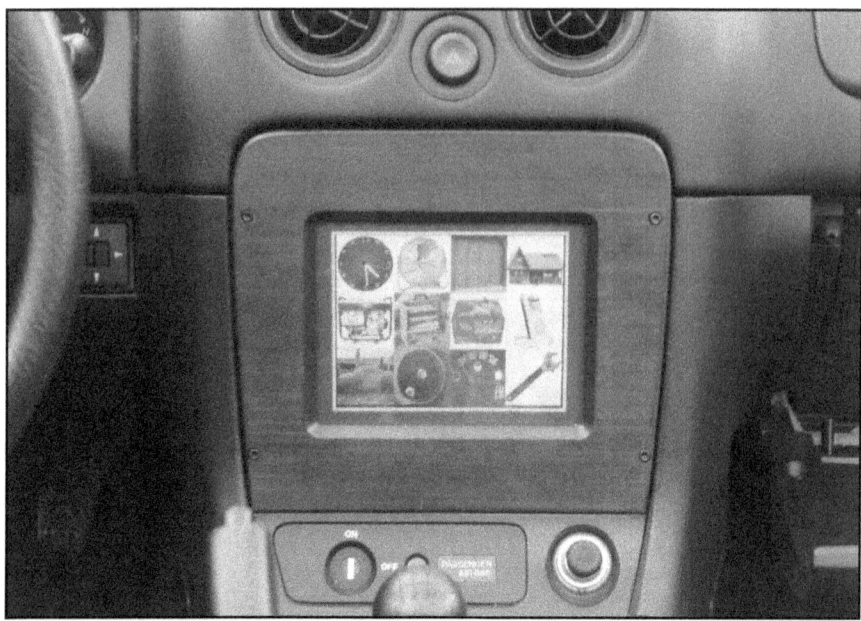

Figure 7.6-3. Touching the home screen brings the user to the main control screen which provides information on the status of the car's electro-mechanical systems. Because the touch screen replaced the stereo and heating system controls, the operation of these items was also added.

Figure 7.6-4. One of the icons on the main control screen is a spanner wrench which turns red in the event of a battery fault condition. Touching the wrench icon brings you to the battery diagnostic screen. Each of the ten batteries is shown in green when its voltage level is within predetermined limits and will turn red and sound an alarm if a fault is detected.

PLC and Metering Overview

I wanted the PLC controller to be mounted in a location where it would be readily accessible both to monitor problems with the vehicle's systems and to demonstrate to people what the brains of the vehicle looked like. We determined that the best location was the vehicle's glove compartment. This allowed easy access to the vehicle firewall to bring signal cables to the unit and also provided a handy location to demonstrate

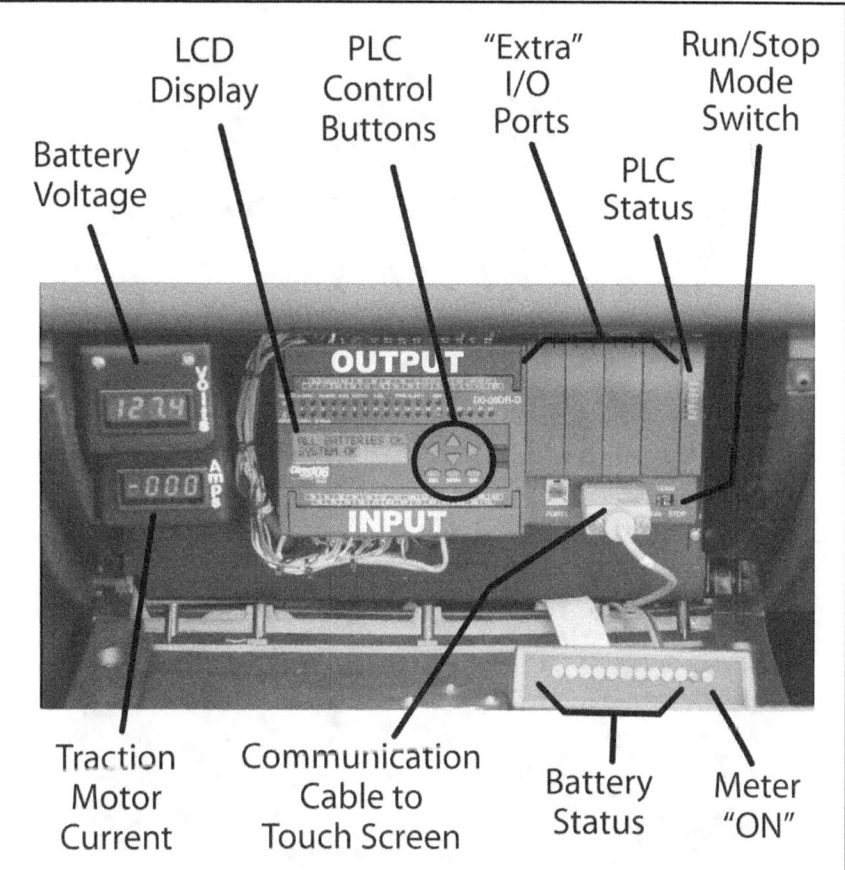

Figure 7.6-5. This view shows the glove compartment where the PLC was ultimately mounted. The tracing arrows identify the various components of the PLC that are described in the text. A standard battery voltage monitor is shown top left, while the traction motor current is monitored below it. A very ingenious battery status monitor is located bottom right. This unit was designed by Ottawa engineer Richard Hatherill, who has kindly provided his design for general distribution (see Appendix D).

typical battery-monitoring methods. When all the viewing and fiddling was completed, the door of the glove compartment was closed, keeping the vehicle's appearance clean and neat. Rick Lane custom-fabricated a sheet aluminum bulkhead plate that mounted at the rear of the glove compartment with a number of holes with grommets to pass wires through.

Figure 7.6-6. Cam Mather is shown pulling some of the 24 conductor cables through the vehicle. They connect the battery voltage monitor and engine control and monitoring equipment to the PLC.

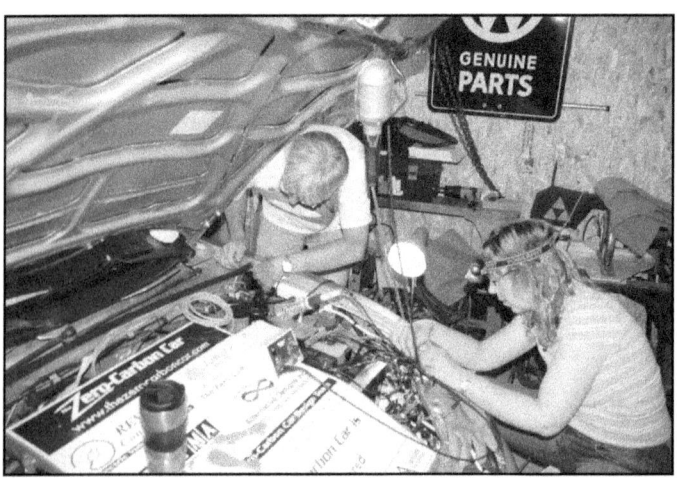

Figure 7.6-7. Rick Lane and Lorraine Kemp are hard at work installing the interface cables that connect the traction batteries to the battery voltage monitor.

Plug-In Hybrid Vehicle Technology

351

Figure 7.6-8. The battery voltage monitor receives voltage-level signals from each of the ten traction batteries. The unit senses when a battery is above or below its normal operating range and signals the PLC.

Figure 7.6-9. The battery charger is connected to a flip-up "twist and lock"-style connector. These units are waterproof and provide a professional-looking finish. The cover is also a great place to add a Bullfrog Power logo.

PLC Inputs and Outputs

Each of the main components of the vehicle is, in some way or another, equipped with an input or output function. For example, the generator system requires a start/stop signal as well as battery voltage level in order to start, stop, and regulate power levels. The battery bank condition must be monitored to ensure that it is not damaged by an overdischarge or defective charging condition. Even the lowly electrical charging plug has a signal related to it, ensuring that no one attempts to drive the vehicle away while it is plugged in.

Each of these signals was reviewed and given a priority it terms of whether or not it should be monitored or controlled by the PLC. In addition to the major control functions, secondary controls were also considered, including the rear window defroster, cabin fans, and wireless garage door opener.

Table 7.6-1 identifies all the PLC input/output structure.

The following information is based on Table 7.6-1 and provides information about each input and output for the PLC:

Battery Low Voltage Detection

Each of the ten traction batteries must be individually monitored to ensure that it operates correctly and is not overly discharged during use. Many electric vehicle users incorrectly suggest that if a vehicle is reaching the end of its driving range or if performance drops off it can be parked for a few minutes to allow the batteries to regenerate before continuing. I also hear remarkable stories about fantastic vehicular range. This is nothing but nonsense. Battery life is compromised if the vehicle is driven under these conditions and not promptly recharged, as we learned in Chapter 7.2 and in Figure 7.2-15. Nor is this problem related only to lead-acid batteries. All battery technologies suffer from this problem to one degree or another.

Another common fallacy is that battery status can be monitored with a simple voltage and current meter such as the model shown in Figure 7.6-5.

The problem with batteries is that no two are created equal: while one unit is operating within its normal operating level, another might well be depleted. To make matters worse, batteries are notorious for disguising which one is the weak link in the chain. Battery voltage is

Input	Description	Output	Description
X0	Battery #1 Low	Y0	Automatic Start/Stop Genset
X1	Battery #2 Low	Y1	Vehicle Interlock
X2	Battery #3 Low	Y2	Activate iPod MP3 Player
X3	Battery #4 Low	Y3	
X4	Battery #5 Low	Y4	Open/Close Garage Door
X5	Battery #6 Low	Y5	
X6	Battery #7 Low	Y6	
X7	Battery #8 Low	Y7	
X10	Battery #9 Low	Y10	Beep Car Horn
X11	Battery #10 Low	Y11	Heater #1 On
X12	Battery Charger Plugged In	Y12	Heater #2 On
X13	Battery Charged	Y13	Rear Window Defogger
X14	Parking Brake On	Y14	Cabin Fan Low Speed
X15	Traction Motor Overheating	Y15	Cabin Fan Medium Speed
X16	not used	Y16	Cabin Fan High Speed
X17	not used	Y17	Manual Override Generator
X16	not used		
X17	not used		
X20	not used		
X21	not used		
X22	not used		
X23	**not used**		

Table 7.6-1. *This table identifies all of the signals that the PLC monitors and controls in the Zero-Carbon Car. The unused channels are available for future enhancements. (Note that PLC input numbering is not continuous).*

very elastic, rising and falling as load levels change. Defective or poorly charged batteries often exhibit normal voltage levels at light loads while quickly falling below safe operating levels under moderate to heavy loads. A volt meter will see the battery bank rising and falling but it cannot tell if the entire bank voltage is fluctuating as a function of load or if a single cell or group of cells has failed altogether.

It is very important to understand this problem, as battery life will be greatly compromised if the problem is not quickly identified and corrected.

One solution is to use a battery condition monitor of the type designed by Ottawa engineer Richard Hatherill (see Figure 7.6-5). Richard has kindly provided the schematics for his circuit, a copy of which is shown in Appendix D. The beauty of this monitor design is its simplicity. A light-emitting diode (LED) is connected through a voltage-monitoring circuit to each traction battery. With a fully charged pack, all ten LEDs are bright and of equal intensity. As the vehicle is driven and load is applied, the battery bank voltage will lower and all of the LEDs should dim in unison. Should an entire battery or a cell within a battery become weak, one of the ten LEDs will dim much more rapidly than the balance, indicating which battery is in need of immediate charging or replacement. The monitor also identifies which of the batteries is giving trouble.

Why not simply stop the vehicle and measure the individual battery voltages? The problem is that voltage elasticity, which causes the battery voltage to spring back to normal as soon as electrical load is removed, makes it nearly impossible to find the culprit.

Richard's design is very effective, but it is only meant for visual monitoring. To provide the same level of battery monitoring quality for the Zero-Carbon Car, I developed a Battery Voltage Monitor (BVM) which performs a similar function, but rather than providing an LED status it sends digital signals directly to the PLC through inputs X0 to X11. (Note that PLC I/O numbering is not continuous.)

In the BVM shown on the next page, there are ten copies of the circuit, one for each battery. For vehicles equipped with different quantities of batteries, the number of monitoring circuits can be adjusted accordingly.

In this design, the battery to be monitored is connected to the input at the top left of the drawing. A set point potentiometer is adjusted to trigger the output relay to come on when the battery voltage reaches no more than an 80% depth of discharge according to the manufacturer's data sheet, typically 10.5 volts under load for a typical lead-acid model.

When the battery voltage drops to the low-voltage cutoff level the relay moves to the opposite state, applying a signal voltage to the ap-

Figure 7.6-10. To provide a high level of battery monitoring quality, the author and George Argiris developed a Battery Voltage Monitor (BVM) similar to Richard Hatherill's design, but rather than providing an LED status it sends digital signals directly to the PLC though inputs X0 to X11. (Note that PLC I/O numbering is not continuous.) This view of the BVM circuit board shows the ten battery voltage signals entering the board at the right. The electronics in the center of the board determine battery status and provide output signals to the PLC through the connectors on the left side of the unit.

propriate input X0 through X11, indicating which battery is in a fault condition.

The entire circuit draws a very tiny amount of current from the traction battery to operate the electronics. The accessory power input is provided by the vehicle's accessory battery (Figure 7.1-3).

Battery Charged Detection

An eleventh battery-monitoring channel is provided in the BVM and differs slightly from the low-voltage detection circuit discussed above. In this circuit, shown in Figure 7.6-12, the voltage detection chip is modified slightly to look for a rising voltage at or above the battery fully charged set point, which is typically 14 volts for a lead-acid model.

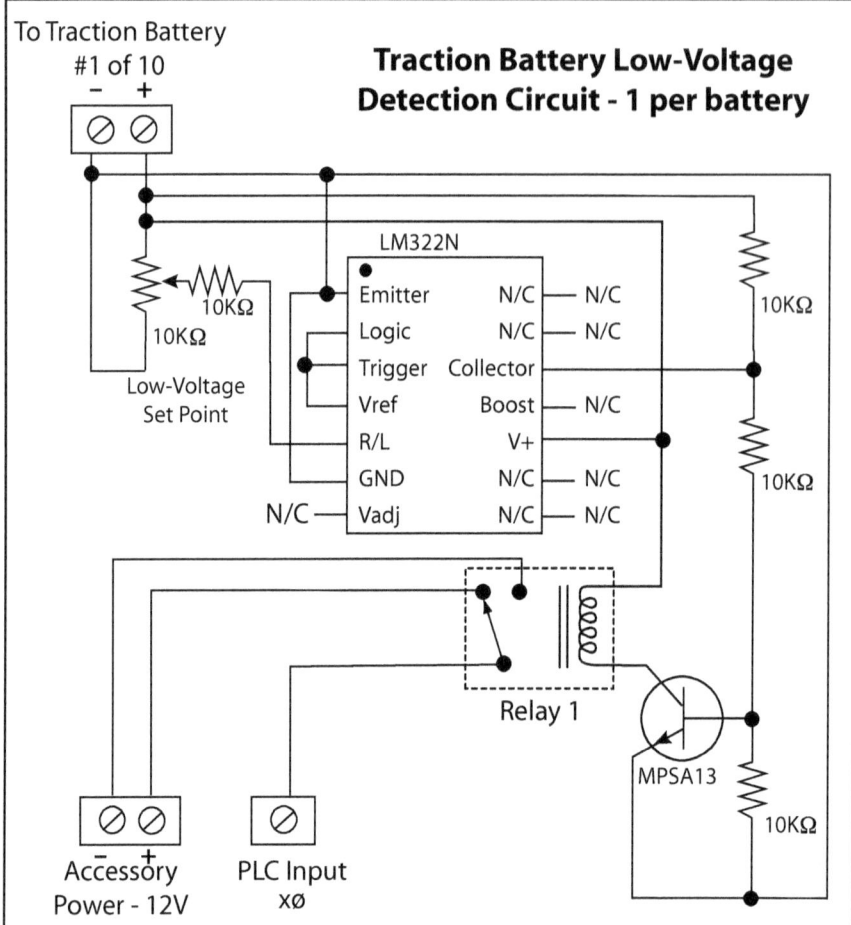

Figure 7.6-11. This schematic diagram shows one channel of the ten that are required to monitor the traction batteries for a low-voltage condition. A discrete electronic version with relay output was chosen for its simplicity and ease of construction. The unit can be used with batteries of six-volt capacity and higher.

The input is arbitrarily set to monitor any one of the batteries, providing an assumption about the charge state of the remaining battery bank.

The output of this circuit is fed to PLC input X13, which indicates that the battery bank is fully charged. Note that the PLC uses this input as a display indication only.

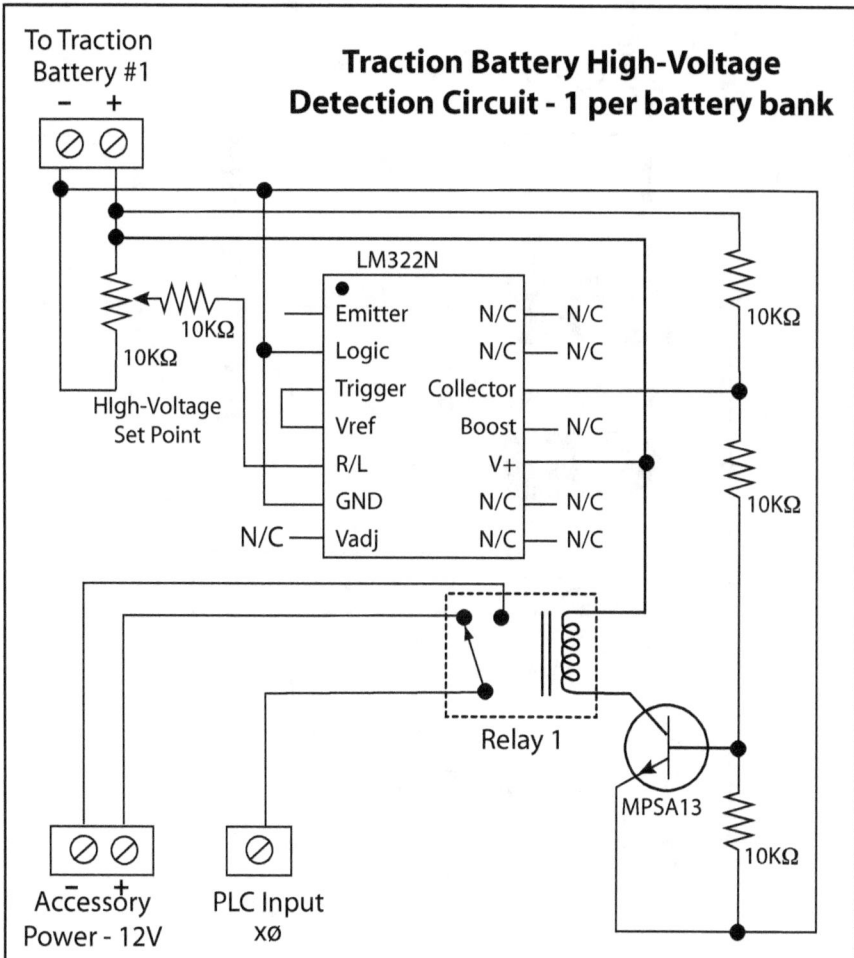

Figure 7.6-12. An eleventh battery monitoring channel is provided in the BVM and differs slightly from the low-voltage detection circuit discussed above. The output of this circuit is fed to PLC input X13, which indicates that the battery bank is fully charged. Note that the PLC uses this input as a display indication only.

Battery Charging Status and AC Mains Schematic

The schematic diagram shown in Figure 7.6-13 describes how the 120-volt AC mains supply is used in the Zero-Carbon Car design, including the battery charging systems and interface to the PLC. On the left side of the drawing, AC mains voltage is applied to the vehicle through the charging plug shown in Figure 7.6-9. Current is fed through a 15-Amp safety circuit breaker which then supplies the following devices:

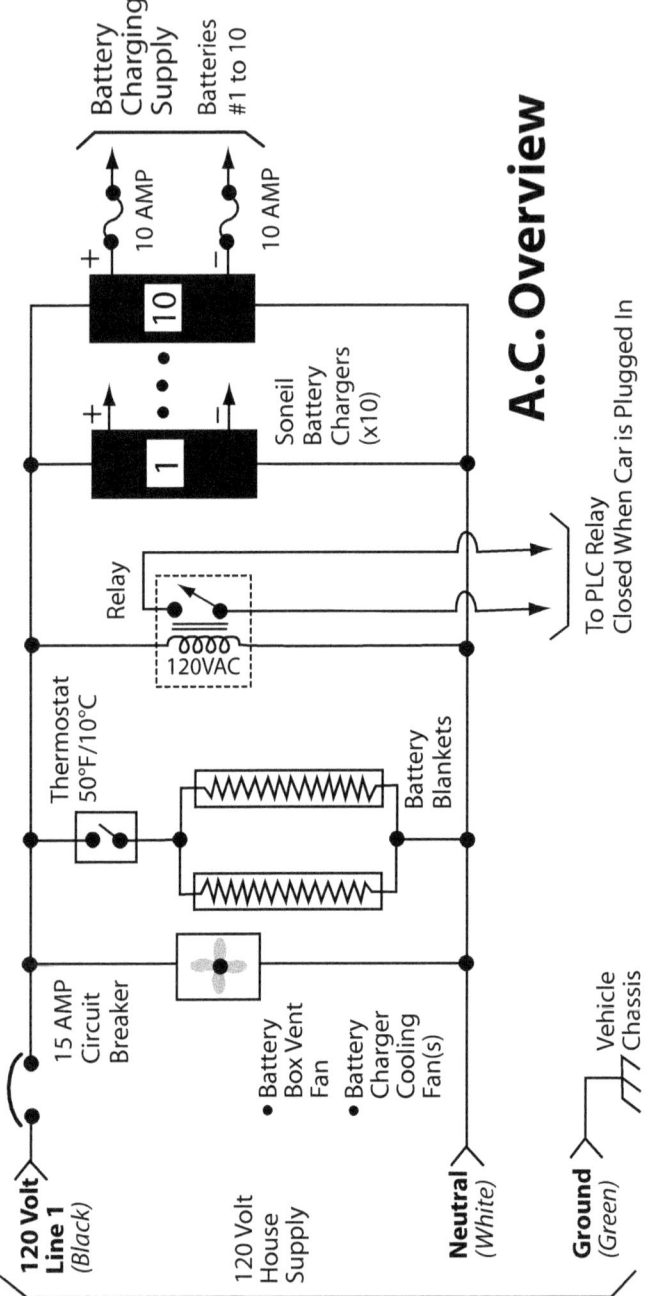

Figure 7.6-13. The schematic diagram shown in Figure 7.6-13 demonstrates how the 120-volt AC mains supply is used in the Zero-Carbon Car design, including the battery charging systems and interface to the PLC.

Battery Box Vent Fan(s)
Battery Charger Cooling Fan(s)

Depending on the operating temperature of the batteries and battery-charging units, it may be necessary to add fan-forced cooling of these devices. The battery chargers are designed to operate any time the unit is plugged in, which leads to battery bank and charger heating. (Battery heating during operation of the genset system may need to be monitored or fans activated on an as-required basis. Generally speaking, the genset should not be providing 100% of the battery-charging capacity and will therefore add little in the way of heat energy.)

Battery Heating Blankets

In areas where the vehicle is subjected to cold winter weather conditions, battery heater blankets may be added to the insulated battery box. A thermostat wired in series with the blankets cycles the heaters on and off as needed to maintain a minimum temperature of 50°F (10°C).

Vehicle Charger Status

A relay with a 120-volt AC coil is connected to the AC mains supply circuit. The relay contact closes any time the vehicle's charging plug is connected to the mains circuit and provides a signal to the PLC at inputs C2/X12. The PLC senses this input and displays an icon on the LCD touch screen indicating that the car is plugged in.

PLC relay Y1 is wired in series with the "clutch interlock switch" shown in Figure 7.1-3 and prevents the vehicle from being driven while the battery charger is plugged in.

Automatic Battery Chargers

Ten Soneil automatic battery chargers are connected to the AC mains supply circuit and provide their charging voltage to each of the ten traction batteries. Note that both the positive and negative charging supply cables are supplied with 10-Amp fuses.

The junction point between the charger output and each 10-Amp fuse is the location where the BVM derives its battery status signal.

AC Mains Ground

The green ground wire of the AC mains is very important for electrical safety reasons. This connection must be firmly bonded to the vehicle's chassis at a point where there is no chance of mechanical or road damage to the wire.

PHEV Operation

Simply put, the Zero-Carbon Car is designed to operate on electrical power only, where possible. For some 85% of the driving done by the average person, this is easily achievable. Electrical energy used to propel the vehicle is cheaper than any form of liquid fuel or hydrogen and can be produced from many renewable or zero-carbon sources. If nonrenewable energy is used, it's possible that, sequestration of emissions and carbon will be developed at some time in the future. Of course, carbon sequestration cannot be accomplished when burning fossil fuels at the vehicle level. This also applies to hydrogen which may be manufactured directly onboard a vehicle using reformation of gasoline, methanol, or natural gas technology as some proponents advocate.

If driving range exceeds the ability of the electric system to supply sufficient energy, the series generator is activated and supplies electrical power for vehicle operation and partial battery charging. It is desirable to have the batteries partially charged when the generator is operating to provide supplementary power for hill-climbing and accelerating. This allows the generator to be as small as possible, with the batteries providing some portion of the vehicle's energy during high-demand periods.

During coasting, braking, or downhill driving, excess energy from the genset or regenerative braking system is directed to battery charging. In vehicles equipped with regenerative braking, batteries must not be fully charged; otherwise the energy recovered during braking cannot be stored.

The PLC interprets this information by monitoring battery bank voltage and relating it to time, as shown in Figure 7.6-14. The graph shows battery bank voltage on the vertical axis with time shown on the horizontal axis. The black line shows how the battery bank voltage might respond during a period of no load, through heavy acceleration, and at cruising speed.

In this example, the battery bank voltage starts out at a normal level of 125 volts while the car is stopped. As soon as heavy acceleration begins, battery voltage drops over time until it reaches a predetermined

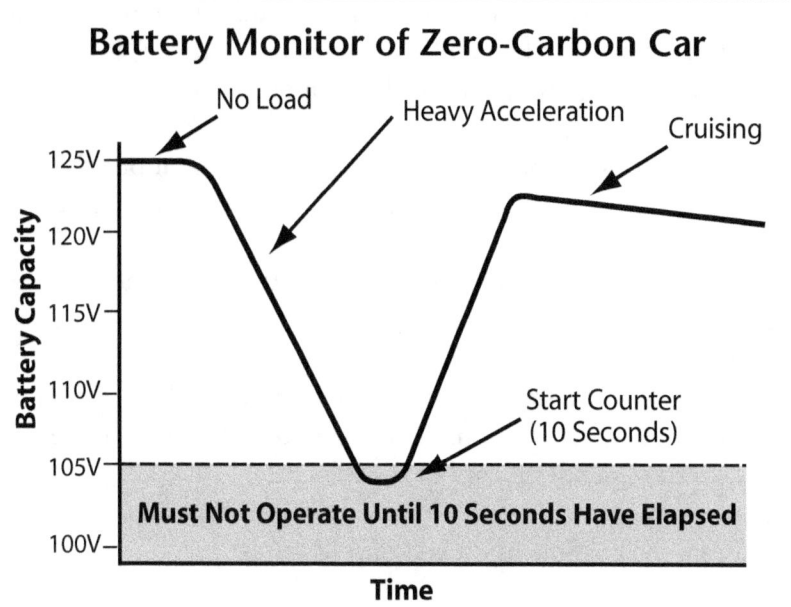

*** The Generator must **NOT** be turned **ON** until the battery capacity has been below the 105V threshold for a period of 10 seconds.

Figure 7.6-14. Simply put, the Zero-Carbon Car is designed to operate on electrical power only, where possible. If driving range exceeds the ability of the electric system to supply sufficient energy, the series generator is activated and supplies electrical power for vehicle operation and battery charging.

"must not operate level" which has been preset to 105 volts. This set point voltage has been fixed by setting the low-voltage detection circuit of the BVM to 10.5 volts for each of the 10 batteries.

If the voltage of any of the ten batteries in the bank drops below this minimum threshold, the PLC is signaled and a ten-second timer is started. If the battery voltage rises above this threshold before the timer has elapsed, the timer is reset and no further action is taken.

If, on the other hand, the battery voltage stays below the minimum threshold for more than ten seconds, the PLC assumes that one or more of the batteries are below the 80% minimum discharge threshold and the generator is started.

The generator then supplies power to the batteries and traction motor until the high-voltage detection circuit indicates that they have reached the upper charge threshold or until one hour of continuous operation has elapsed. (The one-hour run time limit was installed as a safety feature during testing and will be removed after a suitable period of time.)

Additionally, if the hand brake, which is sensed at input X14 (indicating the vehicle is parked), is applied the PLC will turn the generator off.

Lastly, if the vehicle is plugged in to the AC mains, the PLC senses this action on input X12 and turns the generator off, allowing the automatic battery chargers to supply power.

Each of the above state changes are indicated on the PLC's integral LCD display as well as via graphic icons on the LCD touch screen display mounted in the dashboard of the vehicle.

The following truth tables describe how the software logic operates and indicate the possible input and output states.

Zero Carbon Car Truth Table #1 (Batteries OK)

Car Plugged In (Input X12)	Battery Charged (Input X13)	Parking Brake ON (Input X14)	PLC Result Action
0	0	0	Display "SYSTEM OK" and do nothing to Y0 in case the generator is already ON.
0	0	1	Display "SYSTEM OK" and do nothing to Y0 in case the generator is already ON.
0	1	0	Display "BATTERY CHARGED" and after 15 minutes display "GENERATOR OFF" and RST Y0 (open contact).
0	1	1	Display "BATTERY CHARGED" and after 15 minutes display "GENERATOR OFF" and RST Y0 (open contact).
1	0	0	Display "BATTERY CHARGER PLUGGED-IN" and RST Y0 (open contact). Warn user that car is still plugged-in.
1	0	1	Display "BATTERY CHARGER PLUGGED-IN" and RST Y0 (open contact).
1	1	0	Display "BATTERY CHARGED" and RST Y0 (open contact). Warn user that car is still plugged-in.
1	1	1	Display "BATTERY CHARGED" and RST Y0 (open contact).

Table 7.6-2. This table describes all of the possible logic sequences that can occur as well as the PLC input and output states when the vehicle is operating with a properly charged battery bank. Note that an input of "0" indicates that the condition is false, while a "1" indicates that the condition is true. For example, "Car Plugged In = 0" is false, meaning that the car is not plugged in. Likewise a "1" would indicate that the statement is true and that the car is plugged in.

Zero Carbon Car Truth Table #2 (Low Battery)

Car Plugged In (Input X12)	Battery Charged (Input X13)	Parking Brake ON (Input X14)	PLC Result Action
0	0	0	Display "GENERATOR ON" and SET Y0 (close contact).
0	0	1	Display "INDOORS?" "YES/NO". If YES then RST Y0 (open contact), if NO then SET Y0 (close contact).
0	1	0	Display "BATTERY FAULT" and RST Y0 (open contact).
0	1	1	Display "BATTERY FAULT" and RST Y0 (open contact).
1	0	0	Display "BATTERY CHARGER PLUGGED-IN" and RST Y0 (open contact). Warn user that car is still plugged-in.
1	0	1	Display "BATTERY CHARGER PLUGGED-IN" and RST Y0 (open contact).
1	1	0	Display "BATTERY FAULT" and RST Y0 (open contact). Warn user that car is still plugged-in.
1	1	1	Display "BATTERY FAULT" and RST Y0 (open contact).

Table 7.6-3. This table describes all of the possible logic sequences that can occur as well as the PLC input and output states when the vehicle is operating with a battery bank that has one or more batteries operating below the minimum voltage threshold.

User Display and Operation

The operation of the Zero-Carbon Car control system is fully automatic. Whenever the vehicle key is turned on, the software contained in the PLC and LCD touch screen is activated and operation commences.

The user does not have to bother with functionality related to the PHEV control system, although the LCD touch screen provides a host of information. As LCD touch screen software designer Ken O'Rielly suggests, "Dude, we are just getting warmed up!"

Figure 7.6-15. The user does not have to bother with functionality related to the PHEV control system, although the LCD touch screen provides a host of information. As LCD touch screen software designer Ken O'Rielly suggests, "Dude, we are just getting warmed up!" Ken is shown in the foreground, while PLC software developer George Argiris watches from behind.

Figure 7.6-16. After power is applied to the PLC and the LCD touch screen, the Home Screen is displayed. It has no other function than to aesthetically hide the Main Control Screen icons. Touching the screen takes the user to the Main Control Screen.

Clock Icon

The clock icon displays the time of day. Touching the icon increases the size of the clock face to the full size of the LCD screen. Additional display buttons provide time and daylight saving time adjustments. Touch the screen to return to the Main Control Screen

Fan Icon

This icon controls the vehicle's cabin fan speed. When the icon is gray, the fan is off. Tapping the icon will cause the fan speed to cycle from low to medium to high and back to off by activating PLC relays Y14, 15, and 16 respectively. The icon changes to indicate successively faster fan speeds.

Garage Door Icon

PLC relay Y4 is connected to a standard wireless remote control garage door opener. Touching this icon causes relay Y4 to close for two seconds, mimicking the pressing of the remote control button. (The relay output is soldered across (in parallel) to a standard battery-operated garage door opener unit.)

Main Control Screen

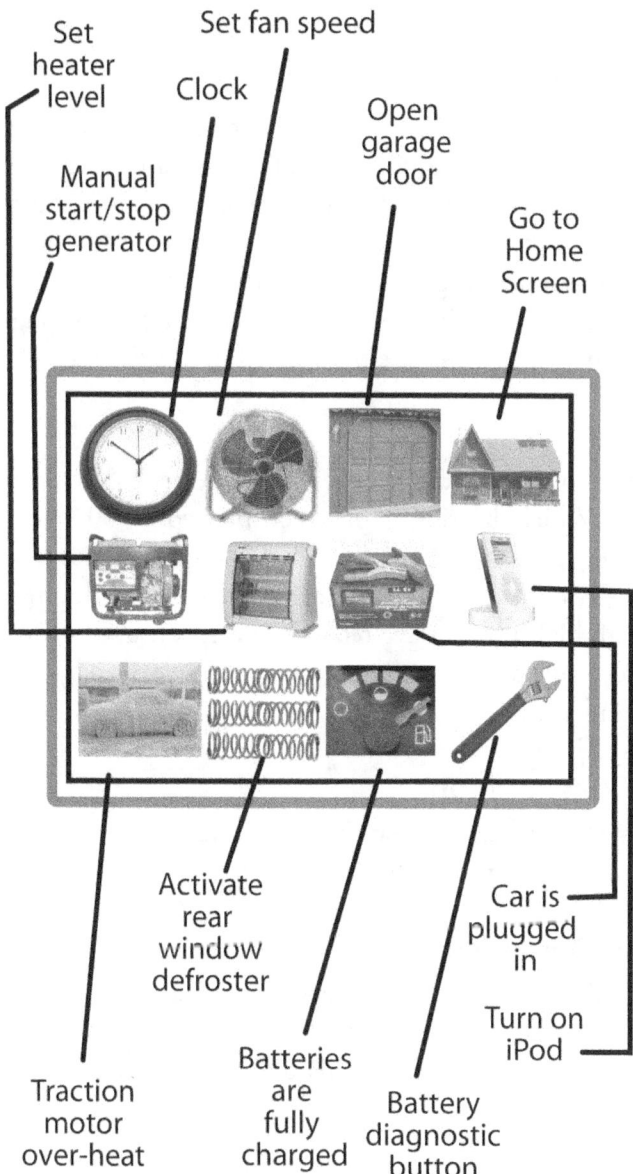

Figure 7.6-17. The Main Control Screen provides access to all of the control and status functions of the PLC system.

Home Icon

Pressing this button causes the LCD touch screen to revert to the Home Screen

Generator Icon

This icon displays the generator status. When it is gray, the generator unit is off. When it is brightly colored, the generator is turned on. The icon is also the manual override button which activates the generator. Pressing the button will cause the PLC to turn on the front-mounted cooling fan and signal the generator to start by closing output relay Y17. (Note that PLC automatic control of the generator is through output relay Y0. Relays Y0 and Y17 are both wired in parallel.)

Heater Icon

Tapping this icon causes the vehicle's cabin heater to cycle from off to low to high. Vehicle heating is provided by two 1,500-watt, 120-volt ceramic heaters installed where the vehicle's cabin radiator heating unit was previously located. The heaters are powered from the traction batteries through high-power slave relays and can consume a total of three kW of power while operating. The slave relays are driven by PLC output relays Y11 and Y12.

Battery Charger Icon

The battery charger icon displays when the vehicle is plugged into the AC mains. The input to the PLC driving this status signal is X12.

iPod MP3 Player Icon

This icon activates PLC output relay Y2, which in turn activates the power supply for a dash-mounted iPod nano and its associated amplifier. Music selection is through the iPod directly.

Icy Car Icon

This icon indicates that the vehicle's traction motor is within normal operating temperature. A thermal switch located in the traction motor activates PLC input X15, signaling that the motor is overheating. When this occurs, the icon shows a picture of a car on fire. The LCD touch screen also beeps to alert the driver.

Defroster Vent Icon

Touching this icon turns on the rear window defroster for ten minutes. When the icon is gray, the defroster is off; when it is brightly colored, the defroster is active. Touching the button a second time will deactivate the defroster. The PLC output relay which drives the defroster slave relay is Y13.

Empty Gas Tank Icon

This icon displays the battery bank state of charge status. When the icon is gray, the batteries are not fully charged. When the icon changes to a brightly colored image of a full fuel tank, the batteries are at the fully charged threshold as defined by PLC input status X13.

Wrench Icon

Pressing this button transfers the user to the Battery Status Screen. When any one of the ten traction batteries is out of tolerance, the wrench icon flashes between gray and red to alert the driver of a potential battery problem.

Further technical details regarding design and construction of the vehicle are available on the Zero-Carbon Car website at www.thezero-carboncar.com.

Battery Status Screen

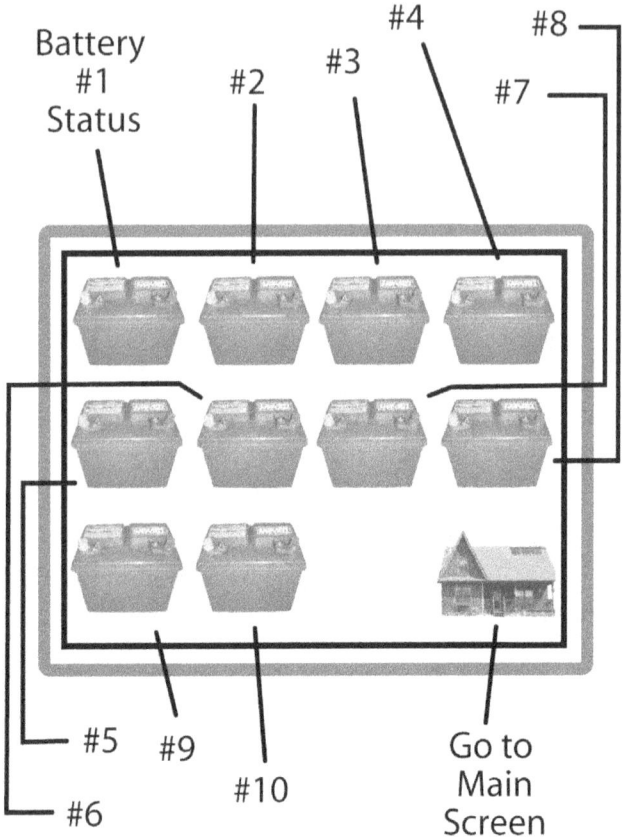

Figure 7.6-18. The Battery Status Screen is shown in this image. Each of the ten battery icons corresponds to the status of one of the ten traction batteries as read by PLC inputs X0 through X11. When the battery is within its normal operating voltage, the battery icon is bright green. When one or more batteries are below this threshold, the icons turn red and cause the LCD touch screen to beep, alerting the user to a problem. The Home icon takes the user back to the Main Control Screen.

7.7 The Test Drive

The time finally came to get the Zero-Carbon Car out of the garage and onto the pavement for some real-world testing and a bit of exercise on a dynamometer. Perhaps the best way to explain the lead-up to the test results is to take a look as the vehicle rolls out under its own power:

Figure 7.7-1. Ken O'Rielly is an admitted computer geek and automotive buff. The Zero-Carbon Car couples both of these passions, which might be reason enough for the development of PHEVs. Ken is shown here installing updated software into the car's computer system.

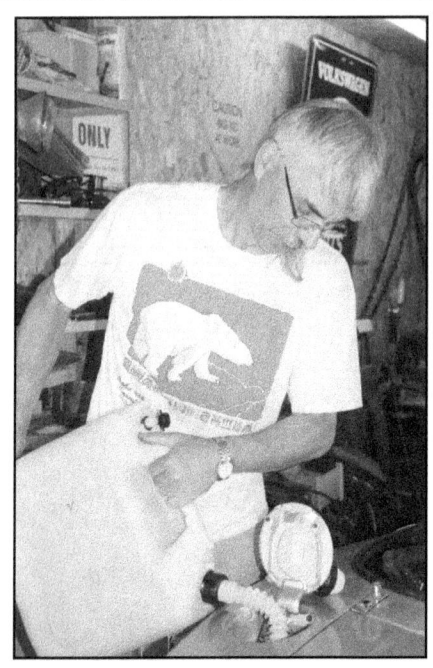

Figure 7.7-2. Rick Lane fills up the car's fuel tank with 5.3 gallons (20 liters) of biodiesel. If you are a typical driver and conform to the norms of driving distance discussed earlier in the book, this should be enough fuel for at least 1,240 miles (2,000 kilometers) of driving!

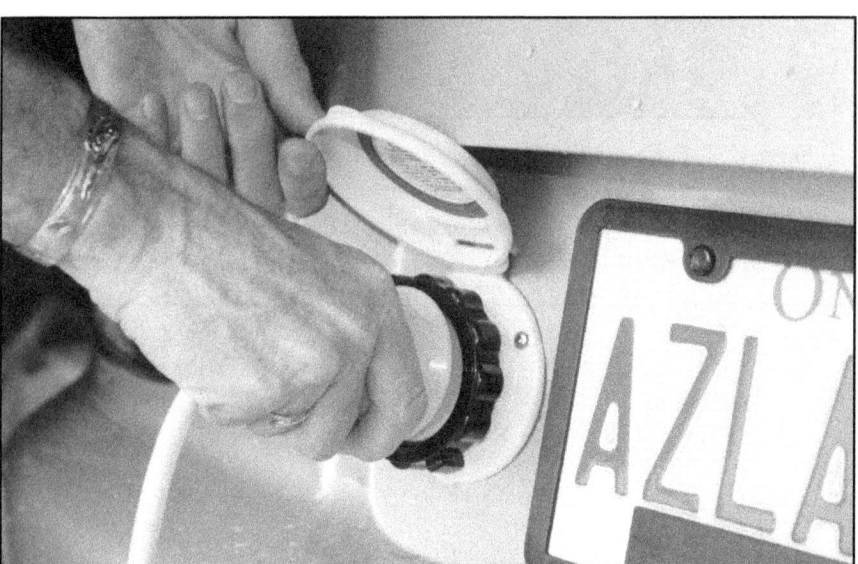

Figure 7.7-3. The batteries have been charging for a few days, so they are ready to roll and it is time to pull the plug. Filling up the battery bank with Bullfrog Power's green energy costs 87 cents and provides enough energy for approximately 12.4 miles (20 kilometers).

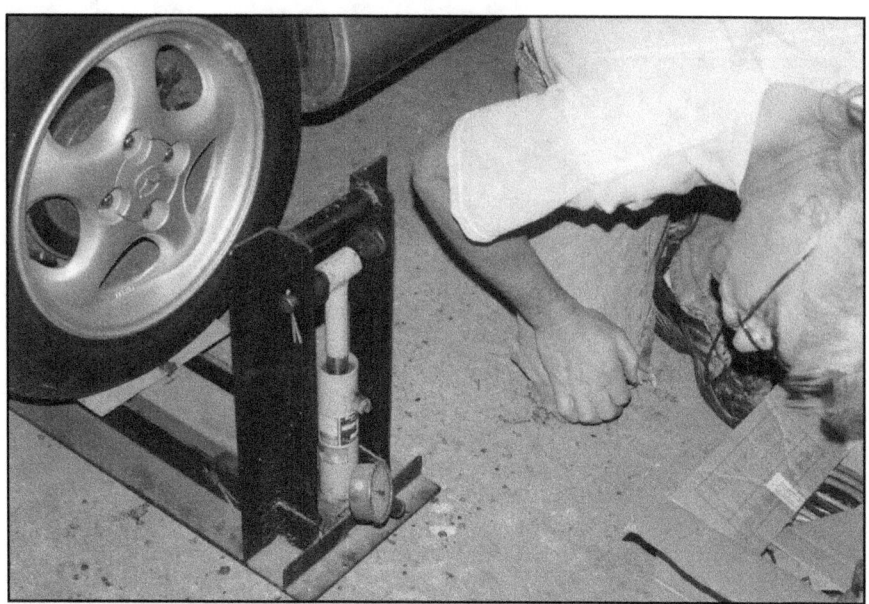

Figure 7.7-4. Rick is shown taking a final reading of the vehicle's weight. We are 480 pounds (218 kilograms) heavier than the stock vehicle, but this additional weight has been corrected for by the addition of supplementary suspension springs.

Figure 7.7-5. The excited looks on the faces of Rick Lane and the author tell the story: the Zero-Carbon Car works!

Figure 7.7-6. After 2 ½ laps around a 5-mile (8 -kilometer) route, the car has used 4.8 kWh of energy to travel 12.4 miles (20 kilometers) before the genset activates. This confirms that the car requires 387 Wh per mile (240 Wh per kilometer) to operate in electric-only mode.

Chassis Dynamometer Testing

After testing the Zero-Carbon Car for general operation and to determine driving range, it was time to test the vehicle's power capacity at the wheels and to review the results.

To perform this test I contacted Edwin Labori of KVR Performance and discussed bringing the Zero-Carbon Car in for testing. KVR Performance sells after-market specialty race parts, manufactures high-performance brake systems, and provides dynamometer tuning on their all-wheel-drive system. Edwin was excited about the opportunity to test an electric-drive vehicle, so one morning Rick and I brought the vehicle over.

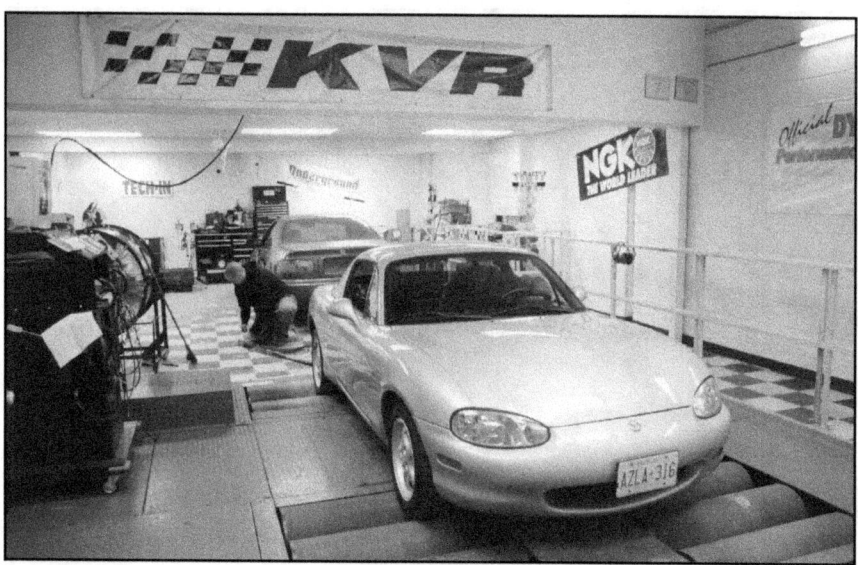

Figure 7.7-7. KVR Performance operates an all-wheel drive chassis dynamometer and provides engine tuning guidance to racing buffs. The Zero-Carbon Car is positioned on the unit and strapped in. "This is pretty cool", exclaims Edwin. "Usually we have to run massive cooling fans, install exhaust plumbing, and wear hearing protection. The dyno makes more noise than the car!"

Figure 7.7-8. The tie-down straps are pulled tight so that the vehicle's suspension system is slightly compressed, ensuring good contact between the tires and the dynamometer rollers.

Figure 7.7-9. Edwin throws his hands up in a show of perplexity. Everything seems to be working backwards, which is exactly how an electric car responds compared to its gasoline counterpart. The KVR staff are amazed that the electric motor seems to be developing maximum torque at a normal stall speed.

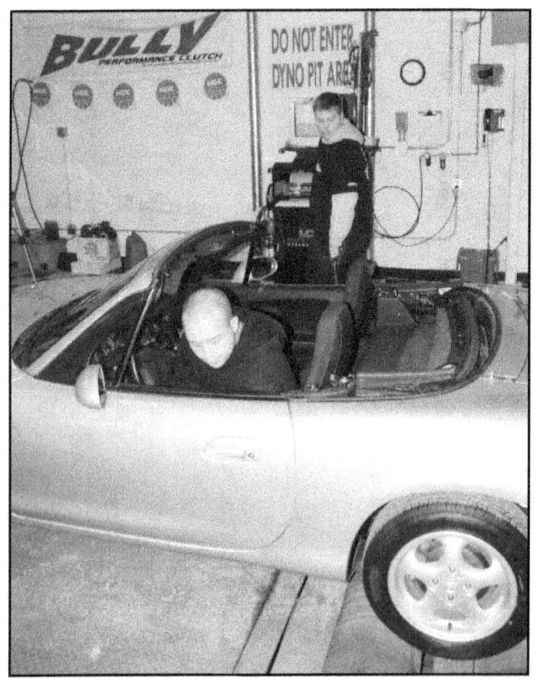

Figure 7.7-10. The other problem Edwin is encountering is that when he steps on the "gas" he can't hear the engine rev up. After years of testing gasoline-powered cars a certain way, he has to adopt a new tactic—watch the wheels spins as he presses the pedal.

Figure 7.7-11. The vehicle runs up to highway speed quickly and without having to shift gears. It is decided that the vehicle has its best performance when starting in third gear and shifting into fourth for highway driving.

Figure 7.7-12. Edwin graphs the results a number of times and performs a number of dynamometer "pulls" to be sure he is recording everything correctly. After a few hours, we have compiled the results shown below.

Reviewing the Results

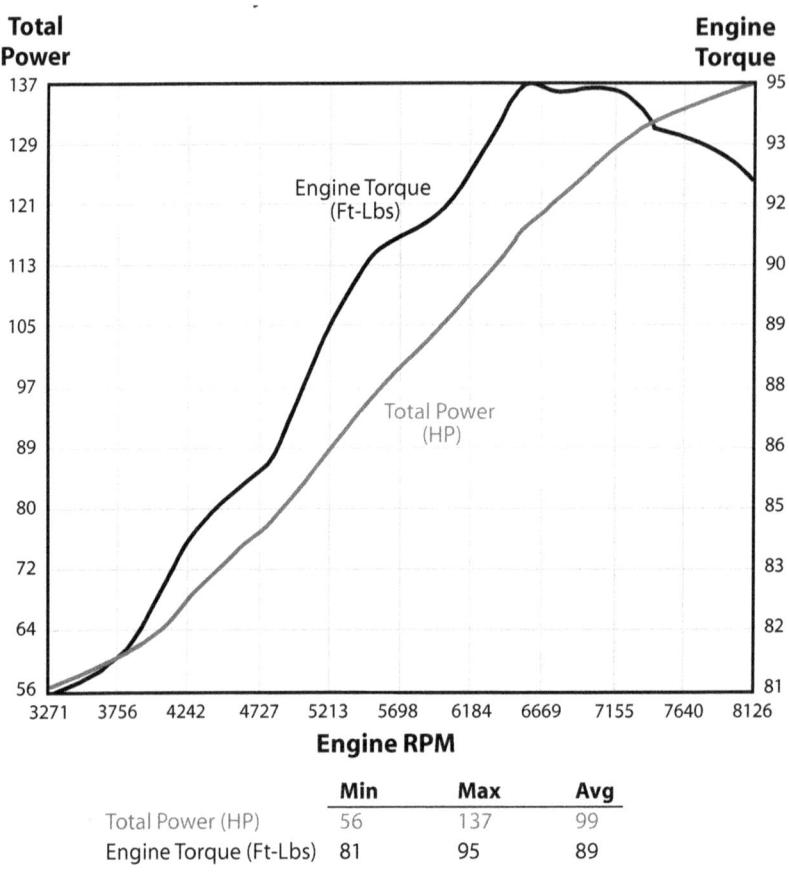

Figure 7.7-13. This graph shows the results for a "race-tuned" Honda Civic with a 1.6-liter gasoline engine. The results indicate that at low-rpm the engine has low torque and horsepower, building to a peak at approximately 7,000 rpm.

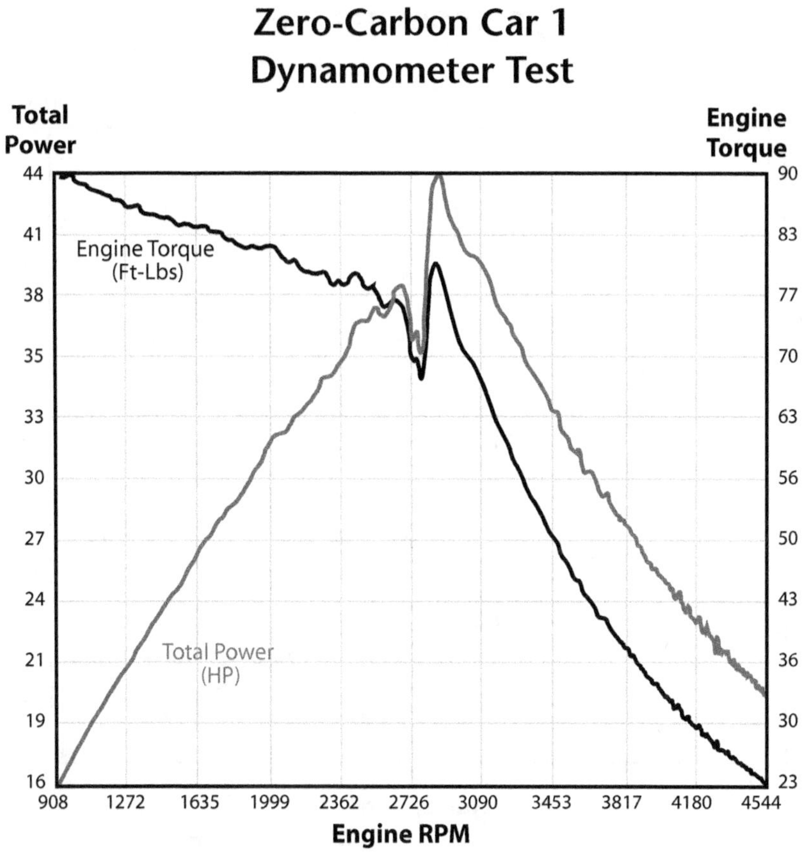

Zero-Carbon Car 1
Dynamometer Test

	Min	Max	Avg
Total Power (HP)	16	44	26
Engine Torque (Ft-Lbs)	23	90	45

Figure 7.7-14. In the series-wound electric motor, torque is highest when the motor is stalled. This gives electric vehicles tremendous acceleration capacity. Torque falls as rpm increases, but horsepower builds as speed increases. The disturbance at the top of the graph is the point where the vehicle was shifted from third to fourth gear. The electric drive system develops as much torque as the race 1.6-liter engine, both vehicles indicating a peak reading of approximately 90 foot-pounds. The Zero-Carbon Car may well be faster off the line than the Honda.

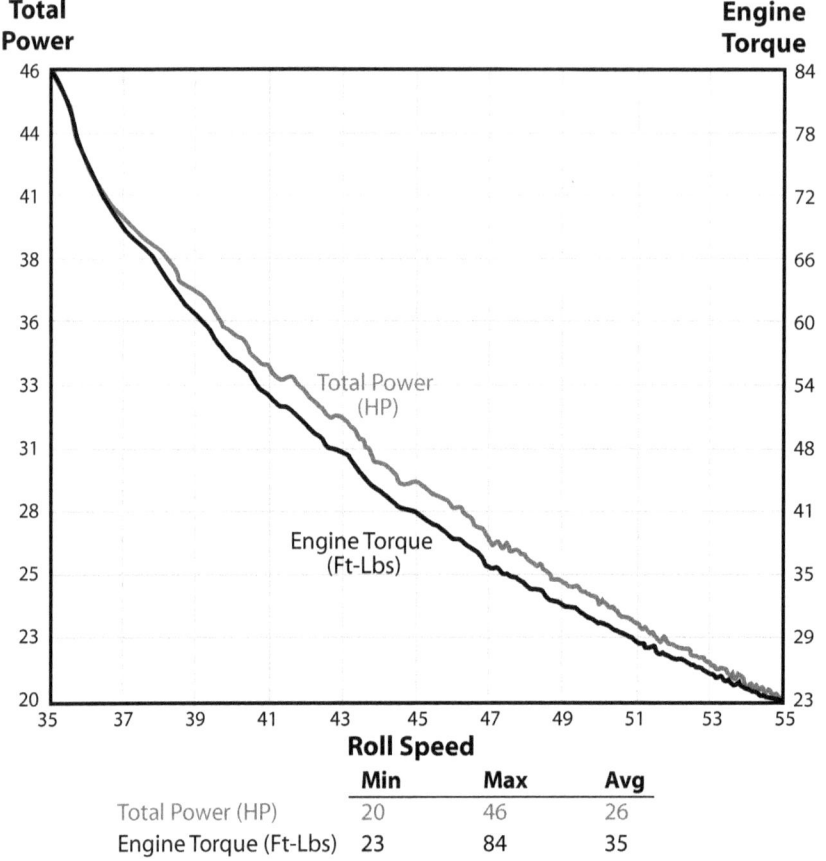

Figure 7.7-15. This graph relates fourth-gear road speed to torque and horsepower. Motor torque is falling off as a function of road speed, indicating that the motor speed is too high to accelerate the vehicle any further. Oddly, gearing down to third gear would have improved vehicular performance.

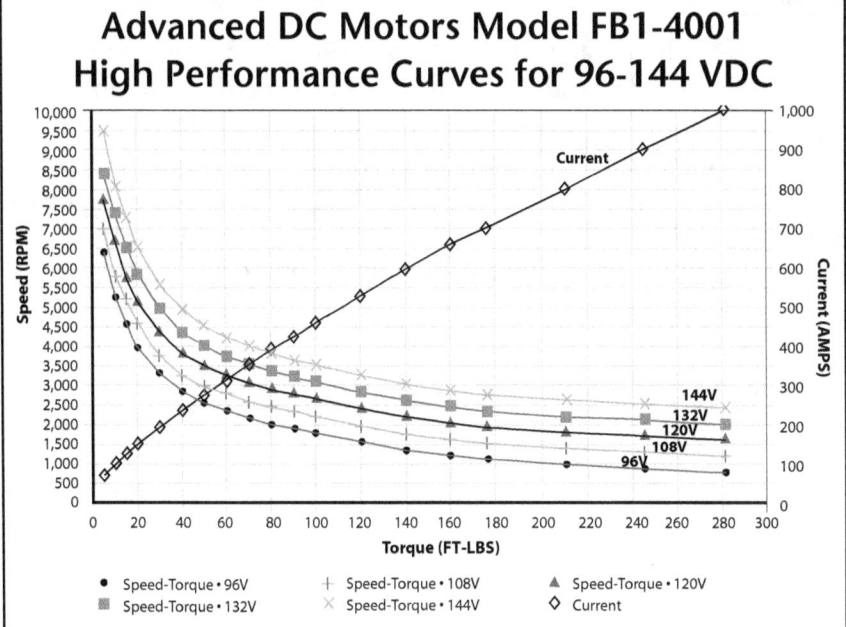

Figure 7.7-16. This graph shows torque versus speed and electrical current for the Advanced DC motor used in the Zero-Carbon Car. Extrapolating the data from the dynamometer reading to this graph confirms the results of the test. (Courtesy KTA Services, www.kta-ev.com)

Speed mph/kph	Voltage (Volts)	Current (Amps)	Power horsepower/kW
3rd Gear			
37/60	110.9	112	16.6 - 12.4
43/70	111.5	128	19 - 14.2
50/80	111.7	170	25.5 - 19
55/90	111.6	204	30.5 - 22.8
4th Gear			
37/60	109.0	110	16 - 12
43/70	108.5	115	17 - 12.5
50/80	108.1	157	23 - 17
55/90	108.1	174	25 - 18.8

Table 7.7-1. This table relates the speed of the Zero-Carbon Car to the battery bank voltage while under a given load, which is shown in the third column in Amps. The last column shows the actual power consumed by the electric motor in both horsepower and kilowatts. It may be hard to believe, but at highway speed the car requires only 25 horsepower (18.8 kW) to operate.

Summary

The Zero-Carbon Car uses less than 50% of the energy used by a typical vehicle and provides more than adequate performance for city or highway driving. The series DC motor and lead-acid battery bank are not ideal for this design, although the vehicle performance tests results prove that the PHEV concept is sound. If test results can be extrapolated to Google.org's fleet of Toyota Prius vehicles modified to PHEVs, it is clear that there are no technical reasons why BEVs and PHEVs could not be on the road today.

Technology is not the problem; society's attitude is.

7.8 Adding Some Bling to the Zero-Carbon Car

After the final road test, a bunch of us were standing back from the car admiring our handiwork when someone commented that it was too bad it looked just like a Mazda Miata. Enough said. After all the work and testing, it was clear that this wasn't just any car; it was the Zero-Carbon Car and everyone had to know.

Cam Mather, graphic artist extraordinaire, came to the rescue with the great idea of creating a big, custom-printed decal similar to a picture printed on a roll of wallpaper. After a few hours on the computer, Cam whipped up some proof images and covered a tiny silver sports car model with them.

Armed with the miniature model and computer files we headed off to Sign It in Vinyl Inc., a company that specializes in custom digital image printing for trade shows, banners, and vehicles. When we explained to production manager Jordan Johnson what we wanted to do, he and Cam worked their magic, creating the layouts and printing off the massive sheets of thin self-adhesive banner material that would adorn the car.

It took a couple of days of wrapping, pulling, tugging, and fitting, but the artwork was finally installed and the Miata was transformed into the Zero-Carbon Car. Perfect!

Great work, and many thanks to the "Sign It" team!

Figure 7.8-1. Armed with the miniature model and computer files we head off to Sign It in Vinyl Inc., a company that specializes in custom digital image printing for trade shows, banners, and vehicles.

Figure 7.8-2. After we explain to production manager Jordan Johnson what we want to do, he and Cam work their magic, creating the layouts and printing off the massive sheets of thin self-adhesive banner material that will adorn the car.

Figure 7.8-3. It took a couple of days of wrapping, pulling, tugging and fitting to install the decals on the Zero-Carbon Car.

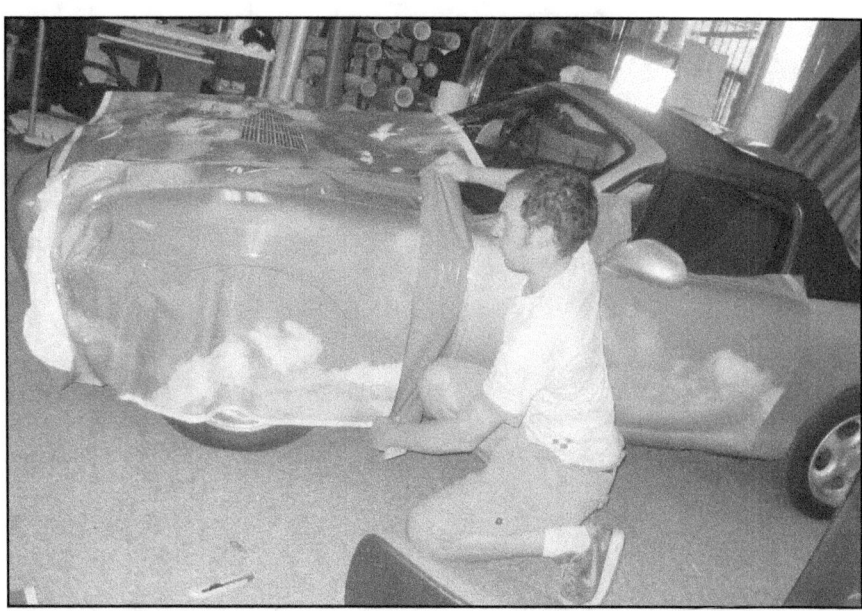

Figure 7.8-4. Getting flat sheets of plastic to fit on curvy bodywork takes an enormous amount of muscle.

Figure 7.8-5. The final detailing of the car was very time consuming, as Sign It owner Emad Memish is demonstrating here.

Figure 7.8-6. Almost done...

Figure 7.8-7. The Sign It team takes a last look at their work, with Jordan doing a few last-minute touch-ups.

Figure 7.8-8. Cam Mather adds one more bit of advertising to the vehicle, with a few words of thanks along with some cautionary notes on the battery cover.

Figure 7.8-9. The Zero-Carbon Car is ready to hit the streets!

8
Zero-Carbon Electricity

Much of the world's energy supply comes from polluting coal-fired power plants, natural gas, and oil generating facilities. All of these sources produce greenhouse gas emissions, and unless they are subjected to carbon taxation an environmental catastrophe will occur. Air pollut-

Figure 8-1. Electricity generated from these fossil-fuel sources generates what are euphemistically referred to as "brown electrons" in order to differentiate them from their renewable or "green electron" cousins.

ants, which lead to respiratory and other ailments, must be curtailed by a change in the energy supply mix or by technological fixes such as exhaust-stack "scrubbers" and the like.

Electricity generated from these fossil-fuel sources generates what are euphemistically termed "brown electrons" in order to differentiate them from their renewable or "green electron" cousins. Electricity sources such as hydroelectric, solar, wind, and biomass are said to be renewable because they will never run out, have no carbon-emission profile, and are therefore sustainable in the long term.

Nuclear power sits somewhere in the middle. Nuclear power is not renewable, yet it emits no products of combustion or greenhouse gases. As carbon taxes would not be applied to nuclear-generated electricity, I will lump its power output into the green side of the balance sheet for purposes of this discussion.

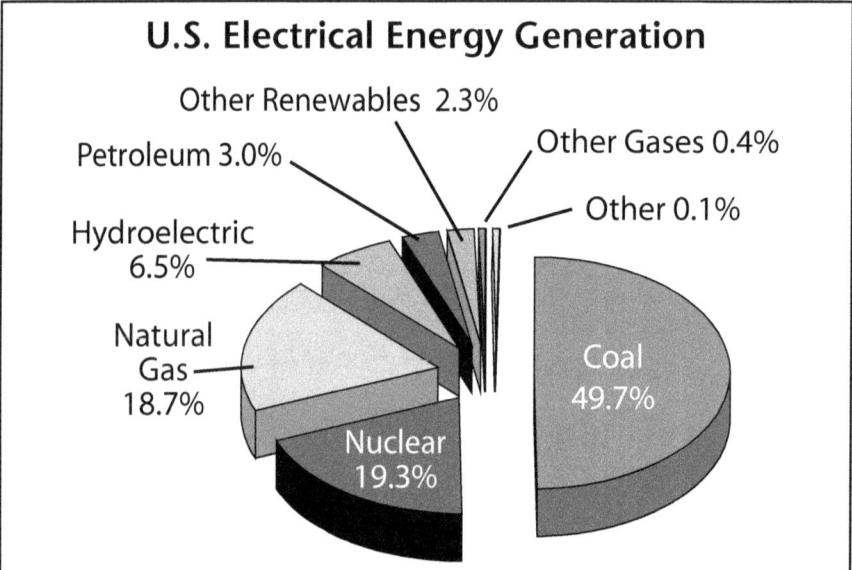

Figure 8-2. This graph shows the distribution of electrical energy generation for the United States. A total of 71.8% of electricity comes from nonrenewable fossil-fuel sources which are major emitters of greenhouse gases. The remaining 28.2% comprises large hydroelectric plants and "new renewable" sources such as wind, biomass, and solar. (Source US Energy Information Administration)

Many environmentalists (correctly) point out that an electric or hydrogen-powered vehicle that is fuelled with brown electrons is not carbon- or emission-free since the source of the emissions is simply moved from the vehicle to the point of electricity production. This is not true of renewable energy sources where the green electrons have no carbon emissions associated with them.

We learned in Chapter 4 (see also Figure 4-19) that a BEV powered by coal-generated electricity has almost the same life cycle carbon profile as a regular gasoline-powered car. It is not until the electricity used to charge the BEV comes from a renewable source such as hydroelectricity that the carbon profile drops to near zero.

Understanding the Terms "Zero-Emission" and "Zero-Carbon"

Zero-emission refers to an electrical generating source that has no atmospheric emissions of any type. This would include solar electric, wind, and sustainable small-hydro facilities.

Zero-carbon refers to generating sources that do not emit any net carbon into the atmosphere. Examples of this category would include landfill and biogas combustion systems, where naturally produced methane is burned to generate electricity. Although there are carbon emissions at source, they are lower than if the original biomass were allowed to rot and vent directly into the atmosphere.

So the next time you see an advertisement suggesting that emissions from coal-fired sources can be lowered by 90%, remember that this refers only to smog-forming compounds. At present, there is no practical way of removing carbon dioxide and other greenhouse gases.

This brings up a number of problems for BEVs and PHEVs, especially for those who wish to operate their vehicles on zero-carbon-emission electricity:
1. Is there enough electricity available to recharge a domestic fleet of electrically powered cars?
2. With only 28.1% of US electrical power coming from renewable or carbon-free sources, where will the electricity come from to power these vehicles?
3. Some areas are supplied almost exclusively with nonrenewable electricity. How can users of BEVs and PHEVs get universal access to clean electricity?

The North American Electrical Power System

As we have discussed earlier, there will be a very slow transition to BEVs and PHEVs under the current carbon and climate change policies in North America. Under these conditions, the electricity supply will have no trouble coping. If the governments of the United States and Canada continue to deny climate change and maintain a head-in-the-sand approach, fuel prices will stay at current low levels (say $3 to $4 per gallon in the United States and $1 to $1.50 per liter in Canada). At this price, automotive manufacturers will have little incentive to build these new-generation vehicles since the public will not be prepared to pay more money for the technology only to save a few dollars in gasoline costs.

However, once a punitive carbon tax is levied on gasoline (for example, $500 per tonne or higher), or when gas supplies are interrupted for geopolitical reasons, vehicle manufacturers will not require any prodding and the models will start to show up on the dealers' lots almost overnight.

The electrical grid will have little problem supplying the energy for this fleet since the vast majority of vehicular charging will occur overnight and during off-peak times. Figure 8-3 depicts the correlation between the typical electrical energy demand and time of day curve of almost all industrialized cities, towns, states, and provinces. The vertical or "Y" axis of the graph shows increasing energy demand in megawatts. The horizontal or "X" axis is the time of day. As would be expected, energy demand is lowest after midnight, when most factories, business-

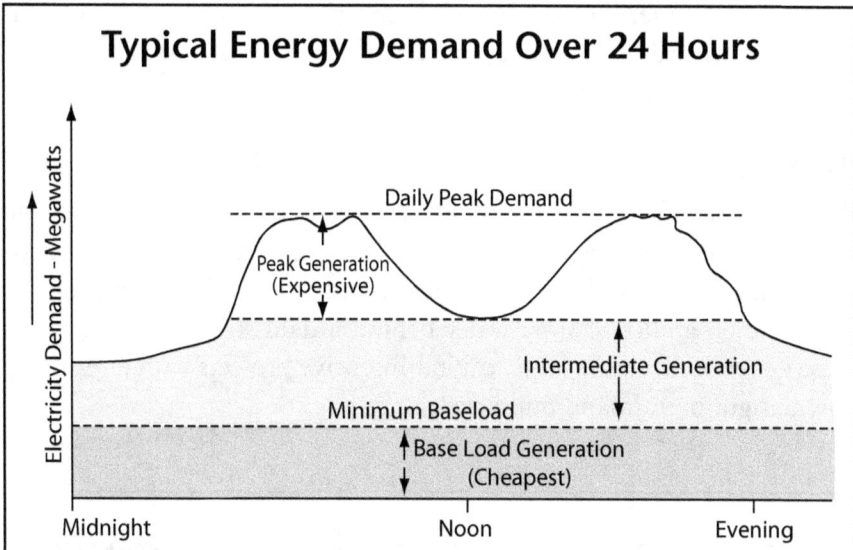

Figure 8-3. The electrical grid will have little problem supplying the energy to charge BEVs and PHEVs as the vast majority of vehicular charging will be overnight and during off-peak times. This chart depicts the correlation between typical electrical energy demand and time of day curve of almost all industrialized cities, towns, states, and provinces. As would be expected, energy demand is lowest after midnight, when most factories, businesses, and homeowners have shut down for the night.

es, and homeowners have shutdown for the night. The period between midnight and 3 a.m. is when the lowest electrical energy consumption occurs; this is known as the minimum baseload level, supplying power for those businesses that remain open, streetlights, and other infrastructure.

Just after dawn, people start waking up, having showers, making coffee and breakfast, and beginning their work day. As a result, electrical power consumption begins to increase. Generally, consumption falls off during the lunch hour and picks back up in the afternoon. Peak power demand is usually reached by 3 p.m., when businesses are at full operation and air conditioning loads are highest. Power levels remain high until after the dinner hour and slowly drop back as people retire for the evening.

Taking a closer look at the graph and you will see that the various levels of electrical power are supplied by three types of generation: baseload, intermediate, and peak.

Baseload Generation

Baseload generation is supplied by those generating facilities that are best suited to operating continuously and at full capacity for long periods of time. This would include nuclear and large hydroelectric sources that have no ability to store water behind a dam structure. In general, these generators require large, capital-intensive projects with a relatively long design, permit, and build cycle.

Figure 8-4. As carbon taxation increases the cost of fossil fuel or as new regulations regarding the purchase of green electricity (electricity generated from renewable energy sources) begins to gather momentum, there will be an increase in the number of small-scale distributed power generating facilities that can operate as baseload sources. One example is this farm-based anaerobic manure digestion system that generates 100 kW of green electricity, providing the farm operator with supplemental income while reducing pathogens and odor in the handling systems.

As carbon taxation increases the cost of fossil fuel or as new regulations regarding the purchase of green electricity (electricity generated from renewable energy sources) begins to gather momentum, there will be an increase in the number of small-scale distributed power generating facilities that can operate as baseload sources. Examples include farm-based anaerobic manure digestion systems and landfill gas power generators. A quick look at Germany shows that so-called electricity feed-in tariffs have allowed the development of over 2,000 farm-based anaerobic digester systems which can generate power 24 hours per day and supply thermal energy to the local area as well. The Province of Ontario, Canada has adopted similar green policies and a proliferation of these carbon-free energy generating facilities are now built, with hundreds more to follow. Expect similar local-scale electricity generating facilities to be built across North America in the coming years.

Intermediate Generation

Intermediate generators are designed to operate for shorter periods of the day, typically up to 16 hours and generally during times of peak demand. These generating facilities are shut down during off-peak periods and include coal- and oil-fired generation, water power, and efficient combined-cycle gas turbine generators.

As more renewable energy sources become part of the electricity mix, wind and solar electric will begin to play a significant role in this class of generation. Although people like to pretend that wind and solar energy cannot be controlled and are therefore of little value, these intermittent generators "integrate" on the grid system, smoothing out customer power demands. Also, as wind energy increases on the grid, other generators will reduce their output, giving wind priority. This strategy allows both coal and hydroelectric generators to "store" energy for use during quiet wind periods.

Peak Generation

Peak generation is designed to operate for short periods of time and is by nature the most expensive form of electrical power generation. These generators are inherently flexible, which further increases their value in supporting the electrical grid during times of high demand. Examples of this generating class include simple gas turbine generators (single-

Figures 8-5a/b. Because small-scale renewable generating facilities such as landfill gas operations are highly flexible in their operating profiles and can generate on demand, they can also be classed as peaking plants. Systems that can guarantee operation during peak periods are paid a premium for this valuable service. The landfill shown here was flaring methane simply to reduce odors, but with the advent of green-power purchase policies, the facility is constructing a power generating station that will be able to run up to 4,000 homes. Without adding another bag of garbage to this site, there is enough gas to run at this level for at least another 50 years.

cycle without heat recovery), combined-cycle gas turbines, and storable water power. Because small-scale renewable generating facilities such as anaerobic digester units and landfill gas operations are highly flexible in their operating profiles and can generate on demand, they can also be classed as peaking plants. Systems that can guarantee operation during peak periods are paid a premium for this valuable service.

Additional sources of peak energy can come from imports from neighboring jurisdictions that have a surplus of power to sell during this period or perhaps from what is known as Vehicle-to-Grid technology (see below).

Vehicle-to-Grid Technology

I suspect that there have been far too many engineers sitting around drinking potent espresso and thinking about all the neat things that future BEVs and PHEVs might be able to do. One of these ideas is known as Vehicle-to-Grid technology or, as it is called by those in the know, V2G technology.

The concept of V2G is that given the vast numbers of electrically powered vehicles that might one day roam the roads, it should be obvious that they will spend the majority of their time plugged in—recharging. As the majority of the recharging will occur at night, when power prices are low, it is conceivable that the same vehicles could sell power during peak times, when prices are higher. This back and forth exchange or arbitrage would occur silently and automatically, courtesy of on-vehicle computing power.

Although the concept is just a thought at this time, several test trials are being conducted to see if the idea has financial merit. Although it should be obvious that you should "buy low and sell high" (which is opposite to what many people do when dealing in the stock market), there are several concerns that must be considered. The main issue is the effect on battery life through all of this energy-exchange process. The second is determining the correct "trigger" prices for the arbitrage to occur and signaling the vehicle that the exchange should happen.

In any event, worrying about V2G at this point in time is of little value, as it will take hundreds of thousands of vehicles to make the technology worthwhile. Better to put the espresso-drinking time into figuring out how to get electrically powered cars on the road in the first place.

System electricity planners respond to numerous factors and time scales. A forecast of abnormally hot weather will ensure that planners have contacted generators to ensure that adequate supply is available, often by adjusting the forward market price upwards to attract suppliers.

Likewise, if government policy implemented a price on carbon, both short- and long-term planning groups would assess electricity supplies to meet the immediate and long-range needs.

Figure 8-6. Baseload generation is supplied by those generating facilities that are best suited to operating continuously and at full capacity for long periods of time. This would typically include nuclear and large hydroelectric sources that have no ability to store water behind a dam structure. These generators typically require large, capital-intensive projects that require a relatively long design, permit, and build cycle, although this dentist-owned 0.5 MW run-of-the-river facility is an exception. North America-wide there is the potential for tens of thousands of megawatts of low-environmental-impact hydroelectric generation.

Figure 8-7. As more renewable-energy sources become part of the electricity mix, wind and solar electric will begin to play a significant role in this class of generation. Although people like to pretend that wind and solar energy cannot be controlled and are therefore of little value, these intermittent generators "integrate" on the grid system, smoothing out customer power demands. Also, as wind energy increases on the grid, other generators will reduce their output, giving wind priority. This will allow both coal and hydroelectric generators to "store" energy for use during quiet wind periods. (Courtesy: Vestas Wind Systems)

Figure 8-8. Fuel cells have numerous applications beyond powering the family vehicle. In fact, so-called stationary power systems that produce both electricity and heat may be a better suited application for fuel cells in the short- to mid-term. In this photograph, the company Fuel Cell Technologies has placed a gas reformer and fuel cell system in the output air stream of a Ford factory paint shop. The highly volatile fumes that were being exhausted into the atmosphere are being captured and used to generate electricity in this unique and ultimately zero-carbon manner. There are as many ways to generate clean electricity as there is demand for this power. (Courtesy: Fuel Cell Technologies)

Demand Time Shifting

In the short term, almost all power consumption patterns look the same as those shown in Figure 8.3, giving an immediate supply of electricity through demand load time shifting. Using the invaluable tool of self-interest, suppliers can adjust (manipulate) electricity markets by lowering night time prices so that people who would have plugged their vehicles in during the day are financially rewarded for plugging in at night. This, together with the fact that the system never operates near full capacity for an entire 24-hour cycle, will cause the baseload portion to rise during the night while daytime peak energy demand will increase only moderately, as noted in the following statement by the US Electric Power Research Institute (EPRI)[1]:

More than 40% of U.S. generating capacity operates at reduced load overnight, and it is during these off-peak hours that most PHEVs would be recharged. Recent studies show that if PHEVs displaced half of all vehicles on the road by year 2050, they would require only an 8% increase in electricity generation (4% increase in capacity).

As more overall energy is required, new renewable plants will be built quickly to fill the need and to take advantage of green-power purchase policies which give preference to low- or zero-carbon generating sources.

Zero-Carbon Coal and Fossil Fuels

Once carbon taxes or carbon emission caps become a reality, everyone will get on the zero-carbon-emission bandwagon, including the worst polluter and greenhouse-gas emitter, coal. Because the coal-power industry is so large and has so much to lose if carbon regulations make

Figure 8-9. As more overall energy is required, new renewable plants will be built quickly to fill the need and to take advantage of green-power purchase policies which give preference to low- or zero-carbon generating sources. The EcoLogo certification program audits green power producers to ensure that they comply with the tenets of low emissions and sustainable environmental footprint. (Courtesy: Bullfrog Power)

coal uneconomical, the industry has been searching for ways to remove carbon dioxide from the exhaust gas stream and sequester it using some form of technology.

The most promising of these schemes is known as carbon capture and sequestration (CCS) technology, where the carbon dioxide gas that is sequestered from the exhaust stream is pumped under pressure into various underground structures including depleted oil and gas reservoirs, coal seams, and deep saline aquifers made of porous rock. In some cases the pressurized carbon dioxide gas can be pumped into mature oil fields to help pressurize the oil and make it easier to recover.

There is a distinct possibility that coal and other "stationary" fossil-fuel processors will be able to use CCS technology to their advantage and actually compete against the oil and gas industry. This surprising scenario is due to the potential ability of coal and other fossil-fuel-powered "stationary" electricity generators to use CCS to make green electrons and thus escape carbon taxation. Since this zero-carbon electrical power is desirable for all aspects of industry, including the recharging of BEVs and PHEVs, it will actually erode the oil and gas industries' grip on personal transportation. For this reason, the electrical-power-generating industry is supporting the development of electrically powered vehicles. Here's what EPRI says on this subject:[2]

> *The EPRI Electric Transportation program helps energy companies increase electricity sales, promote economic development and customer retention, improve customer satisfaction, and contribute to a cleaner environment by accelerating market penetration of electric-drive vehicles.*
>
> *We can build upon these successes:*
> - *Plug-in hybrid vehicles are a common recognized phrase, even written in the 2005 Energy Policy Act*
> - *You and we together have an alliance with an OEM manufacturer to build and test plug-in hybrid vehicles with a path to commercialization*
> - *We have a PHEV project with Eaton Corporation, a major supplier to the heavy duty transportation industry*
> - *We have expanded our competency within our advisors and EPRI technical team*
> - *We have expanded our outreach to universities, ports, airports, and the environmental community*

We are part of a growing national interest in electrifying transportation. We look forward to continuing to grow and collaborate with you in 2007!

With the right carbon policies, coal-fired electricity using CCS to charge electrically powered vehicles could be cheaper than using gasoline to run standard internal combustion engines.

Taking this thinking just a step or two further, it should also be possible for the oil patch industries to see a golden lining as well, once the tired old men who currently run this sector leave and the next generation of more sustainable thinkers come on board.

Currently, the oil industry sells gasoline that is placed in an automobile and burned with very low efficiency, spewing all sorts of smog-forming emissions and greenhouse gases out the tailpipe. The automobile industry adds evermore complex and heavy technologies to its vehicles (further lowering overall vehicular efficiency) to deal with the toxic stew. These technologies include catalytic converters, advanced engine management systems, oxygen sensors, and exhaust-gas recirculation components.

The refining of crude oil into gasoline and diesel fuel also requires vast amounts of energy and in turn spews out even more chemical compounds and carbon into the atmosphere. Delivering the fuel in a tanker truck to the 121,446 retail stations in the United States also consumes a considerable amount of energy, releasing more atmospheric emissions and damaging road surfaces.[3]

Now imagine for a moment that the total world oil-refining process were scaled back from the production of gasoline and other refined fossil fuels to one grade such as #6 Heavy Fuel Oil (HFO) (also known as "Bunker C") that would be used to generate electricity to power society and electric vehicles. HFO is an inherently dirty fuel, but could be refined sufficiently to strip out sulfur and other compounds to a lower level. The refining process would consume considerably less energy and emit far fewer emissions, producing the low-sulfur HFO. Naturally, any carbon emissions from the refining process would be subjected to CCS.

The HFO would then be burned at a centralized power generating facility that would have suitable emission control equipment to capture or neutralize any remaining smog-forming compounds left over from the initial refining process. The emissions from the power generation system would be subjected to CCS, allowing the electricity generated by this system to be green.

Current thermal power generating facilities including coal, oil, and single-cycle natural gas are just about as inefficient as automobile engines; nearly two-thirds of the energy consumed is lost to heat which is either exhausted into the atmosphere or pumped into lakes and rivers in proximity to the power station. To put this in perspective, for every coal or nuclear power generating facility that generates electricity, the equivalent of 2 power stations worth of heat energy are wastefully discarded! Capturing this energy has not been a priority of the power generating industry because of the cost of the heat-recovery equipment and the historically low price of thermal energy. In a more sustainable model, this thermal energy would be captured and used rather than wasted. At the centralized power generation scale, for example, this energy could be used by close-proximity greenhouses, industrial laundry facilities, or any other high-demand thermal energy user.

If the size of thermal generating facilities were reduced it would be easy to place them closer to areas of electrical demand load in downtown cores or suburbs. At such a small scale, CCS technology would not be available locally, but there is no reason why the carbon emissions could not be pipelined to a centralized facility for storage or possible sale as a useful product. With the generating source located close to centers of demand, the waste heat that is normally thrown away could be captured and sold to industry, homes, and other users of heat, which is to say everyone.

Of course none of this is done today because of concern over cost, but if the proper carbon policies were put in place, nothing that I have noted would be impossible. Likewise, if the technology to permanently store carbon dioxide gas were available and cost effective as a result of carbon taxation policies, CCS and the cogeneration of electricity and heat could become widespread and turn the black electrons of coal and other fossil-fuel-based energy sources shiny and green.

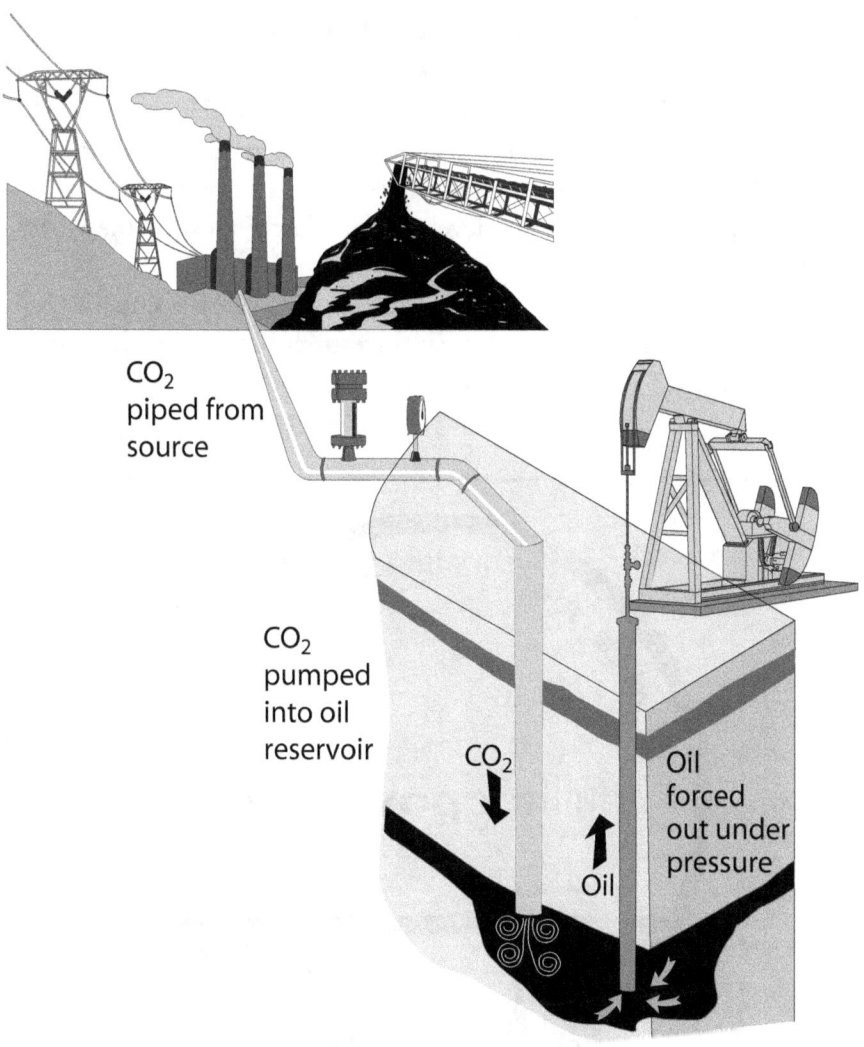

Figure 8-10. Carbon capture and sequestration (CCS) technology strips carbon dioxide from the combustion exhaust stream and pumps it under pressure into various underground structures including depleted oil and gas reservoirs, some coal seams, and deep saline aquifers made of porous rock. In some cases the pressurized carbon dioxide gas can be pumped into mature oil fields to help pressurize the oil and make it easier to recover.

Moving the Charging Plug Around Town

In the previous section we learned that there is plenty of nonrenewable electricity available in the United States to recharge BEVs and PHEVs. A more detailed study would likely show similar results in other countries of the world and if planning were to start in earnest, electrical system strategies could easily be developed to power most of the world's transportation fleet.

But the transition to 100% zero-carbon electricity production is a long way off and still requires massive investment in renewables and the development of CCS technology as well as appropriate carbon regulations.

Figure 8-11. The transition to 100% zero-carbon electricity production is a long way off and still requires massive investment in renewables and the development of CCS technology as well as appropriate carbon regulations. Bullfrog Power identified this problem and found a unique solution for those people who wanted to do something about it. "Bullfrog Power opted for a more direct approach to selling green power, one that could be easily verified and just as easy for the non-technical person to understand." (Courtesy: Bullfrog Power)

The question of what to do in the meantime is one that had been puzzling me at the beginning of this project. Sure, some overzealous people like me could (and would) install photovoltaic panels or wind turbines to charge the battery banks of our BEVs and PHEVs, but what of the rest of society?

The answer turned out to be rather simple: use the financial markets to help create a ready "pool" of carbon-free electricity that anyone can access, and then make the power pool mobile. Simple, really.

The story starts with a discussion with Greg Kiessling, Executive Chairman of Bullfrog Power Inc. (www.bullfrogpower.com), a green-energy retailer based in Toronto, Ontario. "The retail electricity market is becoming aware that not all electrons are created equally," Greg states. "People are becoming polarized, aware that they don't want anything to do with coal or nuclear electricity but feeling powerless to do anything about it."

Greg goes on to explain how Bullfrog Power identified this problem and considered various solutions to offer these people. "Bullfrog discussed the concept of using carbon-offset credits, but people did not seem to understand the process and were sceptical of it. We opted for a more direct approach to selling green power, one that could be easily verified and just as easy for the non-technical person to understand."

The carbon offset program that Greg did not accept has people pay a fee for a given amount of carbon used for powering their houses or flying. The seller of the credits estimates how much carbon these activities generate, and a fee is charged based on an established market price for carbon. The majority of the carbon charge is paid to developers of supposedly environmentally beneficial projects such as tree planting in the Amazon or initiatives to offset coal power generation in China. Although this scheme can actually work, there is some question of the quality and longevity of the carbon abatement projects that people are investing their money in.

"We use the direct connection of the markets to put a premium on clean-energy technologies, using the fees we generate to build even more clean-tech projects," Greg explains. Bullfrog Power does this by purchasing "blocks" of energy from EcoLogo-certified, low-environmental-impact power producers and then reselling these blocks to retail consumers. In this manner, a homeowner who currently receives a blend of electricity from nuclear, fossil-fuel, and renewable sources would in effect be getting 100% clean energy supplied, for example, by a wind turbine in the province.

In this example, the wind turbine can only produce so much energy, which is metered and supplied to the grid. Bullfrog Power subscribes to purchase some or all of the yearly generation of the power source and resells this amount to retail customers. The transaction is direct, clear, and verified by the electrical utility charged with metering the generation and load customer accounts.

In reality, the electrons from the wind turbine do not actually flow to the load customer's home but rather mix with all electrons on the electrical system. It is the act of purchasing the "green attributes" that actually counts, assuring customers that their actions have a positive effect on the environment.

Obviously, it is possible for homeowners to subscribe to Bullfrog Power and be assured that their homes are powered with 100% renewable energy. Any vehicles that are recharged from the house circuits are also supplied with clean energy. But what if the car is plugged in at the office, a friend's place, or some other location that is not currently serviced by Bullfrog Power?

I presented one possible solution to Greg. Why not "sell" the electrical energy blocks to the vehicle rather than the home? To support the Zero-Carbon Car (ZCC) project, I suggested that I purchase a block of 1MWh of electricity (1,000 kWh), provide the attributes to the vehicle, and retire a same-size block of energy from an Ontario-based wind turbine. Using this approach, it would not matter where I plugged the car in. As long as the energy consumed by the vehicle equalled the block of energy retired from the clean technology source, the system would remain balanced.

After discussing this proposal with his colleagues, Greg enthusiastically reported back that Bullfrog Power's retail electricity licence would allow this transaction to proceed. "For Bullfrog, this is a great fit. It allows us to provide retail green electricity not just for lights and fridges but for transportation as well, and it could be a huge market for us in the coming years." Greg went on to explain that he was a member of Plug-In Partners (www.pluginpartners.org), a grassroots initiative to demonstrate to automakers that a market for flexible-fuel PHEVs exists today.

"The concept of mobile clean energy is really the icing on the cake in the advancement of zero-carbon vehicle technology."

Bullfrog Power presented me with the certificate shown in Figure 8-12 which states that 1 MWh of green electrical energy supplied by wind and low-impact hydroelectricity has been attributed the Zero-Carbon Car. Exactly what does this 1 MWh provide?

This is mere pocket change really, when you consider the value of saving the environment. Consider the Bullfrog Power fee as cheap insurance; the energy price is so reasonable, that the vehicle still saves money even after you purchase the more expensive green electricity.

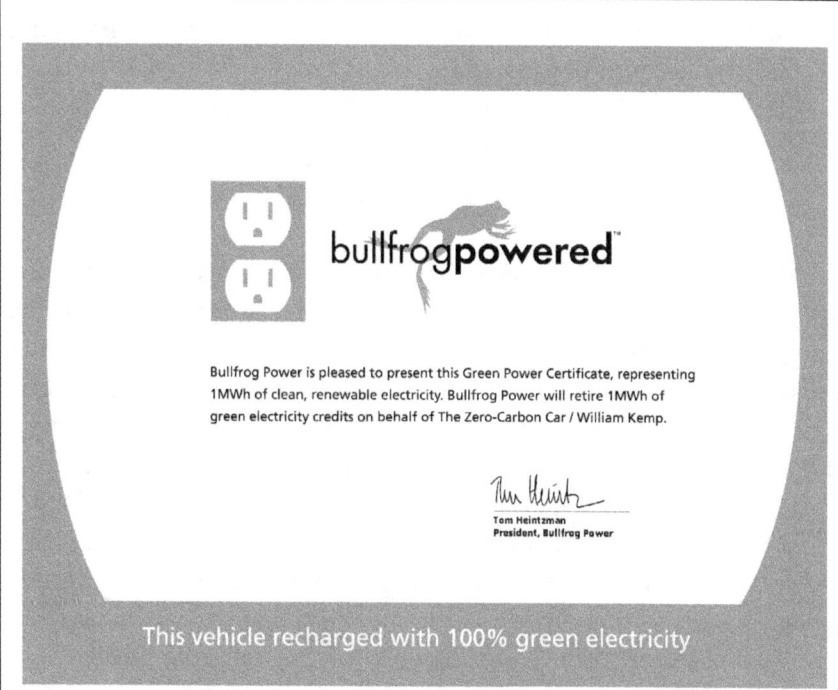

Figure 8-12. Bullfrog Power presented me with this certificate which recognizes that 1 MWh (1,000 kWh) of green electrical energy supplied by wind and low-impact hydroelectricity has been attributed the Zero-Carbon Car. To put the 1MWh certificate in perspective, for the grand sum of $30 added to a regular electricity bill, I can drive carbon free for 2,584 miles, (4,159 kilometers). (Courtesy: Bullfrog Power)

The Value of Zero-Carbon Electricity

My version of the Zero-Carbon Car is a relatively inefficient design compared with factory-built, megabudget concept cars and modified PHEV Toyota Prius vehicles, yet even with the "poor" energy conversion rate the outcome is nothing less than stunning.

The ZCC has a gross battery capacity of approximately 6,000 watt-hours capacity (120 volts x 50 amp-hours). Driving the car on electricity alone to an 80% battery discharge rate occurs at the 12.4 mile (20 kilometer) mark. Therefore, the vehicle consumes:

(6,000 Wh capacity x 80% discharge) ÷ 12.4 miles = 387 Wh per mile
(6,000 Wh capacity x 80% discharge) ÷ 20 kms = 240 Wh per km)

So, to put the 1MWh Bullfrog Power certificate in perspective, for the grand sum of $30 on top of a regular electricity bill ($115 per MWh), I can drive carbon free for the following distances:

1,000 kWh ÷ 0.387 kWh per mile = 2,584 miles
(1,000 kWh ÷ 0.240 kWh per km = 4,167 kilometers)

This is mere pocket change really, when you consider the value of saving the environment. Consider the Bullfrog Power fee as cheap environmental insurance. The energy price is so reasonable, that the vehicle still saves money even after you purchase the more expensive green electricity, as you will see by comparing the operating costs for BEVs and PHEVs to those of standard internal-combustion-powered vehicles.

387 Wh equates to 0.387 kWh. The current price of electricity in Ontario, delivered to a retail customer, including all taxes and distribution fees, is approximately 11.5 cents per kWh. Bullfrog Power adds a further 3 cents to this total ($30 per MWh) to supply the home or car with carbon-free energy. The total cost for the delivered green energy is therefore 14.5 cents per kWh. Accordingly, each mile that the ZCC drives on electricity costs:

14.5 cents per kWh x 0.387 kWh per mile = 5.6 cents per mile
(14.5 cent per kWh x 0.240 kWh per km = 3.5 cents per km)

To put this in perspective, a typical gasoline car based on the United States fleet average economy of 19.8 miles per gallon (8.4 kilometers per liter) consuming gasoline at $3.00 per gallon (79 cents per liter) has the following operating cost:

$3.00 per gallon ÷ 19.8 miles per gallon = 15 cents per mile
($0.79 per kilometer ÷ 8.4 km per liter = 9.4 cents per kilometer)

The greenhouse-gas-spewing gasoline-powered car that has been engineered by the multibillion dollar auto industry is nearly three times more expensive to operate on a per-mile basis than the Zero-Carbon Car developed by a bunch of amateurs! Furthermore, it would take only 200 miles (320 kilometers) driving distance to pay for the carbon attributes provided by Bullfrog Power. Any additional driving and you would begin to accrue savings compared to a standard gasoline-powered vehicle. Imagine what might happen if the professional car builders got involved.

To consider this possibility, let's look at the "fleet" energy efficiency study compiled by Google.org during its plug-in hybrid tests. According to Google, its modified Toyota Prius models using the Hymotion technology discussed earlier have averaged 124.5 Wh/mile during the study. Recompiling the numbers based on these findings we see that operating costs are:

14.5 cents per kWh x 0.1245 kWh per mile = 1.8 cents per mile
(14.5 cents per kWh x 0.077 kWh per km = 1.1 cents per kilometer)

The factory-built Toyota Prius modified to operate as a PHEV is over eight times cheaper to operate on a per-mile basis, and this does not take into account a value for the savings in carbon emissions.

If we take this example just a bit further and add in what I consider to be a modest carbon valuation of $550 per ton ($500 per tonne), we find that the price for gasoline would almost equal the current retail price in the UK and that the operating costs for the gasoline-powered car would look something like this:

$7.80 per gallon ÷ 19.8 miles per gallon = 39 cents per mile
($2.05 per liter ÷ 8.4 kilometers per liter = 24 cents per kilometer)

This is nearly 21 times more expensive that driving a BEV or PHEV.

Of course this is a simple equation and does not take into account maintenance and regular service charges for the gasoline-powered vehicle; nor does it take into account battery-life issues for whatever version the ZCC comes in. However, it can reasonably be said that internal combustion engines will always have relatively high maintenance charges, while BEV and PHEV battery technology should improve to the point where the battery bank will last the life of the car. (Remember that my Honda Civic Hybrid has an 8-year battery warranty.)

Summary

One has to wonder why these types of high-efficiency vehicles are not for sale in the UK today, where gasoline prices actually are at these levels. And of course this is exactly the problem. A carbon tax of $550 per ton ($500 per tonne) or carbon tax equivalent as is currently levied in most of Europe is simply not punitive enough for people to demand these kinds of vehicle-efficiency improvements. In my opinion, increasing the carbon tax on retail gasoline and diesel fuel to the $1,100-dollar-per-ton ($1,000 per tonne) range should just about do it.

One final option to consider is that BEV or PHEV vehicles be "factory equipped" with their own certification of having sufficient green power attributes to last the life of the vehicle. Assuming the vehicle in question were a PHEV Toyota Prius and the life expectancy of the vehicle were 155,000 miles (250,000 kilometers), at current retail prices the amount of green electricity required would be:

155,000 miles x 124.5 Wh per mile = 19.3 MWh

As each retail carbon certificate issued by Bullfrog Power costs $30.00 per MWh, the total cost to offset the carbon contained in the electricity portion of the vehicle would be:

19.3 MWh x $30 per MWh = $579

Of course, these prices are subject to market forces and with inflation pushing these prices upward over time the cost for full carbon abatement could be locked in over the life of the vehicle by including the offset certificate in the purchase price of the vehicle. And if governments or the market dictated a very high value for carbon, Bullfrog Power could command a much higher price for its green energy.

Given that some 80% to 85% of all driving would be done on green electricity and that the environment might just be saved, the investment in BEVs and PHEVs as well as green electricity marketing could be the bargain of the century.

Chapter 9
Zero-Carbon Liquid Fuels

While the world waits breathlessly for hydrogen-powered fuel cells, developments in liquid biofuels are continuing at a surprising, perhaps reckless rate aided by pork-barrel politics and a general lack of common sense.

Almost everyone has heard of ethanol and some enlightened souls may be familiar with biodiesel. While these technologies are not nearly as sexy as hydrogen and fuel cells, they are available now and can offer numerous advantages over oil while also making it unnecessary to replace trillions of dollars' worth of existing fuel infrastructure systems.

At the same time, if biofuel supplies and regulations are not properly developed, they can also provide an equally bewildering series of problems ranging from increased carbon emissions to creating massive damage to the world's rain forests, food supplies and prices.

This rather extensive chapter will attempt to provide an overview of the issues, pro and con, related to biofuels. It will also show what place there is for a sustainable biofuel industry when used in conjunction with high-efficiency PHEVs.

Finally, I will provide an overview of the small-scale production of zero-carbon biodiesel that will be used as the liquid fuel source for the Zero-Carbon Car. The following is a brief examination of biodiesel and its production. For a more thorough discussion, please refer to my previous book *"Biodiesel: Basics and Beyond"* (Aztext Press, 2006).

An Introduction to Biofuels

Modern internal combustion engines including those used in transportation vehicles do not have to run on gasoline. In fact, early automotive pioneers did not have access to refined gasoline and consequently used peanut oil and alcohol for fuel. However, the discovery of large volumes of easy-to-access and cheap fossil fuels spelled the end of these early alternatives.

Ethanol from Food – A Non-Starter

The consumption of fossilized fuels is not the only means of extracting energy from plant life. Fermentation of grapes and apples has been fuelling binges for as long as man can remember. The naturally occurring sugars in the fruit produce wine and cider with a maximum alcohol content of approximately 14%. Applying heat to these beverages, in a process known as distillation, allows extraction of the ethanol alcohol at up to 100% concentration.

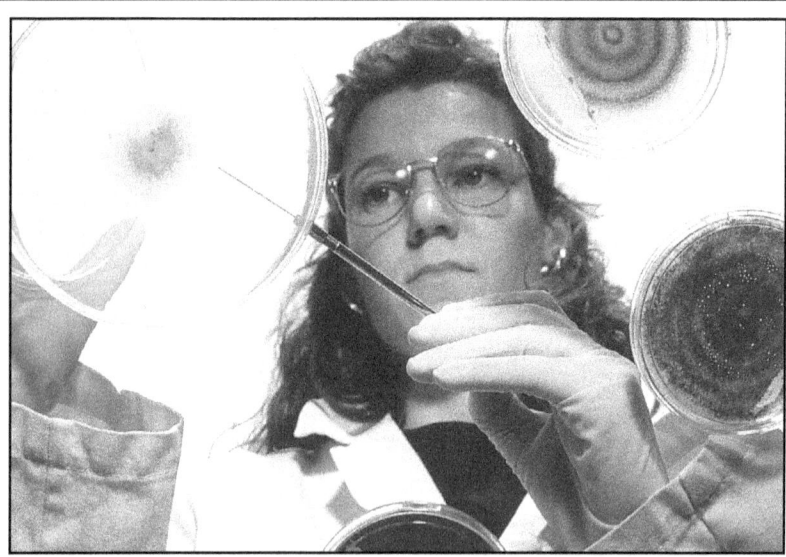

Figure 9-1. Plant grains and fiber can be converted to sugar, fermented into ethyl alcohol (ethanol), and used as a blending ingredient with gasoline or as the main fuel. High concentrations of ethanol can reduce greenhouse gas emissions by up to 80% relative to gasoline. (Courtesy: Iogen Corporation)

The primary source of plant sugars can vary, as automotive-grade ethanol can be produced from food grains such as corn, wheat, and barley. Recent advances in enzymatic processes even allow the conversion of plant waste in the form of straw and agricultural residue into sugars which can in turn be fermented into ethanol.

Conventional grain-derived and cellulose-based ethanols (those made from farming and timber byproducts) are the same and can be easily integrated into the existing gasoline supply chain. Ethanol may be used as a blending agent or as the main fuel source. Gasoline blends with up to 10% ethanol can be used in any vehicle manufactured after 1970.[1] High-level ethanol concentrations of between 60% and 85% can be used in special «flex-fuel vehicles.»[2] At the time of writing Ford and General Motors warranties allowed up to 10% ethanol blends in their

Figure 9-2. The majority of ethanol is derived from corn. The production of ethanol fuels strengthens agricultural regions and creates new and more stable markets for the corn-cropper provided that government subsidies remain in place. At the same time, market pressure for more ethanol plants has caused large swings in the price of corn and other grains, causing hardship in poor regions of Mexico as well as domestically. In addition, the environmental benefits of growing crops to produce road fuels are negligible at best. The long-term prospects for food-crop-based ethanol are very poor and will immediately fade as soon as carbon taxation starts or government subsidies end.

Figure 9-3. North American ethanol production currently stands at 3.36 billion gallons (12.72 billion liters) and is expected to increase by 28% before the end of the decade. This is just a drop in the bucket compared with fossil-fuel gasoline usage, but with good demand management (through CAFE and properly managed carbon taxation) it could go a long way toward reducing greenhouse gas emissions. (Courtesy: Iogen Corporation)

North American vehicles and have begun to introduce models that operate on high-ethanol blends as well.[3]

Adding ethanol to gasoline increases octane, reducing engine knock and providing cleaner and more complete combustion, which is good for the environment. Ethanol produces lower greenhouse gas emissions than gasoline: a 10% ethanol blend with gasoline (known as E10) may reduce GHG emissions by 4% for grain-produced ethanol and 8% for cellulose-based feedstocks. At concentrations of E85, GHG emissions are reduced by up to 80% when using cellulosic ethanol.[4]

Proponents argue that ethanol is produced from domestic renewable agricultural resources, thereby reducing our dependence on imported oil and providing a major source of economic diversity for rural farming economies.

In the United States, 77 ethanol plants produce over 3.3 billion gallons (12.5 billion liters) of ethanol per year. Canadian production currently stands at 62.6 million US gallons (237 million liters) per year. Of the 77 plants in the United States, 62 use corn as the feedstock. The remainder use a variety of seed corn, corn and barley, corn and beverage waste, cheese whey, brewery waste, corn and wheat starch, sugars, corn and milo, and potato waste. In the United States there are 55 proposed new plants and 11 currently under construction. Canada currently has 6 plants with 1 under construction and 8 new proposals on the drawing board.[5]

There is no question that using properly managed, domestically grown feedstocks from renewable fuel crops such as switchgrass or from waste biomass sources is better than importing fossil fuel from the Middle East. However, there is a great deal of controversy about diverting food grain stocks into fuel feedstocks.

At current rates of gasoline consumption it will be impossible to use ethanol as a full replacement for gasoline used in internal combustion engines; there is simply not enough farmland, irrigation water, or petrochemical feedstocks for fertilizer and pesticides available. Even if 100% of the US corn supply were distilled into ethanol, it would supply only a small fraction of total domestic gasoline demand. In his 2007 State of the Union address, President George Bush called for 20% biofuel content in gasoline and diesel fuel. This will require an eightfold increase in US biorefinery capacity and will supply approximately 15% of US motor fuel requirements. In addition, a massive increase in natural gas imports into the United States will be required to distil the corn into alcohol. North American natural gas supplies are in a free fall, so this option will be costly if not impossible.

According to Alberta economics professor Kurt Klein, "this level of biofuel production would be devastating for the livestock industry and the food supply. Costs of all forms of livestock feed would increase in price and with a horrendous dislocation of agricultural production. There would be no more exports of corn and wheat from the U.S., and remember that Canada actually imports about 20 percent of its corn requirements from the U.S. now."[6]

I also worry about the long-term effects on the farming sector directly. During this massive boom in ethanol production, corncroppers are ramping up production as never before. This effort requires increased land use for monocropping corn, which comes at the expense of all other crops. Farmers must invest more aggressively in land leases and equipment purchases to ensure that they get their share of the rewards, all the while leveraging assets to make this happen.

During the boom cycle, ethanol and hence corn prices will rise, affecting everyone who buys corn and corn products. This is, indeed, everyone. The corn cropper is hurting his dairy neighbor down the road because protein meal prices are rising as a result of the increased pressure placed on all animal feeds. And everyone has no doubt heard of the "tortilla riots" in Mexico when tens of thousands of Mexicans swarmed the streets of Mexico City in a protest against the rising price of tortillas. These flat corn breads are the main source of calories for many of Mexico's poor, who saw prices rise by over 400% in early 2007 partially as result of US-based ethanol production.

Is it acceptable to cause financial hardship or starve people in distant lands in order to allow the middle and upper classes to drive around in their SUVs with a supposedly clear conscience?

And what happens when the subsidies stops flowing and people come to their senses? Corn ethanol will cease to exist as a transportation fuel and farmers and related suppliers will go bankrupt by the thousands.

Add to this the dubious greenhouse gas emission reductions and negligible forward energy-balance issues and food-grade ethanol will not be a viable long-term solution to our energy woes.

Cellulosic Ethanol – The Better Choice

Cellulose-based ethanol eliminates the diversion of food crop to fuel conversion and offers an advanced new transportation fuel feedstock which has some advantages over grain-based ethanol:

- Unlike grain-based ethanol production, the manufacturing process does not consume fossil fuels for distillation and thus reduces greenhouse gas emissions.
- Cellulose ethanol is derived from nonfood renewable sources such as straw and corn stover. Fast-growing perennial plants such as switchgrass can be grown with little assistance from petroleum-based pesticides and fertilizers and can be cultivated on marginal lands.
- There is a potential for large-scale production since it is made from agricultural residues which are produced in large quantities and would otherwise be destroyed by burning.

Cellulose-based ethanol has, in the past, been very expensive as a result of the inefficient processes required to produce it. The Iogen Corporation, working in conjunction with the Government of Canada and Shell Corporation, has recently launched its most cost-effective method for producing what the company refers to as EcoEthanol™. Once the first phase of production, using straw and corn stover, is under way it won't be long before other feedstocks can be used, including hay, fast-growing switchgrass, and wood-processing byproducts.

The Downside of Ethanol

There are very few downsides to ethanol as a fuel. Because ethanol contains oxygen it permits cleaner and more complete combustion. Ethanol contains approximately 47% less energy than gasoline on a volume basis. At high-concentration levels, it takes more ethanol than gasoline to drive a given distance; however, this will be much less of a problem when driving PHEVs that use very little liquid fuel in normal driving. In any event, the current higher price per unit volume of ethanol compared to gasoline will eventually be offset by carbon valuation policies that will favor sustainably produced renewable fuels.

Biodiesel as a Source of Green Fuel

Internal combustion engines ignite fuel using one of two methods. A spark ignition engine produces power through the combustion of the

Figure 9-4. Iogen Corporation founder Brian Foody Sr. stands in a field of straw grass, which may become the twenty-first-century equivalent of a Saudi Arabian oil field. (Courtesy: Iogen Corporation)

Fuel Source	Volumetric Energy Density
Ethanol:	89 MJ/Gallon
	23.4 MJ/Litre
Gasoline:	131 MJ/Gallon
	34.6 MJ/Litre

Table 9-1. This table illustrates the volumetric energy density of ethanol compared to gasoline. A typical gasoline-powered car will travel approximately one-third further than a vehicle using the same volume of ethanol. The current higher price per unit volume of ethanol compared to gasoline will eventually be offset by carbon valuation policies that will favor sustainably produced renewable fuels.

Figure 9-5. Loading straw into an ethanol plant is a lot cleaner and much more sustainable than extracting oil from the North Sea. Waste grasses, straw, and wood products are "carbon-neutral" fuels that are available domestically and in endless supply. Think of these sources as "stored sunshine." (Courtesy: Iogen Corporation)

gasoline and air mixture contained within the cylinders. An electric spark which jumps across the gap of an electrode ignites this volatile mixture.

The compression ignition or diesel engine, as it is more commonly known, uses heat developed during the compression cycle. (Have you ever noticed how hot a bicycle air pump becomes after a few strokes?) With a compression ratio of 18:1 or higher, sufficient heat is developed to cause diesel fuel sprayed into the cylinder to self-ignite. The higher energy content of diesel fuel and the oxygen-rich combustion process of the compression ignition engine contribute to improved fuel economy, power, and reduced CO_2 emissions, all desirable features for fleet and other high-mileage vehicles.

Biodiesel, like its cousin ethanol, is a domestic, relatively clean-burning renewable fuel source for diesel engines and oil-based heating appliances. It is derived from virgin or recycled vegetable oils and ani-

Figure 9-6. Biodiesel fuel, a renewable, clean-burning fuel source for diesel engines and oil-based heating appliances, is produced from virgin vegetable oils as well as waste vegetable or animal fats. Dr. Martin Reaney, Chair of Lipid Utilization and Quality at the University of Saskatchewan, studies the canola oilseed plant used extensively in the production of biodiesel. He is also researching ways of using so-called "industrial mustards" that may produce high oil yields and grow on marginal soils.

mal-fat residues from rendering and fish-processing facilities. Biodiesel is produced by chemically reacting these vegetable oils or animal fats with alcohol and a catalyst to produce compounds known as fatty acid methyl esters (FAME or biodiesel) and the coproduct of the chemical production process, glycerine.

Biodiesel that is destined for use as transportation fuel must meet the requirements of the American Society of Testing and Materials (ASTM) Standard D6751 for pure or "neat" fuel graded B100.[7] Fossil-fuel diesel (or petrodiesel) must meet its own similar requirements within the ASTM standard. Biodiesel and petrodiesel can be blended at any desired rate, with a blend of 5% biodiesel and 95% petrodiesel denoted as B5, for example. The testing and certification of any transportation fuel is a requirement of automotive and engine manufacturers and is implemented to minimize the risk of damage and related warranty costs.

From an agricultural viewpoint, biodiesel offers many of the same advantages and disadvantages to the farming community as ethanol feedstocks. In addition, the process of making biodiesel is *relatively* simple and low cost, which may lead to cooperative and rural ownership of processing and production facilities, further increasing farming income and risk diversity. Even the coproduct of the biodiesel manufacturing process, glycerine, has a market in food, cosmetic, and other industries.

A primary advantage of the biodiesel production process is that two streams of product are generated from the oilseed input. After crushing, the protein meal is sold for livestock feed as well as all manner of products from tofu to baking additives. The oil stream can likewise be used in the food industry or provide the input feedstock for biodiesel.

From an energy and therefore carbon balance standpoint, biodiesel produced from soybeans is superior to corn-based ethanol, yielding 3.2 units of energy output for every unit of energy used in the growing and production cycle.[8]

In Canada, it is estimated that some 0.5 million tonnes of canola seed become distressed as a result of weather or storage problems. Although the oil that is extracted from these seeds is not suitable for a food product, there is no reason why it could not be used as a fuel input.

Likewise, the restaurant industry disposes of or recycles millions of tonnes of used cooking oils known as yellow grease, which can also be diverted into the biodiesel production process as is beginning to happen in North America.

Figure 9-7. Every year approximately 5% or more of the total Canadian canola crop is declared "off-specification" and in effect wasted. Generating a special commodity price for "fuel-grade" canola (or any oilseed) would put more money into farmers' pockets and help alleviate the losses incurred in a horticultural roll of the dice with nature.

At the back end of the restaurant industry are millions of grease traps that capture cooking oils and food products contained in the wash water. Rather than allowing them to travel into municipal sewage facilities, these grease traps are evacuated by recovery companies and the grease is disposed of. The extracted effluent is known as brown grease and has a very high energy content. Because of nonexistent carbon regulations, this material has traditionally been landfilled, allowing the material to decay and release immense quantities of carbon dioxide and methane into the atmosphere. Current environmental regulations are tightening up and are beginning to prevent the material from being landfilled. Studies have shown that this material can be converted into biodiesel, although it would not be competitively priced with gasoline. Allow carbon taxation to step in and the inequity in pricing would be solved in an instant.

Biodiesel in the Transportation Sector

The modern diesel engine is a far cry from the smoky, anaemic model of the 1970s. Fast forward to today. Gone are the smelly, smoky, lumbering diesels of old. Witness the new Mercedes E320 family of "common rail,

Figure 9-8. As the price of petrodiesel continues to rise and biodiesel production costs fall, rural communities will produce their own democratic energy to fuel their part of the economy. (Courtesy: Lyle Estill/Piedmont Biofuels)

turbo-diesel" engines that offer no "dieseling" noise, smoke, or vibration, achieve superb mileage, and have better acceleration than the same model car equipped with a gasoline engine.[9]

Biodiesel offers some distinct advantages as an automotive fuel:[10]

- It can be substituted (according to vehicle manufacturer blending limits) for diesel fuel in all modern automobiles. Although B100 may cause failure of fuel system components such as hoses, o-rings, and gaskets that are made with natural rubber, most manufacturers stopped using natural rubber in favour of synthetic materials in the early 1990s. According to the US Department of Energy, B20 blends minimize all of these problems.
- Performance is not compromised using biodiesel. According to a 3 ½-year test conducted by the US Department of Energy in 1998, using low blends of canola-based biodiesel provides a small increase in fuel economy. Numerous lab and field trials have shown that biodiesel offers the same horsepower, torque, and haulage rates as petrodiesel.
- Lubricity (the capacity to reduce engine wear from friction) is considerably higher with biodiesel. Even at very low concentration lev-

Figure 9-9. The simplicity of a biodiesel production facility is shown in this aerial view of a continuous-production plant. Soybeans are delivered to the processing plant for conversion to food-grade oil. Alcohol and a catalyst are added to the oil to produce biodiesel and glycerine. The biodiesel is stored in the tank farm. The glycerine byproduct is sold and the process water and alcohol are recycled, creating an environmentally friendly processing cycle, with the biodiesel having a much higher energy balance than corn-based ethanol. (Courtesy: West Central)

els, lubricity is markedly improved. Reductions in the sulfur levels of petrodiesel to meet new, stringent emissions regulations have, at the same time, reduced lubricity levels in petrodiesel dramatically. Biodiesel blending is currently being considered by the petrodiesel industry as a means of circumventing this problem.

- Because of biodiesel's higher cetane (ignition) rating, engine noise and ignition knocking (the broken motor sound when older diesels are idling) are reduced.
- Carbon lifecycle emissions can be dramatically reduced or even eliminated based on careful selection of oil feedstocks and process chemicals.

For the above reasons, biodiesel was selected as the liquid fuel to power the Zero-Carbon Car. Later discussion will show how this fuel can be made to have a zero-carbon emission profile

Biodiesel Composition

The concept of using plant matter to operate internal combustion engines is older than the gasoline and diesel fuels that are so ubiquitous in our lives today. Rudolf Diesel developed the *compression ignition* engine and demonstrated it at the Paris World Exhibition in 1900. His fuel of choice for powering the new engine: peanut oil.

Although the concept of using plant matter to operate internal combustion engines has been revisited numerous times since Diesel's early experiments, the discovery of cheap fossil oils delayed any significant development of biofuels.

The development of the diesel engine and fuel system progressed very quickly after its first demonstrations to the public owing to its increased efficiency compared with that of the steam engine, its relative portability (paving the way for personal transportion, farming, and industrial uses), and access to cheap and convenient diesel fuel oil. Engine development continued for the next 80 years using low viscosity petrodiesel fuel, while the much higher viscosity plant oils were left behind on the grocers' shelves for baking, salad dressing, and french fries.

Figure 9-10. Over 200 public and private fleets in the United States and Canada currently use biodiesel, and the number is increasing rapidly. Environmental stewardship regarding climate change, urban smog, and air quality is creating mass-market acceptance of biodiesel fuel. (Courtesy: Saskatchewan Canola Development Commission)

Figure 9-11. The discovery of cheap fossil oils delayed any significant development of biofuels. Given the massive financial subsidies and military support lavished on the fossil fuel industry as well as the environmental damage it has caused, perhaps biofuels should have been with us from the beginning.

With the first worldwide oil "shortages" in the 1970s, researchers began working in earnest in an attempt to develop the biofuel market. The many shortcomings related to the direct use of plant oils and their **total** incompatibility with petrodiesel fuel and existing distribution infrastructure[11] pushed researchers in the direction of chemically modified forms of plant oils and animal fats known as biodiesel.

Biodiesel is a renewable, relatively clean-burning, carbon-neutral fuel that can be obtained from a variety of oilseed plants, waste oils, and rendered animal fats. These unprocessed materials (collectively referred to as feedstock "oils") can be converted into a petrodiesel-compatible fuel using a process known as chemical transesterification.

The properties of rendered animal fats and plant oil vary widely from those of petroleum diesel fuel, primarily in the areas of *viscosity*, *atomization*, and the *coking* of engine components. All plant and animal oils have essentially the same chemical structure, consisting of *triglycerides*, which are chemical compounds formed from one molecule of glycerol and three fatty acids.

Glycerol (common name glycerin) is an alcohol that can combine with up to three fatty acids to form mono-, di-, and triglycerides.

Figure 9-12. Biodiesel is a renewable, relatively clean-burning, carbon-neutral fuel that can be obtained from a variety of oilseed plants, waste oils, and rendered animal fats.

Fatty acids are chains of *hydrocarbons* that vary in carbon length depending on the oil feedstock. If each carbon atom has two associated hydrogen atoms, the fatty acid is known to be *saturated*. If two carbon atoms are double bonded, having less hydrogen, the fatty acid is *unsaturated*. Likewise if more than two carbon atoms are unsaturated, the fatty acid is said to be *polyunsaturated*.

Triglycerides are the main compounds or components of animal fat and vegetable oils. They have a lower density than water and will therefore float on it. If the oil is solid at room temperature the triglycerides are known as "fats"; if they are liquid they are called "oils." As a general rule, triglycerides that are liquid at room temperature are unsaturated, which is a desirable property for engine fuels.

Plant oils have viscosities that can be as much as 20 times higher than that of fossil diesel fuel, while chicken fat, yellow grease, lard, and tallow remain stubbornly solid and unusable in their unaltered state. The problem of the high and variable viscosity of feedstock oils can be corrected by adapting the engine to the fuel or vice versa. This chapter focuses on the latter concept: adapting the fuel to the millions of engines that are now operating or will be produced for many years to come, in-

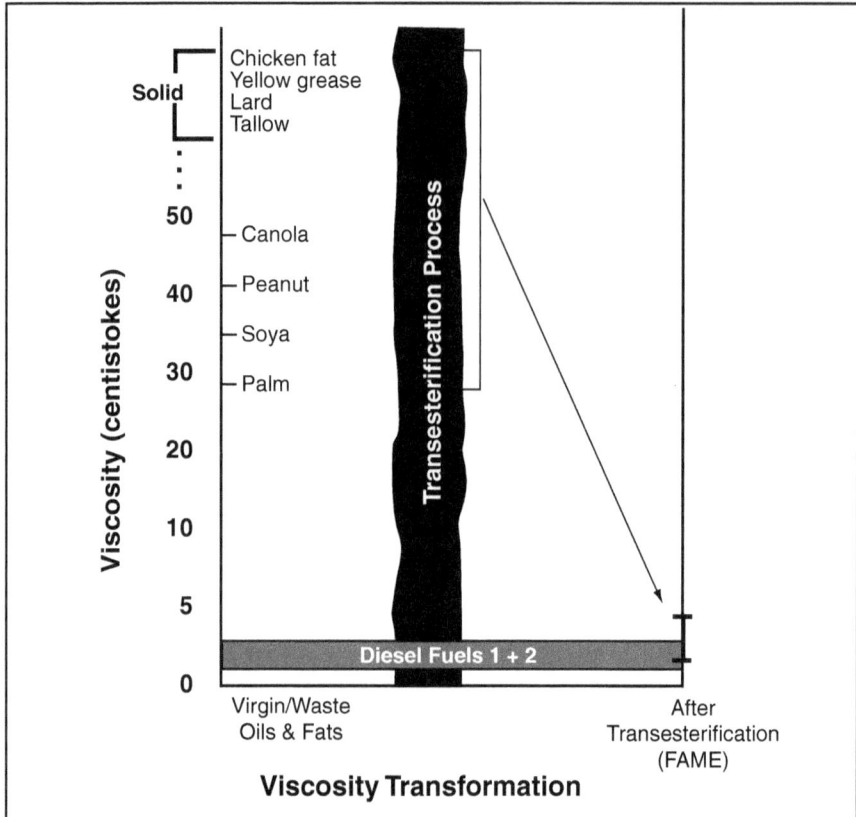

Figure 9-13. Plant oils have viscosities that can be as much as 20 times higher than that of fossil diesel fuel, while chicken fat, yellow grease, lard, and tallow remain stubbornly solid and unusable in their unaltered state. The problem of the high and variable viscosity of feedstock oils can be corrected by adapting the engine to the fuel or vice versa.

cluding the hybrid power plant used in the Zero-Carbon Car.

Chemical transesterification of feedstock oils is a well-known process which solves the problem of feedstock viscosity as demonstrated in Figure 9-13. The process was first described in 1852[12] when it was originally used as a means of producing high-quality soaps, and with a bit of retooling it was found to work wonders in the production of biodiesel. Simply stated, biodiesel is produced by the reaction of feedstock oils with an alcohol in the presence of a catalyst to produce fatty acid methyl esters (FAME) or biodiesel. The typical process is:

100 kg feedstock oil + 10 kg methanol ≈ 100 kg FAME + 10 kg glycerol

The resulting FAME is known to be chemically contaminated with numerous compounds resulting from the esterification process, requiring further downstream processing to ensure a fuel quality compatible with ASTM Standard D6751 for biodiesel, a fuel comprised of the mono-alkyl esters of fatty acids derived from vegetable oils or animal fats. Hereinafter, FAME that is directly taken from the reaction process will be referred to as "*raw FAME*" and when it meets the fuel quality standard it will be referred to as *FAME*.

The first reference to FAME production was in 1937,[13] and within the next year a bus fuelled with palm-oil-based biodiesel ran between Brussels and Louvain.[14] However, at that point further scientific research and production ground to a halt.

Some forty years later, Professor Martin Mittelbach, Ph.D., University of Graz, Austria, and his team of researchers were producing rapeseed-oil-based biodiesel and testing its feasibility as a diesel fuel substitute.

Figure 9-14. Professor Martin Mittelbach, Ph.D., University of Graz, Austria, and his team of researchers have long been acknowledged as the founders of the modern biodiesel industry. Dr. Mittelbach is the author of numerous technical papers on the subject and the author of Biodiesel: The Comprehensive Handbook.

When prodded to discuss the origins of the modern biodiesel industry, Dr. Mittelbach modestly admits that he has been "involved since the beginning. We (University of Graz) were the first to produce biodiesel in Europe, more than 20 years ago, although I am not exactly sure what moved me into this program!...Because of my research on carbon-based compounds, a discussion ensued with an agricultural group in Austria that had some experience using straight vegetable oils mixed with 50% fossil fuels in tractors. The farmers found that after a period of time running on this mixture, total engine breakdowns would occur and they had to stop this practice. We looked into the problems, examined the prior research, and I guess you could say the rest is history."[15]

Within the next decade hundreds of research programs sprang up around the world as interest in clean and renewable fuels started to take hold.

In the United States, the demand for soybean meal (the residual husk of the bean after crushing) was greater than the demand for oil, causing an imbalance in supply and depressed soy oil prices and galvanizing the United Soybean Board into action. It voted to promote biodiesel production using soy oil, which ultimately led to the current National Biodiesel Board (NBB), headed by its executive director Joe Jobe.

The Canadian Renewable Fuels Association (CRFA), which merged with the Biodiesel Association of Canada, develops market strategy and educational data for both the ethanol and the biodiesel industries.

As a result of lobbying efforts and continued research, biodiesel has received the support of numerous federal, state, provincial, and local governments which see it as a means of reducing greenhouse gas and smog-forming emissions, supporting local agribusiness, and helping to reduce North American dependency on foreign oil.

The Pros of Biodiesel

Blending
One primary advantage of biodiesel is its ability to fit almost seamlessly into the existing fuel distribution and retail sales system while other alternative "fuels" such as hydrogen require the complete rebuilding of distribution technology at a cost of trillions of dollars.

Biodiesel can be used in all modern diesel engines and oil-fired heating systems with minor (if any) modifications, with the following note

Figure 9-15. A primary advantage of biodiesel is its ability to fit almost seamlessly into the existing fuel distribution and retail sales system, while other alternative "fuels" such as hydrogen require the complete rebuilding of distribution technology at a cost of trillions of dollars.

of caution: FAME *may* cause long-term degradation of natural rubber hoses and gaskets and some paints, and replacing natural rubber hoses, "O" rings, and gaskets with polymeric (synthetic) versions such as Viton® may be necessary. However, experience has shown that blends of 20% biodiesel or less seldom cause any problems at all, and in any event late-model vehicles seldom use natural rubber components.

Biodiesel fuels are in commercial use in many European countries including Austria, the Czech Republic, Germany, France, Italy, and Spain. Biodiesel can be used in either its pure or neat form or as blends mixed with fossil diesel fuels. Neat biodiesel is designated B100, while blends are marked "BXX" where "XX" represents the percentage of biodiesel in the fuel mixture. Further, because biodiesel is stable at any concentration, users are free to choose the blending level they prefer based on availability, desired operating temperature, or price.[16]

Biodiesel Concentration

Germany, Austria, and Sweden market neat biodiesel, although blends of 5%-20% are the preferred concentration. The "European Directive for the Promotion of the Use of Biofuels," published in 2003, mandates that all member states ensure minimum market shares of biofuels (ethanol, ethanol derivatives, and biodiesel). Market share of biofuels is to reach 5.75% by 2010.[17]

For several reasons, such as the maximum available production of biodiesel, fuel quality and stability, and the political realities of displacing fossil diesel from the market, North American industry proponents consider a blend level of B20 to be the upper limit concentration. There is also a popular myth that 100% concentrations of biodiesel will damage engines, leading to expensive repairs. This is simply nonsense and is nothing more than posturing by engine and automobile manufacturers for two very obvious reasons:

1. Approving a new fuel type will not increase vehicle sales but will add complexity to servicing and warranty issues. From the auto industry's point of view, why take on additional risk if there is no downstream financial benefit?

Figure 9-16. Biodiesel is readily biodegradable and nontoxic, making it the ideal fuel choice when used in environmentally sensitive areas such as parklands, marine habitats, or ski resorts. It is known to be less toxic than table salt and is as biodegradable as sugar.

2. Approving the use of 100% biodiesel will give legitimacy to the multitude of biodiesel home-brewers, who might use substandard fuel in the vehicle.

Biodegradability and Nontoxicity

Biodiesel is readily biodegradable and nontoxic, making it the ideal fuel choice when used in environmentally sensitive areas such as parklands or marine habitats. It is known to be less toxic than table salt and is as biodegradable as sugar.[18]

High Cetane Value

The cetane value is a rating of the relative ignition quality of diesel and biodiesel fuels, with higher ratings offering improved ignition performance. As the cetane value increases, fuel ignition will be smoother and more complete, improving combustion and reducing emissions from unburned fuel. Virtually all biodiesel fuels have cetane values several percentage points higher than that of petroleum diesel fuel.

High Lubricity

Biodiesel, which contains no sulfur, has excellent lubricating properties, far in excess of those of petrodiesel, which help to reduce fuel system and engine wear. As petroleum diesel fuel sulfur levels continue to be legislated downwards, its lubricity will decline to the point where additives will be necessary. The addition of 1% biodiesel to low-sulfur petrodiesel will improve the fuel blend lubricity to within specification.

Low Emissions

As a renewable fuel source, biodiesel operates on a closed-carbon cycle, which reduces CO_2 production by 2.2 kg for every liter of fossil fuel displaced.[19] This is because of the regenerative (biological) nature of all energy sources that absorb CO_2 from the atmosphere during their growing phase, only to release the same compound during fuel combustion. Additionally, the FAME molecule contains 11% oxygen, which leads to improved combustion and significant reductions in particulate matter (PM) or soot.

As part of the Clean Air Act Amendments enacted by the US Congress, the Environmental Protection Agency (EPA) was directed to ensure that any new commercially available motor vehicle fuel or fuel ad-

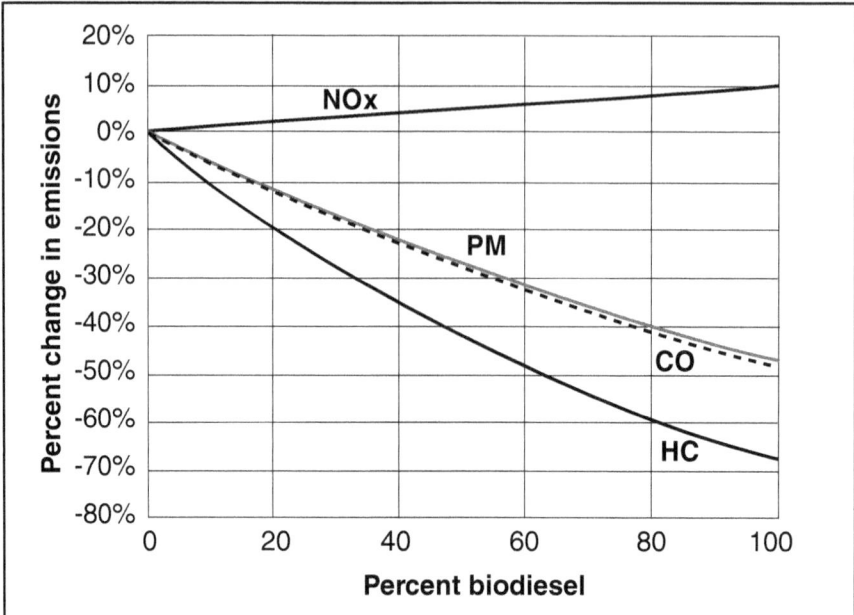

Figure 9-17 Average emission impacts of soybean-based biodiesel for heavy-duty highway engines. NOx = Nitrogen Oxides; PM = Particulate Matter; CO = Carbon Monoxide; HC = Hydrocarbons. (Source: U.S. EPA Report EPA420-P-02-001, October, 2002)

ditive would not present an increased health risk to the public. Under this directive, the EPA established a registration program and testing protocols which are outlined in CFR Title 40 Part 79 as part of Tier I and Tier II emissions testing.

The EPA completed a major study of the impact of various concentrations of soybean-based biodiesel in the operation of heavy-duty highway-based vehicles. The results of the study are shown in Figure 9-17 and clearly demonstrate the superior emissions reductions of FAME fuels.

Nitrogen oxides (NOx) do increase as a result of high engine combustion temperatures and a variety of new technologies have been developed to reduce this air toxin.

Renewability

The US Department of Energy (DOE) and the US Department of Agriculture (USDA) in 1998 completed a thorough study of the energy balance of biodiesel and found that for every unit of fossil energy used in the entire biodiesel production cycle, 3.2 units of energy were delivered when the fuel was consumed.[20]

Readers who are interested in learning more about the life cycle energy requirements for the production of soybean-based biodiesel are encouraged to read the entire 286 pages of analysis completed by the US Departments of Energy and Agriculture.[21]

Low Sulfur

In order for petroleum diesel fuel to be given a rating of "low" or "ultra-low" sulfur content, it is necessary to subject the fuel to an energy-intensive refining process that generates additional carbon dioxide emissions. The resulting fuel will have reduced lubricity levels that must be supplemented with lubricating additives.

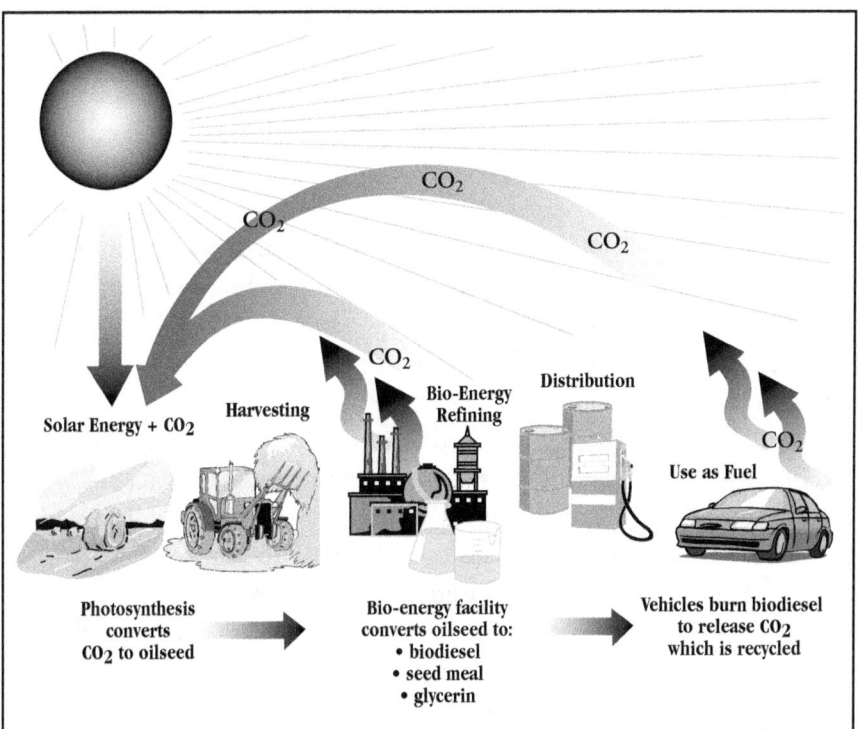

Figure 9-18. Biodiesel is able to provide a positive net energy balance because growing plants absorb massive amounts of energy from the sun. Driving this energy absorption is the process of photosynthesis, which recycles atmospheric carbon dioxide, making biodiesel a carbon-neutral fuel source.

Biodiesel by contrast retains its excellent lubricity while being intrinsically free of sulfur. Having a virtually zero-sulfur level allows the optimum use of oxidation catalytic converters in the exhaust system.

The Cons of Biodiesel

So far, biodiesel sounds like the perfect fuel. Unfortunately, there are a number of issues which must be considered on the negative side of the balance sheet.

Oxidation and Bacteriological Stability

Biodiesel is biodegradable, which is an excellent environmental benefit, but it creates long-term storage and fuel stability issues. When it is exposed to high temperatures, oxygen, or sunlight, or placed in contact with nonferrous metals, FAME will deteriorate, resulting in polymerisation (fuel thickening) which leads to plugged filters and "glazing" within the fuel injection system. To prevent storage degradation, antioxidant additives may be added to extend storage life. Of course, the simplest corrective action is to simply limit the storage time of FAME fuels through rapid consumption and rotation.

Although one study says there is no problem with bacteria,[22] nevertheless there is evidence that there is a problem that water content can destabilize FAME as well as create an active growing medium for microorganisms. Biodiesel manufacturers produce fuels that have very low water content, but biodiesel is hygroscopic and actually attracts water much more readily than petrodiesel does.

Biodiesel will become saturated at water levels above approximately 1,000 ppm, and if water ingress continues unchecked, water no longer remains bonded and collects at the bottom of the storage tank, leading to the very condition that promotes microbial growth.

Edward English II, Vice President and Technical Director of Fuel Quality Services, comments that "Biodiesel is a very different product from fossil diesel fuel and there needs to be better education surrounding handling and storage of these fuels. B100 may well leave the factory meeting ASTM standards, but each time it is pumped, transferred, stored, blended, and dispensed, it will pick up ever-increasing amounts of water. This stuff is like a sponge, and the only way neat or blended biodiesel will remain stable and microbial free is through proper handling, monitoring, and remedial procedures."[23]

Nitrogen Oxide Emissions

Numerous studies have confirmed that overall emissions from the combustion of biodiesel are low but show slightly elevated levels of nitrogen oxides (NOx). This increase is regarded as a problem related to higher combustion cylinder temperatures and is not inherently a fuel-related issue. Manufacturers believe that improvements in engine sensor and management technology as well as NOx catalytic reduction are just around the corner.

Unfortunately, this issue of NOx emissions is placing a damper on the entire diesel engine industry. Diesel proponents believe that gasoline-based exhaust emission strategies are strangling the potential of the diesel market and that each engine technology should receive its own emission profile, as is done in Europe, in light of the considerable fuel and greenhouse gas emissions savings of diesel engines.

Not every automotive manufacturer is worried. Mercedes-Benz is well known for its innovative, quality automobiles featuring state-of-the-art engineering. The company marked the epitome of its technological prowess by showcasing its leading edge BLUETEC technology, launching the diesel power train of the future. "Vision has therefore

Figure 9-19. The release of Mercedes-Benz BLUETEC technology coincided with the US release of low-sulfur diesel fuel in the autumn of 2006. Restricting sulfur content to a maximum of 15 ppm permits the use of particulate filters and efficient nitrogen oxide exhaust treatment. But the question remains: will the vehicles using BLUETEC technology be certified for use with biodiesel fuels? (Courtesy Mercedes-Benz Canada).

become reality as the extremely economical Mercedes-Benz CDI models are the cleanest diesel in the world in every category and consume between 20 and 40 percent less fuel than the gasoline counterparts," states a January 2006 corporate press release. Mercedes-Benz certainly recognizes the importance of emissions reduction technology, as over 50% of their total production volume is now captured by diesel engines.

Cold Flow Issues

No. 2 diesel fuel suffers from a thickening condition known as "waxing" or "gelling" when temperatures drop below the cloud point of the fuel. Should this occur within the fuel system of a vehicle, expensive cleaning becomes necessary.

By contrast, biodiesel suffers from a similar but reversible problem. Should low-temperature fuel gelling occur, causing loss of engine power or complete fuel starvation, the problem can be remedied by simply

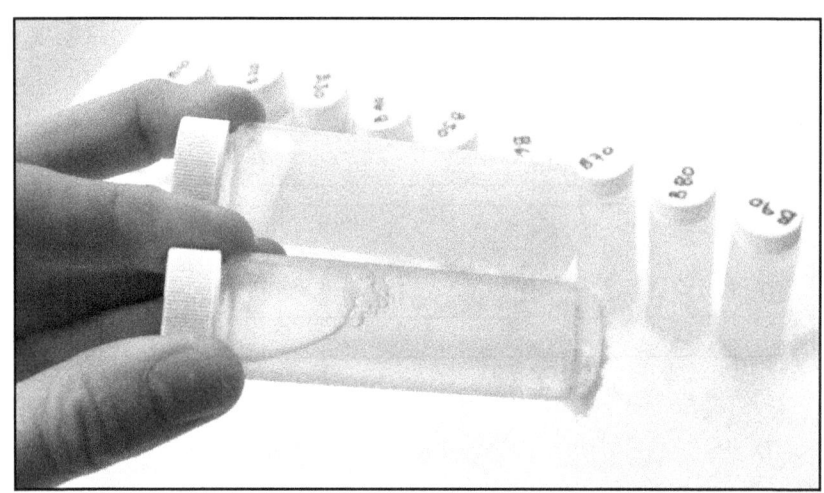

Figure 9-20. Both No. 2 diesel fuel (bottom view) and biodiesel can be "winterized" through the addition of so-called "pour point enhancers," which may be as simple as blending No.1-D into the fuel blend. In this image, one vial of No. 2-D as well as vials of biodiesel in concentrations ranging from B10 through B100 are allowed to sit outdoors. B100 (top vial) is butter-solid at 14°F (-10°C), while No. 2-D as well as B10, 20, 30, and 40 all remain functional, above their respective cloud point ratings. Using this simple arrangement of test vials will provide the biodiesel user with a method of determining maximum winter blending limits.

moving the vehicle to a warm location such as a parking garage until fuel temperatures moderate.

Both No. 2 diesel fuel and biodiesel can be "winterized" by the addition of so called "pour point enhancers," which may be as simple as blending No.1-D into the fuel blend.

No. 2 diesel fuel is treated by fuel refineries to meet the expected minimum temperatures within a given geographical location. To a retail consumer, the transition from "summer" to "winter" diesel fuel is completely transparent.

To counter the problem of cold flow, fuel-additive companies have developed a range of products to improve the cold weather performance of petrodiesel and biodiesel fuels. Primrose Oil Company Inc. offers the Flow-Master® winter diesel fuel treatment product, claiming it is more cost effective than using No. 1-D blended fuels.

The reason fuel gels at cold temperatures is that waxes inherent in the fuel begin to form microscopic crystals. If untreated, these crystals will immediately agglomerate (combine) with one another to form a gel and eventually solidify, blocking fuel lines and filters. Pour point enhancers limit the ability of wax crystals to grow large enough to agglomerate. Primrose indicates that its product will improve the cold flow rating of any untreated fuel by a minimum of 20°F to 30°F (11°C to 17°C).

It is virtually impossible to determine the exact concentration of biodiesel that can be used at a given geographical location without knowledge of the fuel's cloud point temperature. There is considerable variability in biodiesel cloud point temperature resulting from the inconsistency of the feedstock oil saturation level, with long-chain compounds displaying poor cold weather properties. Tallow, lard, palm oil, and yellow greases used to make biodiesel may cause the fuel to remain solid or semisolid just below room temperature, requiring great care in storage and use.

Some researchers have taken a different approach to the problem by attempting to modify the FAME chemical structure through the use of alternate alcohols which have shown improved cold weather performance. Unfortunately, these alcohols have a higher cost and this process is of limited value in the current marketplace.

Repeated cooling and filtration of crystal growth within the FAME has also been attempted with varying rates of effectiveness. However, this process requires considerable amounts of energy and removes valuable esters that are lost during the filtration process, lowering overall biodiesel yields.

As carbon emissions become valued, there will be more impetus for researchers to solve this issue, using the most effective technological means at their disposal.

OEM Warranty Issues

One of the popular misconceptions about biodiesel is that it will not affect engine and fuel system warranties provided the fuel meets applicable specifications.[24] Statements such as these are not only misleading; they are simply wrong.

All major engine, vehicle, and fuel injection equipment manufacturers have clearly stated guidelines regarding the use of biodiesel fuels. Without hesitation, all manufacturers state that biodiesel that is used within the blend limits of their warranty statements must meet the appropriate national and/or international fuel standards.

The Volkswagen policy statement stipulates that "vehicle damage that results from misfueling or from the usage of substandard or unapproved fuels cannot be covered under our vehicle warranties," clearly placing the burden of proof regarding **any** fuel-related damage, fossil- or FAME-based, on the customer.

Figure 9-21. Volkswagen of America has determined that diesel fuel containing up to 5% biodiesel certified to ASTM Standard D6751 meets the technical specifications for Volkswagen vehicles equipped with TDI engines imported into the United States.

The Diesel Engine

There are two types of engine predominantly used to power road-based vehicles. The spark ignition engine operates on gasoline or less frequently on liquid petroleum gas (LPG). The compression ignition engine uses diesel fuel and is named after Rudolf Diesel, who patented his heavy oil engine in 1892.

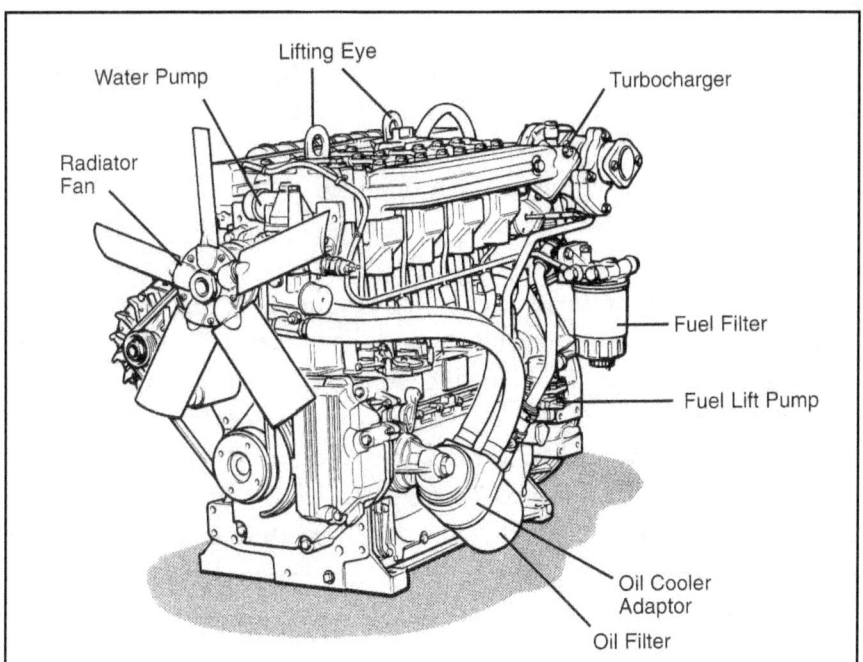

Figure 9-22. A typical light-duty diesel engine has between 4 and 8 cylinders, arranged with a series of intake and exhaust valves which control the air admission, compression, ignition, power, and exhaust phases. Because of their high volumetric efficiency, diesel engines use 30% to 60% less fuel than gasoline engines of similar power. (Courtesy: Lister Petter)

Diesel engines are very popular in Europe, with demand expected to rise from a current level of approximately 30% of total car sales to an anticipated 40% of the market. Surprisingly, the greatest demand for diesel technology comes not from the penny-pinching small-car market but from the luxury car buyers, with 44% of all luxury cars sold in Europe powered by diesel engines. Luxury diesel sales represent a very large percentage of specific markets: Belgium, 87%; France, 82%; Austria, 77%; and Italy, 70%.[25]

Most North Americans have a complete disdain for diesel engines, thinking of them as slow, noisy, polluting, and generally uninspired. While this may have been the case with grandpa's old smoker, advances in technology have placed the diesel *ahead* of the gasoline-powered engine in several key areas:[26]

Fuel Economy: Because of their high volumetric efficiency, diesel engines use 30% to 60% less fuel than gasoline engines of similar power.

Power: Diesels produce more torque and power at lower engine speed than gasoline engines of similar displacement.

Durability: Diesel engines are designed to last well in excess of 300,000 miles (500,000 km) and require less maintenance than gasoline engines.

Greenhouse Gas Emissions: Diesel engines have higher thermal efficiency cycles than gasoline engines and diesel fuel contains more energy per gallon than gasoline, allowing a diesel engine to burn less fuel for a given power output and produce significantly lower CO_2 emissions.

Noise and Smoke: Using the latest "Common Rail Direct Injection" (CDI) and lean-burn technology as well as particulate traps and catalytic converters, today's diesel vehicles have none of the smoke or

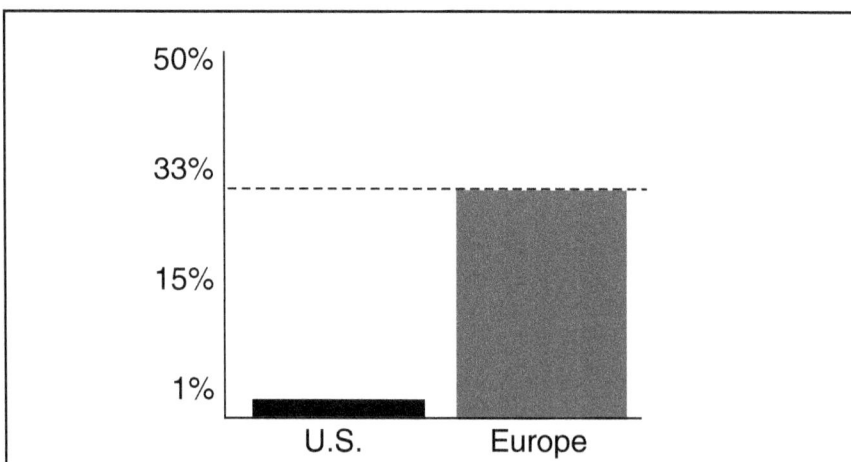

Figure 9-23. Current sales of diesel-powered vehicles have soared in the European Union owing to the numerous advantages diesels have over gasoline engines. In the luxury car market, France has seen 82% of market share go to diesel engines.

Figure 9-24. The prestigious Mercedes-Benz E320 CDI offers a 25% improvement in fuel efficiency compared to its gasoline-powered counterpart, while outperforming it by a full second in 0 to 60 mph acceleration, according to company data. Motor Trend's Frank Markus is even more direct: "Your eyes, ears, and nose will have trouble detecting that this is a diesel, while your backside will detect some serious pressure just at idle." In 2006 the company introduced its BLUETEC powertrain system, which meets all North American emission standards. (Courtesy: Mercedes-Benz Canada)

noise common with older designs.

As a result, diesel-powered vehicles offer a cleaner, quieter, and more powerful alternative to identical automobiles and trucks equipped with less efficient gasoline-powered engines. Taking this one step further, diesel technology is an ideal candidate for advanced PHEV vehicles with their long life, fuel efficiency, and ability to operate on zero-carbon biodiesel.

The light-duty diesel market in North America is practically non-existent, while the Europeans have created huge fuel and GHG savings by developing advanced engine technologies and emissions standards. Concurrently, the United States is fostering the Tier 2 emissions regulations that are a detriment to the development of diesel engine technology here, denying consumers access to the obvious benefits and national fuel consumption reductions that diesel technology would bring.

Figure 9-25. With "lean-burn" technology as well as particulate traps and catalytic converters, there is no need for modern diesel-powered vehicles to have any of the smoke or noise common in older designs. The City of Ottawa, Canada "Ecobus" features these technological advances.

In order to get the benefits of diesel technology in North America, regulators may have to develop a compromise and recognize that different engine technologies have different emission profiles. US emissions standards accommodate gasoline engines which emit relatively little NOx but much more CO, HC, and CO_2 than their diesel counterparts. By placing the emphasis on smog-forming NOx and particulate matter (of greater concern in diesels) instead of CO_2, the US standards discourage the development and marketing of diesel engines.

A more balanced approach to matching technical developments with emissions standards would encourage the development and marketing of diesels in North America.[27]

Engine Technology Overview

All internal combustion engines include a cylinder block which houses the major components of the running gear. An eccentric crankshaft runs along the length of the engine, providing support and a means of transferring the linear motion of the piston to the rotary motion required to turn the vehicle's drive wheels, or in the case of the Zero-Carbon Car, the electrical generating unit.

Each piston is fitted into the cylinder machined in the engine block. A series of piston rings installed on the outer diameter of the piston and placed in contact with the cylinder wall ensures a gas-tight seal.

An intake and exhaust valve (or more commonly multiple intake and exhaust valves per cylinder) are operated from the camshaft, which forms part of the running gear of the engine and is synchronized with the rotation of the crankshaft. As the crankshaft rotates, the camshaft sequentially opens and closes the intake and exhaust valves in accordance with the four-stroke operating theory described below.

In the spark ignition engine, a properly balanced mixture of fuel and air is admitted into the cylinder and compressed, creating a volatile, explosive mixture. An electric spark is activated at the correct timing sequence, igniting the mixture and providing a downward power force on the piston which is then transferred to the crankshaft and ultimately to the drive wheels.

In contrast, the diesel engine eliminates the spark plug and related components and instead uses the heat of compression to perform the ignition function. The four-stroke timing sequence diagrams shown on the following pages illustrate this process.

In the diesel engine, air alone is admitted into the cylinder during the first 180° rotation of the crankshaft, creating the intake cycle shown in Figure 9-26a. The intake valve (left side) is opened by the camshaft (not shown) as the crankshaft rotates in a clockwise direction, forcing the piston down. The downward motion creates a vacuum which draws air into the cylinder. At the lowest point of piston travel in the cylinder, the intake valve closes, creating an airtight chamber.

The next 180° rotation cycle is known as the compression stroke, detailed in Figure 9-26b. The piston is now travelling upwards, compressing the air inside the cylinder. In diesel engines, the compression ratio (ratio of cylinder volume at bottom compared to volume at top) may range as high as 21:1, causing pressures in excess of 500 pounds per square inch (3.4 MPa) and temperatures in excess of 1,000° F (538° C).

When the piston reaches the top of the cylinder at the conclusion of the intake cycle, the power stroke begins, as shown in Figure 9-26c. Diesel fuel is supplied to the fuel injector under very high pressure. Using mechanical or, more commonly, computer/electronic control, the

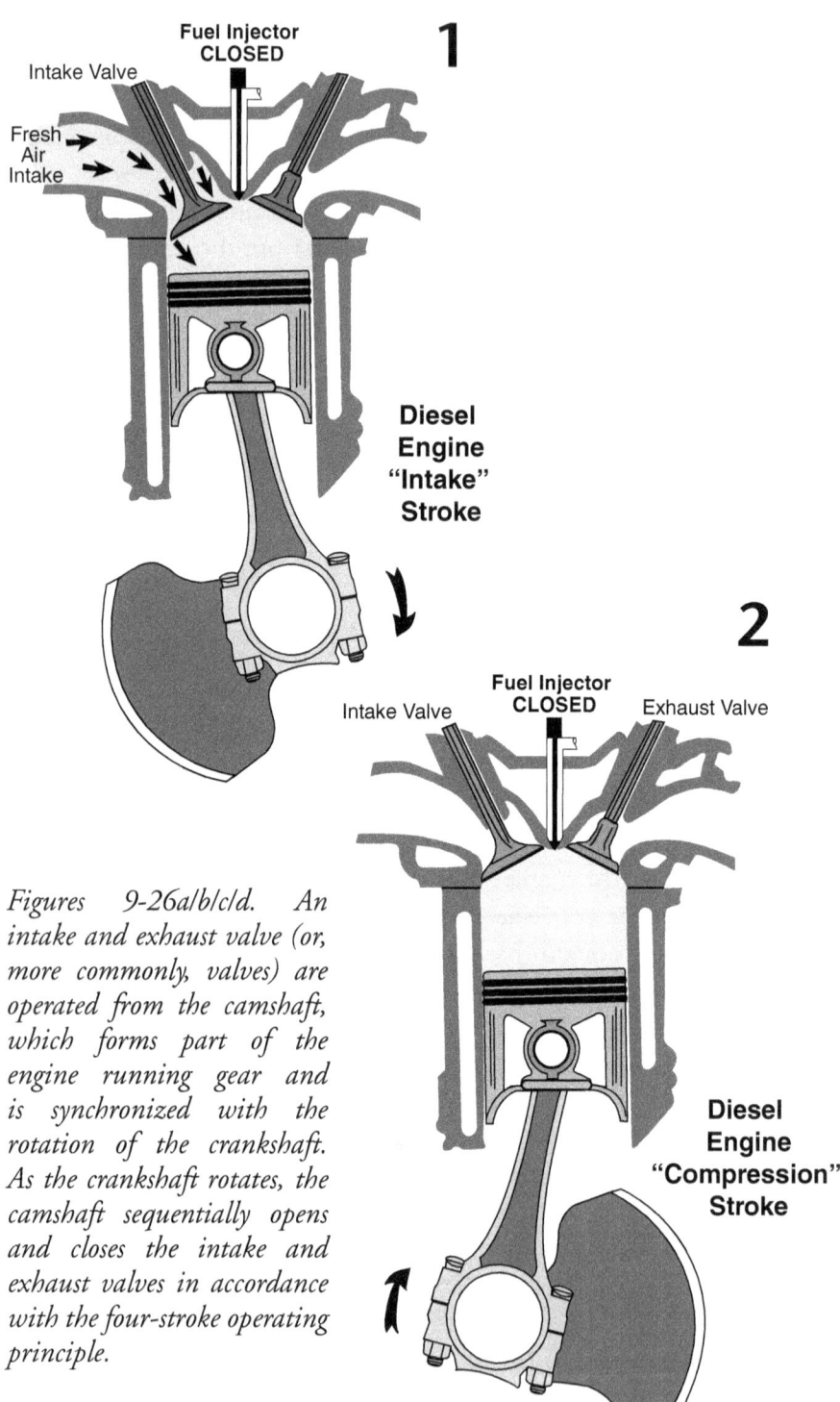

Figures 9-26a/b/c/d. An intake and exhaust valve (or, more commonly, valves) are operated from the camshaft, which forms part of the engine running gear and is synchronized with the rotation of the crankshaft. As the crankshaft rotates, the camshaft sequentially opens and closes the intake and exhaust valves in accordance with the four-stroke operating principle.

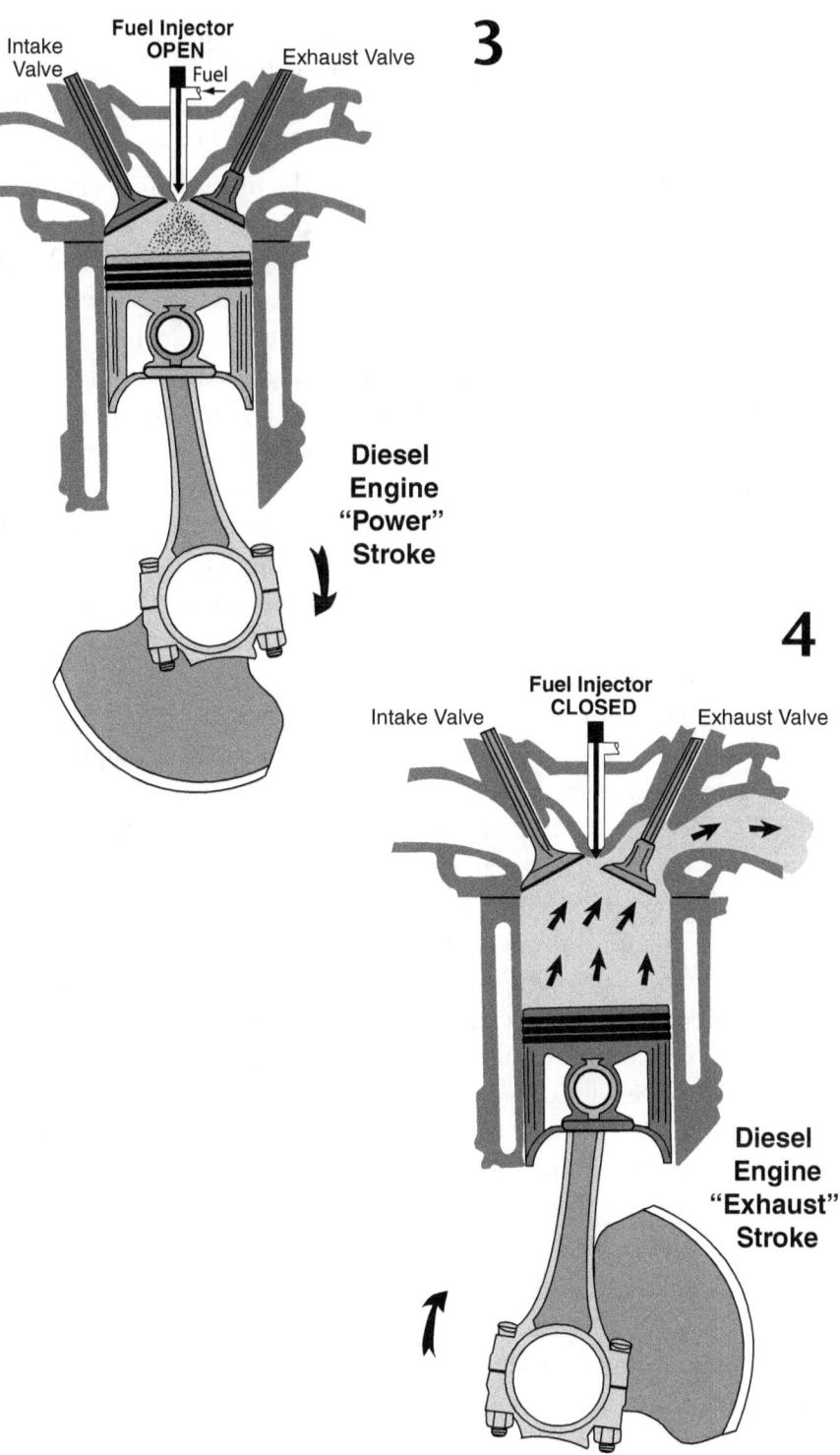

injector is opened and fuel is sprayed in a fine mist (*atomized*) into the cylinder. A fraction of a second later, the fuel will ignite, causing massive expansion of the burning gases and forcing the piston downward, applying power to the piston rod, crankshaft, and vehicle drive train.

The time delay between the opening of the fuel injector (signifying the initial fuel spray into the cylinder) and ignition of the fuel is known as *ignition delay* and is determined by the *cetane* rating of the diesel fuel. The ability of a fuel to combust in the presence of heat is known as the auto-ignition ability of fuel and is a key property of diesel fuel.

The final 180° cycle is the exhaust stroke, when the camshaft opens the exhaust valve as seen in Figure 9-26d. As the piston sweeps upward, combusted fuel (exhaust) components are forced out of the cylinder and sent on their way to the exhaust system, noise reduction muffler, and tailpipe. Advanced engines may also contain particulate matter traps to capture the dusty particles caused by the incomplete combustion of diesel fuel, NOx adsorbers to capture and treat nitrogen oxides in the exhaust stream, and catalytic converters which utilize the high temperatures to catalyze or neutralize exhaust gas components.

It requires two full rotations of the crankshaft to obtain one half-cycle (180°) power stroke from the engine. In order to improve the smoothness and power of the engine, multiple pistons are fitted into the cylinder block and connected to the crankshaft. Each piston is "timed" to produce its power stroke at slightly different times, allowing almost continuous overlap in the power-generation cycle of the engine, in turn reducing mechanical vibration and noise.

Automobiles such as the Mercedes-Benz Smart™ car have only three cylinders and a total engine displacement (net cylinder volume x number of cylinders) of only 52 cubic inches (850 cc). By contrast, a large diesel-powered pickup truck may have an engine with eight cylinders and a displacement of 458 cubic inches (7,500 cc). Of course industrial engines can have much larger displacements than this.

Fuel Injection Systems

At the risk of oversimplifying diesel engine fuel injection technology, I am going to make the generalization that there are two important classes of diesel engine that are relevant to our discussion: engines using "basic" fuel injection and those using "common rail" fuel injection.

Basic Fuel Injection

Filtered fuel enters a high-pressure injection pump which pressurizes the fuel, feeding it into the fuel injector located in the top end of the cylinder. The injection pump is provided with multiple fuel supply lines, one for each cylinder and fuel injector. As diesel fuel is incompressible, an excess fuel return line constantly recirculates fuel when an internal pressure relief valve reaches fuel operating conditions. Pressurizing any liquid will cause its temperature to rise, thereby increasing the temperature of fuel in the tank and distribution system.

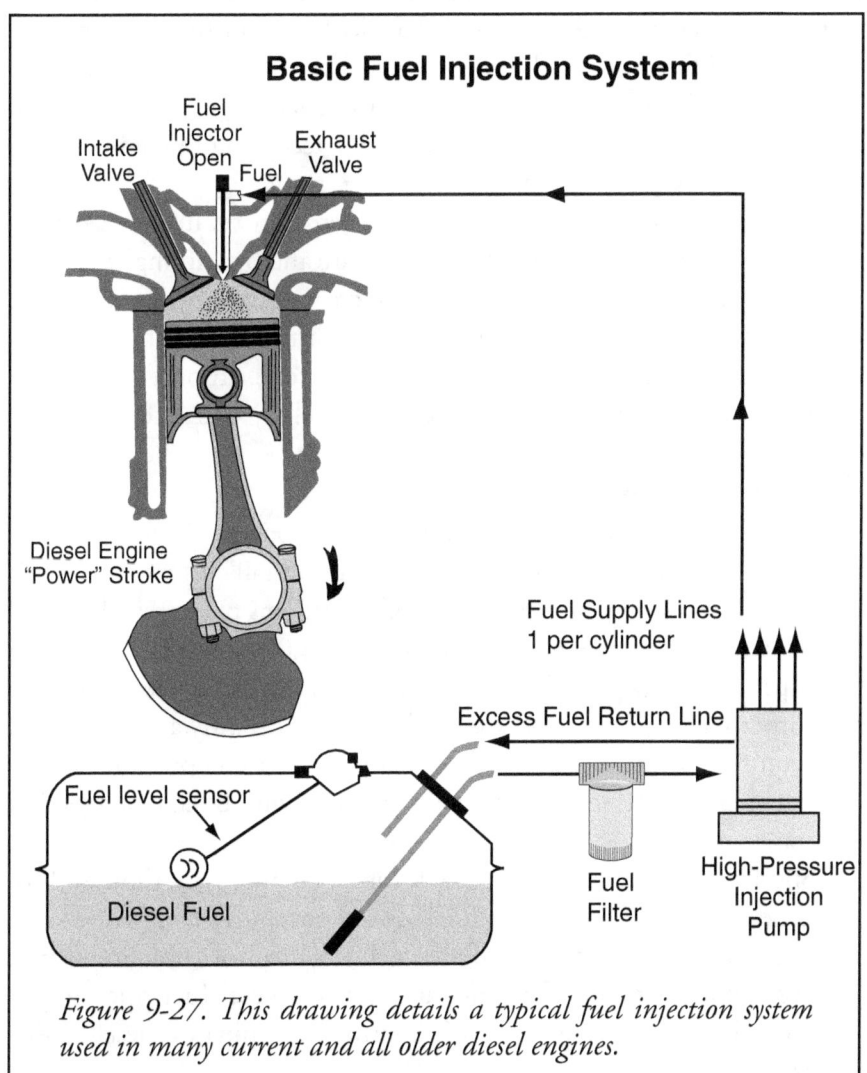

Figure 9-27. This drawing details a typical fuel injection system used in many current and all older diesel engines.

Figure 9-27 shows an overview of a typical fuel injection system used in many current and all older diesel vehicles. Fuel is drawn from the vehicle's storage tank and filtered to remove any debris and water that may be present in the tank. The filter is manufactured from tightly woven cellulose material which is able to stop particles larger than 20 microns (0.0008 inches) in size, preventing abrasive material from damaging the fuel system components.

When the piston reaches the top of the cylinder during the compression stroke, air in the cylinder will be heated above the ignition temperature of the fuel. Depending on the model and age of the engine, fuel will be atomized by either mechanical or electrical control of the injector nozzle. The "fineness" of the resulting fuel spray will determine combustion completeness, engine noise, and efficiency. A disadvantage of this system is that fuel pump pressure varies according to engine speed, resulting in varying fuel spray patterns.

Researchers have found that this problem can be minimized with the use of increased and constant fuel pressure and by "timing" the spray pattern by modulating the fuel injector to create "microbursts" of fuel in the cylinder as the piston is sweeping through the end of the compression stroke and the beginning of the power stroke. In order to accomplish this, much higher pressures and faster fuel injectors are required.

Common Rail Direct Injection

Common Rail Direct Injection (CDI) systems (Figure 9-28) use an ultrahigh pressure injection pump and a "common rail" or pressure manifold to ensure high, even fuel pressure. CDI systems pressurize the diesel fuel to enormous levels, often in excess of 22,000 pounds per square inch (psi) ($\approx$150 Mpa). The common rail manifold is able to act as a pressure storage reservoir and ensure that fuel is instantly available when it is required by the fuel injectors. Note that higher fuel pressures generate higher excess fuel temperatures as compared with basic fuel injection systems.

Special piezoelectric fuel injectors can open and close thousands of times per second, allowing the fuel control computer to provide multiple "bursts" of fuel, which offers precise control of the combustion process. Pilot injection of minute amounts of diesel fuel prior to the main combustion injection initiation virtually eliminates diesel engine "clacking" noises.

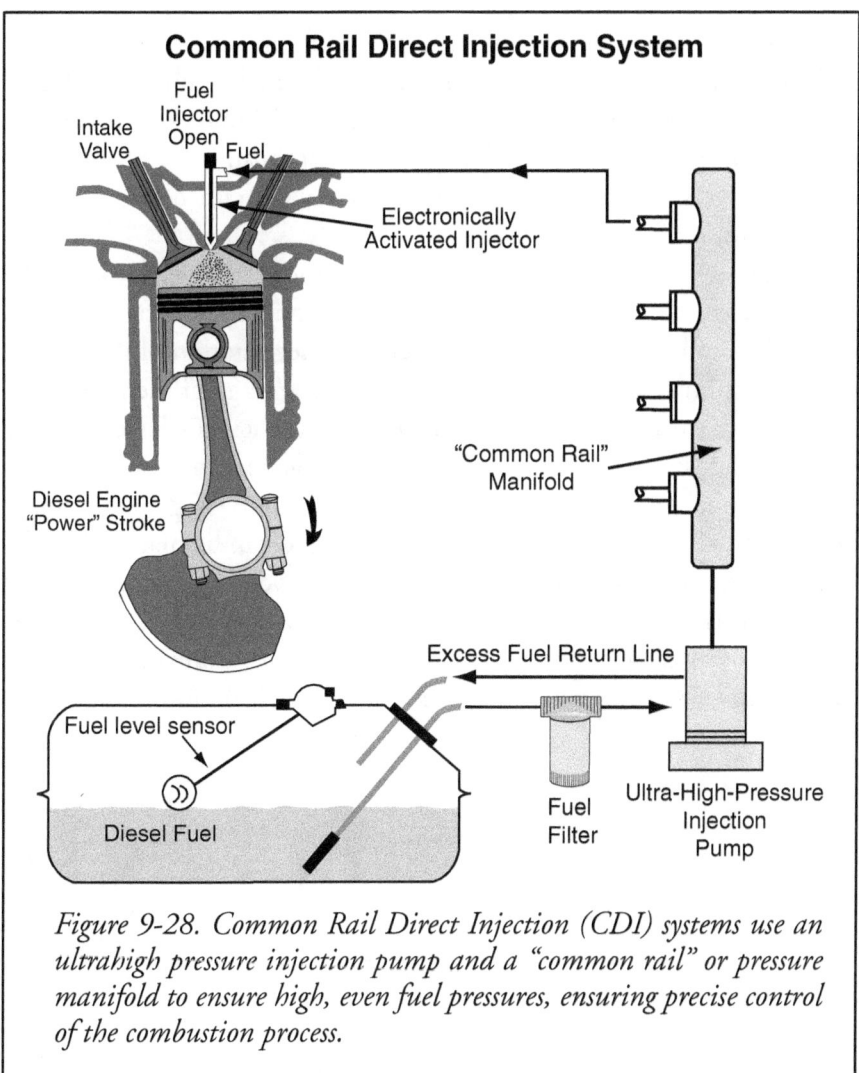

Figure 9-28. Common Rail Direct Injection (CDI) systems use an ultrahigh pressure injection pump and a "common rail" or pressure manifold to ensure high, even fuel pressures, ensuring precise control of the combustion process.

A 4-valve-per-cylinder, 2-intake and 2-exhaust arrangement increases power, fuel efficiency, and responsiveness by allowing the fuel injector to be placed in a central location, creating a symmetrical fuel spray pattern and best fuel/air mixing.

Engine and Vehicle Efficiency

Although the internal combustion engine and fossil-fuel-powered vehicle have been around for a long, long time, they are not the most energy-efficient members of the mechanical engineering team. Gasoline and diesel fuel contain a given amount of energy per unit of volume. No. 2-D, for example, contains approximately 129,000 *BTU* per gallon (40.9 MJ/liter). Due to inefficiency of the combustion-to-mechanical-energy conversion process, most of the energy contained in a quantity of fuel is wasted.

Only 18% of energy input remains to produce power to move the vehicle. Up to 10% of this total is used to overcome the rolling resistance of the tires and air resistance. Movement and acceleration of the vehicle may be attributed to as little as 8% of the total energy of the fuel.

In the Zero-Carbon Car, the diesel engine does not have as many cumulative losses when driving the electrical generating unit. Because of the unique design of the generator and its ability to generate direct rather than alternating current, the engine speed can be matched to the diesel engine, allowing it to operate at its peak efficiency point. The total losses of this system are much lower than they those involved in mechanically driving a vehicle and are estimated to be in the order of 69% for the total system compared with 92% for a typical motor vehicle.

The Biodiesel Production Process

The simplest method used to produce FAME is based on the single reaction tank batch process. An updated version using two reaction tanks is used in many smaller commercial facilities (< 4 million liters per year / ≈ 1 million gallons per year). The advantage of the two-tank system that it ensures completeness of the reaction process and guarantees that *total glycerin* levels are within the ASTM specification of less than 0.24%. An example of such a system is shown in Figure 9-29.

- Vegetable oil or animal fat is loaded into a storage tank and heated.
- An alcohol and catalyst (methoxide) reactant is loaded into a second tank.
- The feedstock oil and reactants are pumped into Batch Tank #1, where they are mixed and heated until the conclusion of the reaction process.
- The materials in Batch Tank #1 separate into two layers or phases: raw FAME and glycerin.
- Glycerin is pumped to a processing section which neutralizes the catalyst, recovers and recycles the methanol, and provides unrefined glycerin (glycerol) for sale.
- The raw FAME is pumped to Batch Tank #2 where it is reprocessed with methoxide reactant to remove trace amounts of bonded glycerin.
- Excess glycerin is pumped off and recovered as in the first reaction step.
- Raw FAME is sent to a washing and catalyst neutralizing tank. Wash waters are processed for disposal.
- The high-quality FAME is then dried and readied for testing and sale.

Vegetable oil or animal fats having a free fatty acid content of less than 2% and preferably less than 1% are loaded into the feedstock tank shown at upper left. A steam-heated exchanger coil heats the oil feedstock to approximately 60° C. It does not make any difference which parent feedstock oil or fat is used to produce the FAME provided that it meets all of the requirements of ASTM D6751 at the completion of the production process. The reader is again reminded that fats and oils that are fully or partially hydrogenated as well as highly saturated will be affected by cold flow issues.

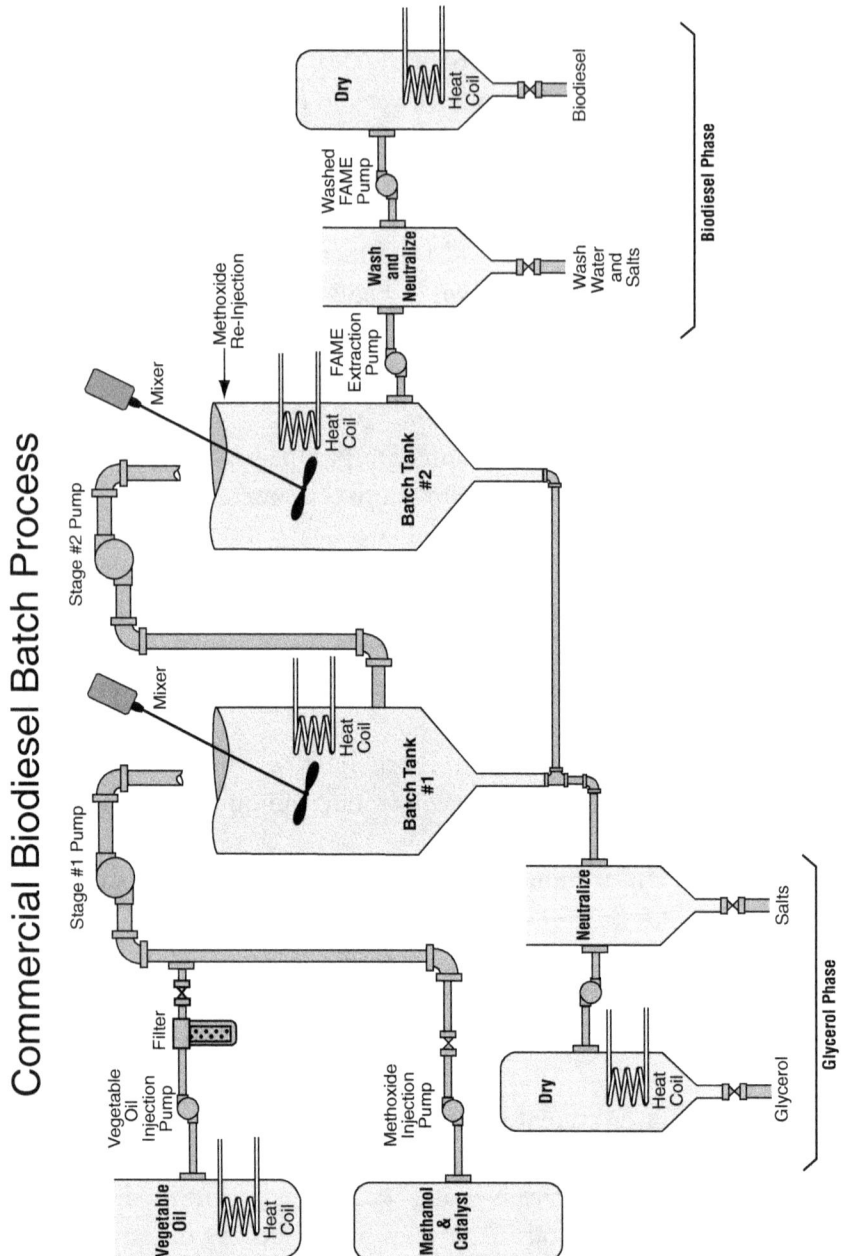

Figure 9-29. The simplest method used to produce FAME is based on the single reaction tank batch process. An updated version using two reaction tanks is used to ensure completeness of the reaction process and to guarantee that total glycerin levels are within the ASTM specification of less than 0.24%.

A second storage tank is filled with a commercial premixed (methoxide) reactant solution comprising methanol (alcohol) and a catalyst of either sodium hydroxide (NaOH) or potassium hydroxide (KOH) which, when blended, is known as sodium or potassium methoxide depending on the chosen catalyst. The sodium or potassium methoxide can also be produced at the factory.

Methanol is the most commonly used alcohol, although other alcohols may be interchanged. Alcohol selection is determined by several factors including cost, toxicity, ease of recycling, and quantities required to complete the reaction process. Although methanol is principally derived from natural gas and is highly toxic in the environment, which lowers the overall "green value" of biodiesel, it meets the litmus test of low cost, ease of use, and ability to be recycled. Cellulosic ethanol is an obvious zero-carbon contender but is currently not used because of the lack of any carbon valuation policy.

Because oil and alcohol do not have an affinity for each other, a catalyst is used to initiate the transesterification reaction. A catalyst may be either alkaline (base or basic) or acidic in nature, with experience showing that acid-catalyzed reactions operate too slowly for cost-effective biodiesel production purposes.

The selection of base catalyst is also a matter of choice, although sodium hydroxide is the most commonly selected compound. The importance of the feedstock selection and free fatty acid level must be considered along with catalyst selection. The combination of base catalysts and FFA levels exceeding approximately 2% will cause the formation of soaps that can hinder the reaction process and create a contaminant in the glycerin phase.

Once the feedstock oil has reached operating temperature, it is pumped and filtered into the batch reactor tank, where it is continuously stirred and heated. The methoxide reactant is then pumped into the reactor tank and vigorous mixing is continued for a period ranging from 30 minutes to one hour, with the close contact of the feedstock oil, alcohol, and catalyst ensuring the completeness of the reaction.

At the end of the mixing period, the mixer and heating coils are deactivated and the solution is allowed to settle, causing it to separate into two phases with the raw glycerin (glycerol) sinking to the bottom of the tank and the raw FAME floating on top as the second phase. Reaction completeness is reportedly in the range of 85% to 94%, which is below the requirements necessary to ensure that bonded glycerin meets the ASTM standard.

The glycerol phase is pumped to a processing unit where methanol is captured and recycled for further use. The glycerol is also washed with acidic water to neutralize the basic catalyst. Wash water is processed and either discarded after meeting required environmental standards or recycled for further use in the plant. The glycerol is then dried and sold to a refining company.

The raw FAME that was produced in the first reaction is sent to a second reaction tank where it is reprocessed with fresh methoxide reactant in a process similar to that of the first reaction. Introducing a second reaction drives the transesterification to 95%+ completeness, ensuring that bonded glycerin levels will now fall within the required standards.

Glycerol is again pumped off and processed as in the first reaction stage.

The raw FAME is subjected to methanol capture before being washed in the same manner as in the glycerol phase. A drying step ensures that the high-quality, washed FAME contains less than the prescribed amount of suspended water before it is sent to storage.

Small-Scale Biodiesel Production System

When Lorraine and I designed our off-grid home we decided to add a small workshop area in the detached garage to house the biodiesel production facility that would be used to fuel the Zero-Carbon Car. This extra space is located behind the door shown in Figure 9-30 and consists of an area of approximately 6 x 18 feet (1.8 x 5.5 meters). The room is well insulated and tightly sealed with standard construction-grade vapor barrier and acoustical gap sealant and is fitted with a ventilation fan that is capable of drawing approximately 500 cubic feet (14 cubic meters) of air from one end of the room to the other per minute of operation. The high-volume air turnover rate is required to keep methanol vapors to an absolute minimum, certainly below the *lower explosive limit* (*LEL*), as well as to ensure that air quality remains high and nonpoisonous.

The production lab is shown in an overview image in Figure 9-33. All of the production equipment is mounted along the left wall, while the "wet laboratory," safety equipment, and storage area are located along

Figure 9-30. When Lorraine and I designed our off-grid home we decided to add a small workshop area in the detached garage to house the biodiesel production facility that would be used to provide fuel for the Zero-Carbon Car as well as our diesel-powered Smart™ car.

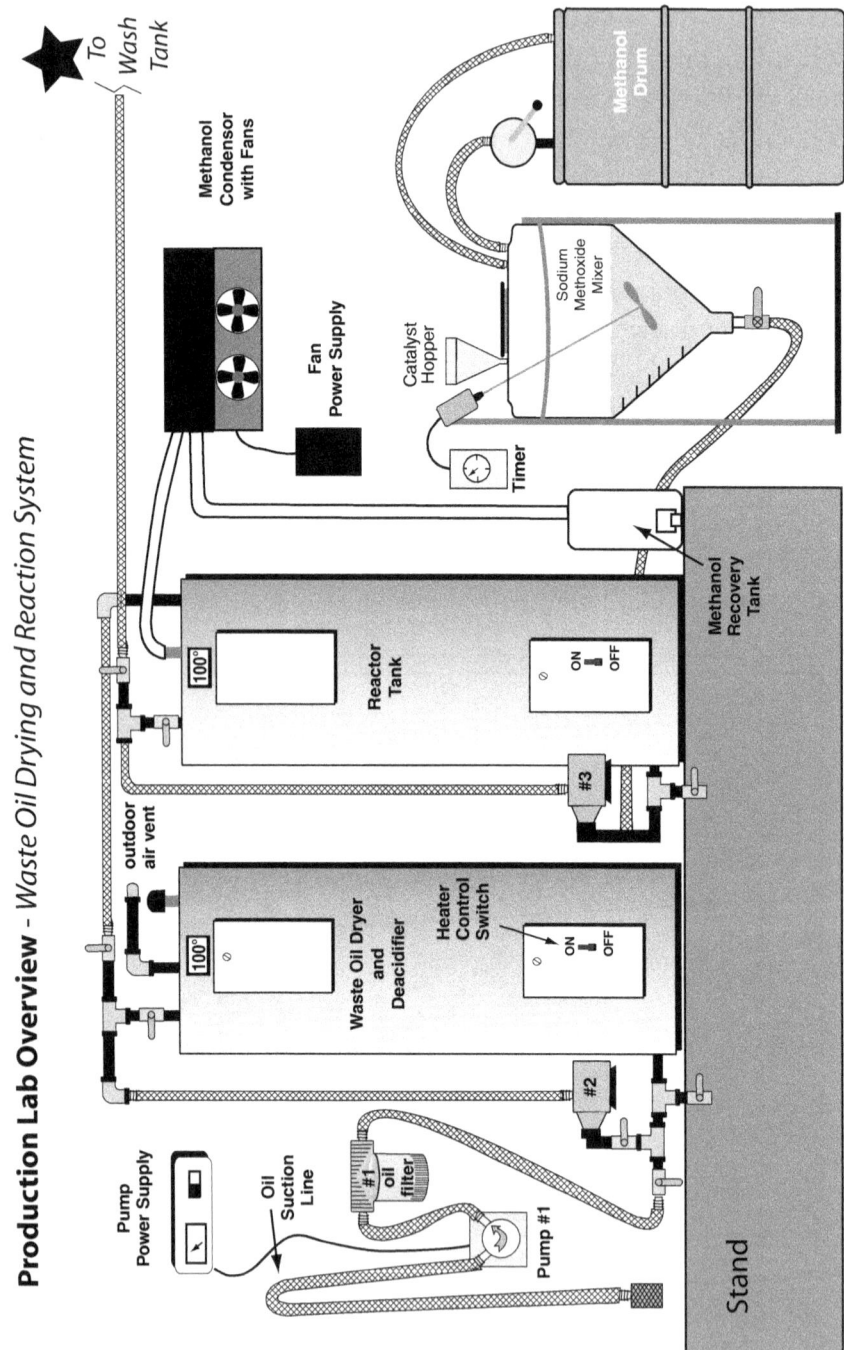

Figure 9-31. Biodiesel Production Lab Overview – Waste Oil Dryer and Reaction System

Plug-In Hybrid Vehicle Technology 463

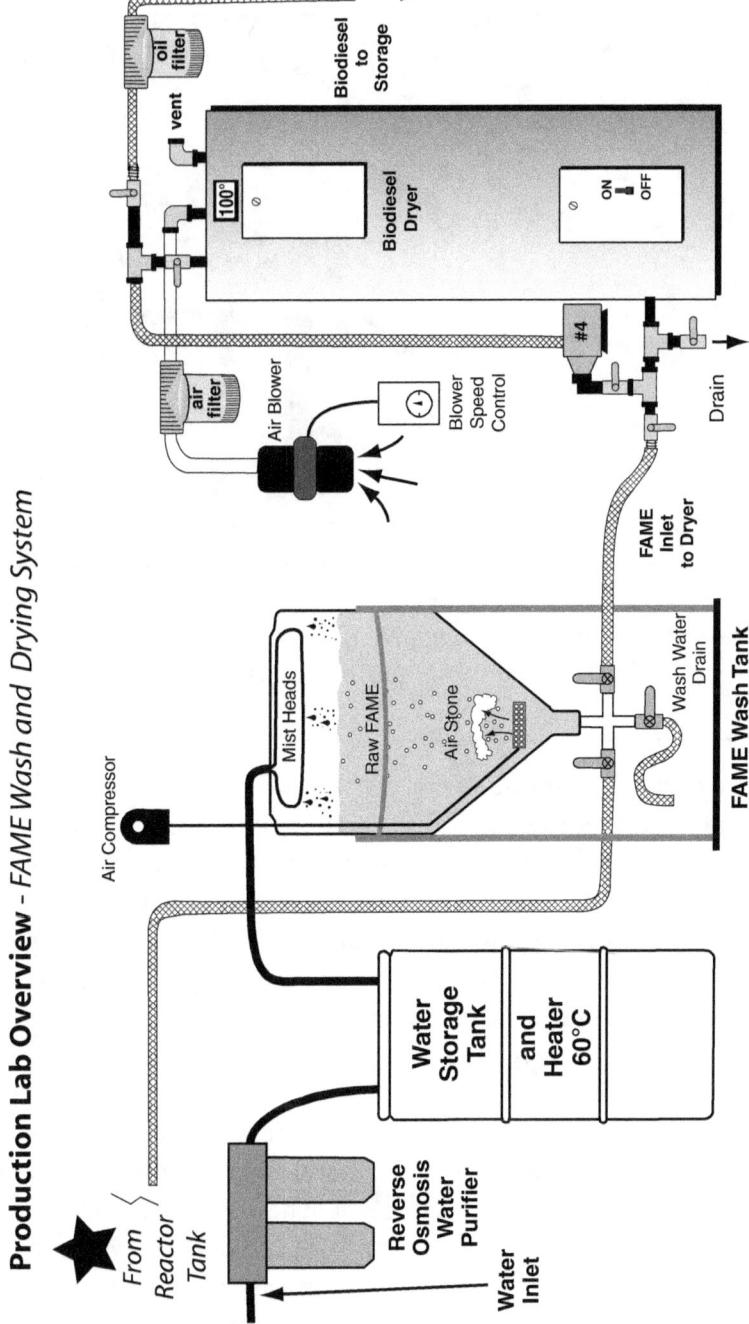

Figure 9-32. Biodiesel Production Lab Overview – FAME Wash and Drying System

Figure 9-33. The production lab is shown in an overview. All of the production equipment is mounted along the left wall, while the "wet laboratory," safety equipment, and storage area are located along the right wall.

the right wall. This image shows, starting at the bottom left corner, the first tank which receives the WVO and is used to filter, dry, and deacidify the oil to desired standards. The second tank is the main reaction tank which performs the chemical conversion (transesterification) of WVO into biodiesel. It is also used to separate the raw biodiesel from glycerol and to recycle the excess methanol prior to washing the biodiesel.

The next small white conical tank is the chemical mixer which combines the methanol stored in the adjacent tank with a powdered catalyst to form sodium (or potassium) methoxide, which is pumped into the main reaction tank to start the conversion of WVO into biodiesel.

The small white tank beside the methanol is the biodiesel wash water storage tank. It contains a submersible heater which warms the wash water to 60° C before it is sprayed into the raw biodiesel which is stored in the large white conical wash tank. The wash tank is fitted with both water spray nozzles and an air bubbling system which are used to remove contaminants from the raw biodiesel.

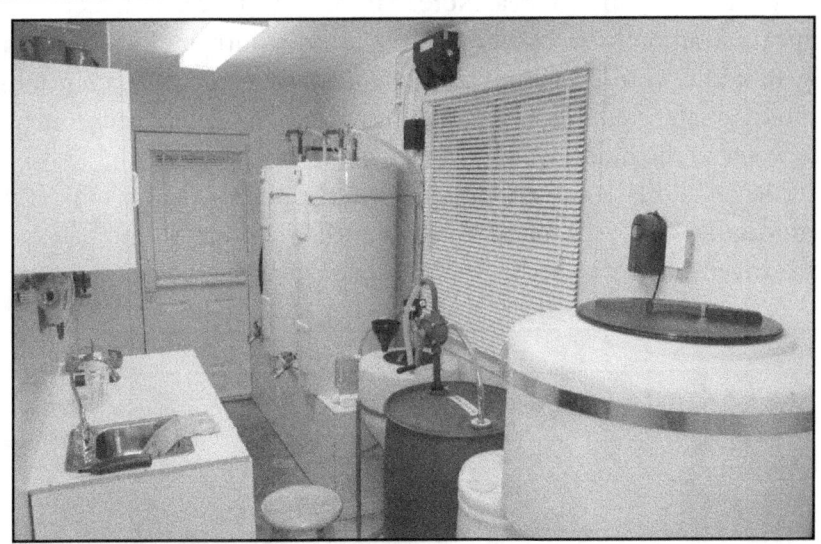

Figure 9-34. This view of the production lab was taken from the rear wall facing the front door. Note that the WVO and main reaction tanks are mounted on a cabinet which raises them off the floor. This arrangement provides additional storage space and makes accessing the valves and controls of the reaction tanks much easier on one's back.

The last tank visible at the far rear of the picture (located behind the large conical wash tank) is the biodiesel dryer. This unit removes any remaining water from the biodiesel before final filtration and storage.

Also visible in this image are the ventilation fan located along the rear wall and the electrical subpanel which provides electrical circuit control and protection for each of the process tanks, pumps, and heaters used in the system.

Running along the right sidewall is the wet lab area which is used to analyze both the WVO and the biodiesel produced by the system. The storage cabinets include all of the necessary process chemicals as well as test equipment, scales, and measuring beakers.

Safety is of primary importance when handling any chemicals and a variety of smocks, eye protection, rubber gloves, and spill cleanup materials as well as an eyewash station and multiple fire extinguishers are provided.

The photograph in Figure 9-34 was taken from the rear of the production lab and shows how all of the equipment and lab facilities fit neatly into this compact area. Note that the WVO receiver tank and the main reaction tank are lifted off the floor on a horizontal storage cabinet. In addition to providing additional storage space, the cabinet also houses a small reverse osmosis water filtration unit which feeds mineral-free water to the biodiesel wash tank.

The small black box mounted above the main reaction tank is the air-to-liquid heat exchanger which is used to condense methanol vapors driven off from the biodiesel reaction process. The unit operates

Figure 9-35. This view of the wet lab shows the ample storage space for process chemicals, gloves, safety equipment, and other materials required in the biodiesel production process.

as a fan-driven cooling unit, causing methanol vapor to condense and drip into the storage tank located next to the reaction tank. Condensed methanol is returned to the methanol storage tank for future use.

Because of the inline design of the production system, it is possible to simultaneously process approximately four 40-gallon ($\approx$ 150 liter) batches of biodiesel at one time, with one batch in each stage:
- drying and deacidification of WVO
- transesterification of WVO into biodiesel along with methanol recovery
- washing of raw biodiesel
- drying of biodiesel prior to testing and storage

Processing of waste stream glycerol and wash water is handled "offline."

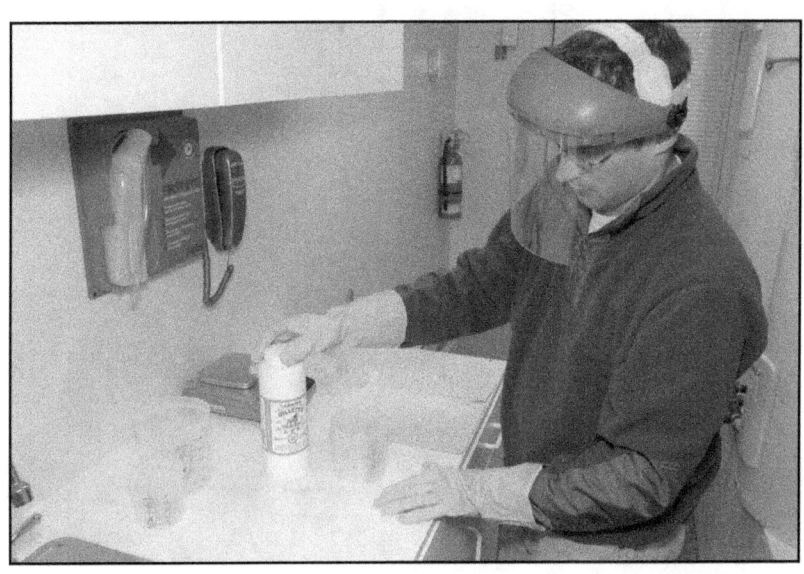

Figure 9-36. Safety is of primary importance when working with any chemicals. The wet lab area has storage for work smocks, gloves, and face shields as well as a telephone, eyewash station, and Class A, B, and C and foam fire extinguishers. Antistatic, ground-connected wrist straps are also provided to prevent the accidental ignition of methanol. The safe production of biodiesel does not happen by chance!

The WVO Receiver/Dryer

WVO is delivered to the facility and immediately transferred to the receiver/dryer tank shown in Figure 9-37. This tank is made from a 60-gallon (227-liter) electric water heater although larger, commercially available process tanks can be used if the system is to be scaled up for cooperative or small-scale commercial use.

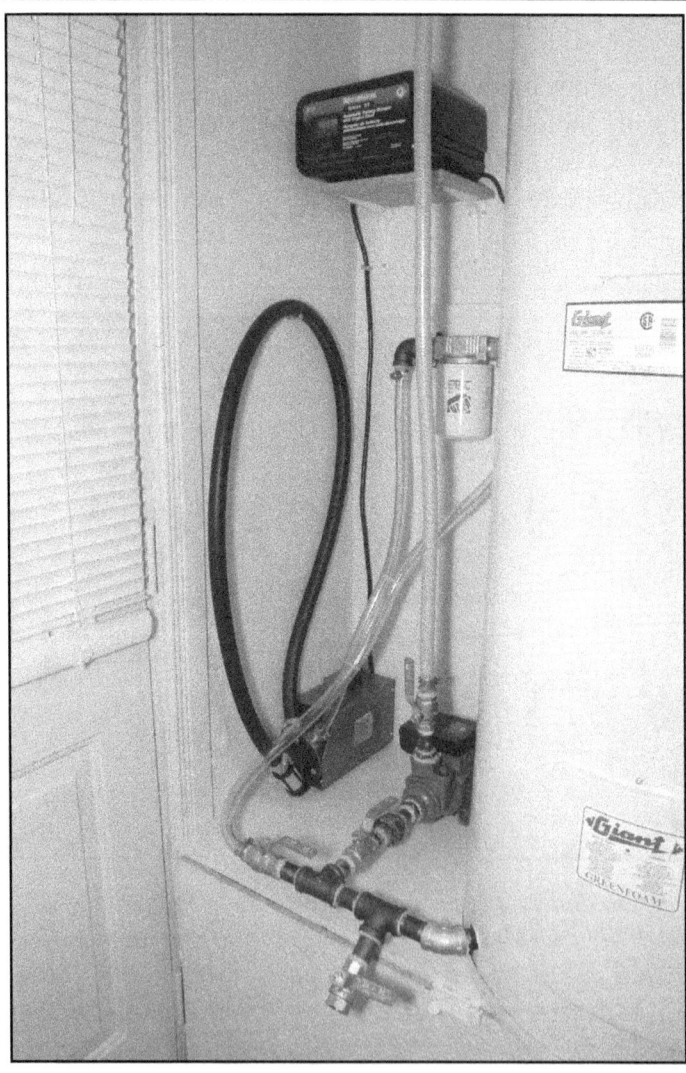

Figure 9-37. WVO is delivered to the facility and immediately transferred to the receiver/dryer tank.

Figure 9-38. A load of WVO recently delivered from a restaurant will be transferred immediately to the receiver/dryer tank.

Figure 9-39. Geoff Shewfelt is shown inserting the suction hose of the receiver/dryer tank into a pail of WVO. The suction hose is equipped with a particle strainer at one end and a 20-micron filter at the outlet of the transfer pump, ensuring that food particles are filtered out and only WVO is sent to the receiver/dryer tank.

Figure 9-40. This detail view shows the suction hose and particle strainer inserted into a pail of cold-pressed canola oil that was recovered from a batch of off-specification oilseeds.

Fellow biodiesel enthusiast Geoff Shewfelt is shown in Figures 9-38 and 9-39 with a load of WVO recently received from a local restaurant and transferred immediately to the receiver/dryer tank, where a fuel transfer pump and suction line draw the WVO from the storage pails (Figure 9-40). The WVO is analyzed to determine its free fatty acid composition and may be subjected to deacidification and heating to remove excess water. WVO absorbs water from the foods that are fried in it, and if sufficient water remains in high-FFA oil the transesterification process may fail, producing a jelly-like gravy rather than biodiesel. This is an important step that most biodiesel processors tend to skip.

Upon completion of drying and/or deacidification, the WVO is transferred to the main biodiesel reaction tank using the circulation pump fitted to the receiver/dryer unit.

The Biodiesel Reaction Tank

The biodiesel reaction tank is configured in a similar manner to the receiver/dryer tank as shown in Figure 9-41. The reaction tank is fitted with a circulation pump and a "sight glass" (Figure 9-42) created from reinforced, braided plastic tubing. This sight glass permits the filling of the reaction tank with an exact amount of WVO and reaction chemicals and also provides a way of monitoring the completeness of the reaction process.

A vapor recovery unit is also installed above the tank to capture the excess methanol driven off during the transesterification process. Excess methanol is used to ensure that the conversion of waste oil to biodiesel is driven to completion, although a large volume of methanol is not required for the transesterification process, thus allowing recycling of the excess. It is much simpler and safer and requires less energy to capture the methanol at the reaction stage than to try and recover it from the wash water.

Leaving methanol in the biodiesel is simply not an option.

Sodium Methoxide System

In order to "crack" WVO into biodiesel and its coproduct glycerol, it is necessary to use an alcohol and catalyst solution such as sodium (or potassium) methoxide. This solution is created by the careful measurement and mixing of methyl alcohol (methanol) and sodium (or potassium) hydroxide. Figure 9-43 shows a 55-gallon (208-liter) drum of methanol that has been delivered by a local fuel supply company. The drum is fitted with a hand-operated chemical pump suitable for methyl alcohol and a vapor recovery tube, both of which are fitted into the bung connections of the drum.

The outlet of the hand pump is connected to the white conical bottom tank (left), and the vapor recovery line is fitted to ensure that methanol vapors are returned to the storage tank. The conical tank is fitted with a screw-top sealing lid and a small funnel and stopper. The funnel is used as a hopper, which allows the addition of sodium hydroxide catalyst to the previously added methanol.

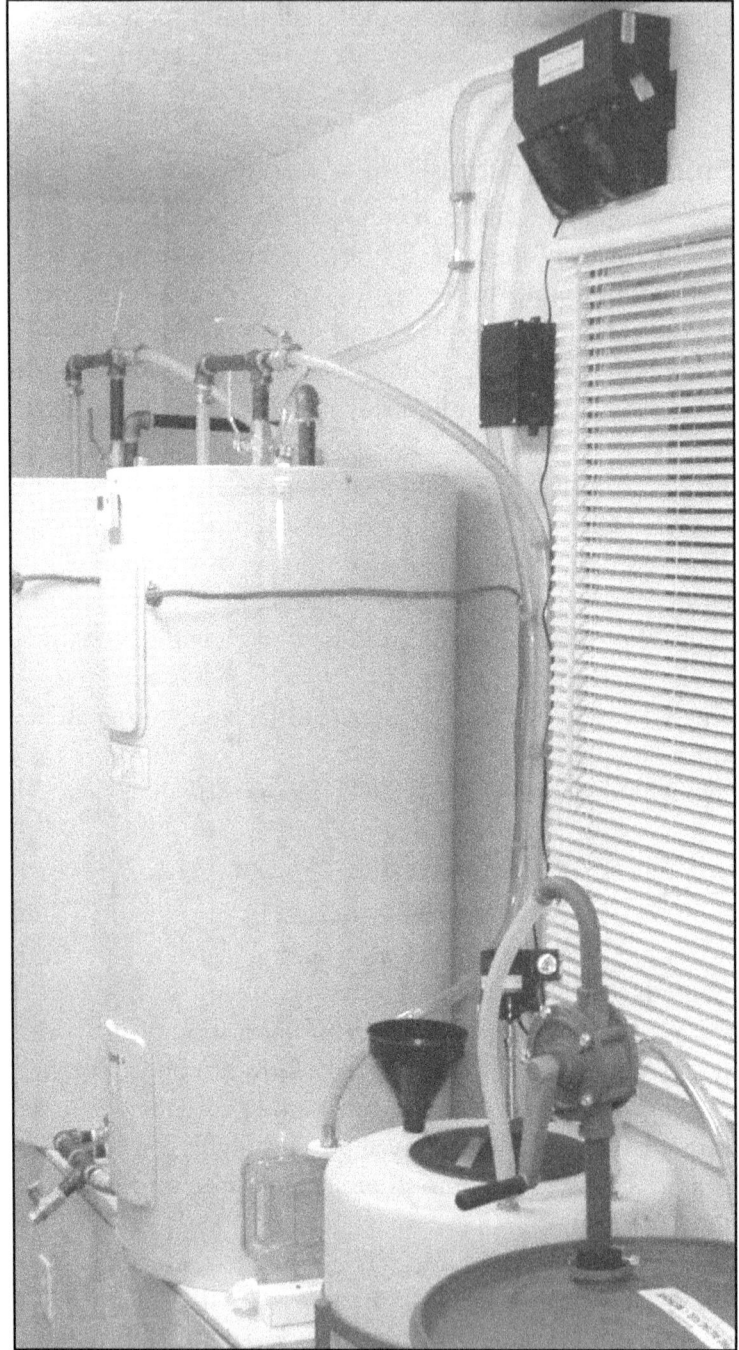

Figure 9-41. The biodiesel reaction tank is configured in a similar manner to the receiver/dryer tank. Both are visible in this view.

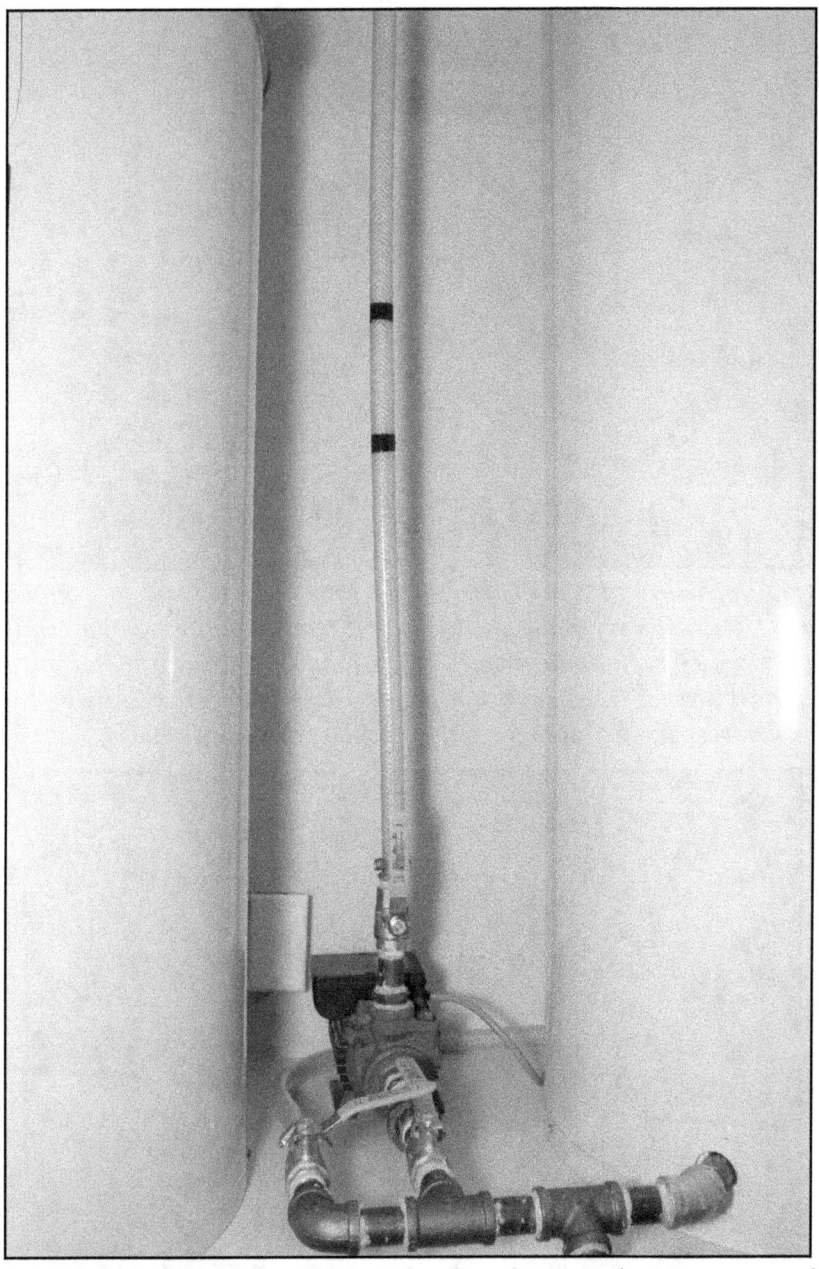

Figure 9-42. The reaction tank is fitted with a circulation pump and a "sight glass" created from reinforced, braided plastic tubing. This sight glass permits the filling of the reaction tank with an exact amount of WVO and reaction chemicals and also provides a way of monitoring the completeness of the reaction process.

Figure 9-43. In order to "crack" WVO into biodiesel and its coproduct glycerol, it is necessary to use an alcohol and catalyst solution such as sodium methoxide. This solution is created by the careful measurement and mixing of methyl alcohol (methanol) and sodium hydroxide in the mixing tank shown above. The barrel to the right is filled with methyl alcohol.

Figure 9-44. A small, spark-proof mixer is fitted to the tank and is also connected to an electrically operated mechanical timer that is used to control the mix timing of the solution, ensuring that the sodium hydroxide completely dissolves in the methanol.

A small, spark-proof mixer is fitted to the tank as shown in Figures 9-43 and 9-44. The mixer is also connected to an electrically operated mechanical timer that is used to control the mix timing of the solution, ensuring that the sodium hydroxide completely dissolves in the methanol.

The outlet at the bottom end of the conical tank is connected to the suction side of the reaction tank through a shutoff valve. Opening this valve causes the sodium methoxide solution to be drawn into the reaction tank containing the WVO, starting the transesterification process.

Upon completion of the transesterification process, two liquid components or *phases* are created. A glycerol phase sinks to the bottom of the reaction tank, while the lower-density raw biodiesel phase floats on top. The glycerol is removed by draining it from the tank and transferring it to a separate refining station.

Upon removal of the glycerol phase, the raw biodiesel is reheated to cause excess methanol to boil off. The methanol may be vented into the atmosphere or preferably directed to a reflux condenser that converts the vapors to liquid. The captured methanol may then be reused, lowering production costs.

After glycerol and methanol have been recovered from the reaction tank, the raw biodiesel is transferred to the wash tank for final processing.

Biodiesel Washing System

Raw biodiesel is transferred from the main reaction tank to the large white conical-bottom washing tank shown in Figure 9-45. This tank is fitted with two washing systems known as mist and bubble wash technologies. Regardless of which washing procedure is used, a small water storage tank located to the immediate left of the washing tank is required. This 10.5-gallon (40-liter) tank receives potable water from a reverse osmosis filtration unit located in the storage cabinet under the reaction tank, which is in turn fed by the household potable water supply. The purpose of the reverse osmosis system is to remove dissolved minerals such as calcium and iron that are contained in the well water in our geographical location. The water storage tank is fitted with a submersible water heating element that heats the wash water to between 120° F and 140° F (50° C and 60° C), greatly improving the wash speed

Figure 9-45. Raw biodiesel is transferred from the main reaction tank to the large white conical-bottom washing tank shown above. This tank is fitted with two washing systems known as mist and bubble wash technologies. The small white tank to the left of the wash tank is the wash water storage tank.

and quality. Prior to starting the wash cycle, a small amount of acetic acid is added to the wash water and a submersible pump is activated.

If the mist washing process is used, the wash water is pumped to a series of mist heads mounted around the perimeter of the wash tank lid, causing a gentle shower of slightly acidic (softened) water to spray over the biodiesel surface (see Figure 9-46). Water has a higher density than oil or biodiesel and therefore falls to the bottom of the tank, absorbing any free glycerin and excess catalyst (sodium hydroxide) along the way.

Washing may be completed using the spray mist method described above or using a "bubble washing" technique in which water is added to the biodiesel directly from the wash tank.

It is necessary to wash biodiesel a number of times in order to ensure compliance with fuel quality standards. At each successive washing stage the wash water contains fewer contaminants, allowing it to be reused for earlier, more heavily contaminated wash stages. This process is known as counter-current washing and greatly reduces the amount of water used in the production process. When the wash water is saturated with con-

Figure 9-46. Prior to starting the wash cycle, a small amount of acetic acid is added to the biodiesel wash water and a submersible pump is activated, sending water to a series of mist heads mounted around the perimeter of the wash tank lid and causing a mist of slightly acidic (softened) water to spray over the biodiesel surface.

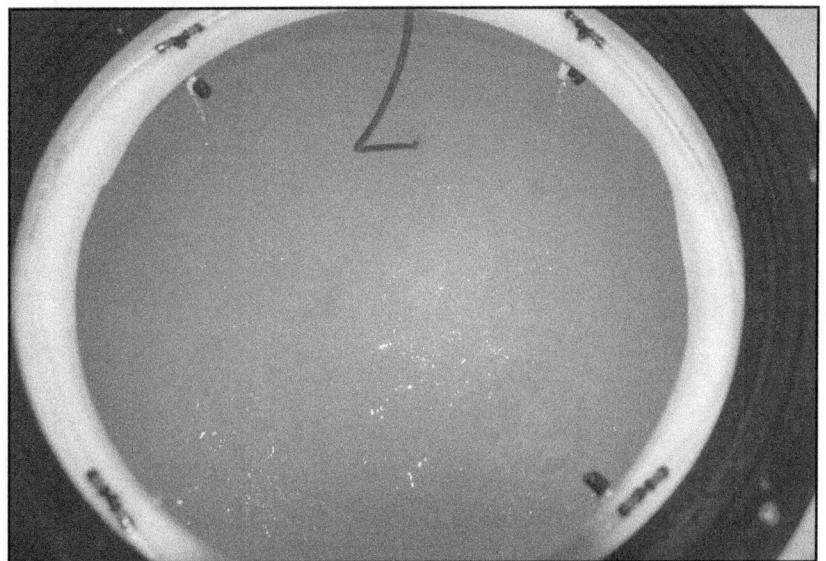

Figure 9-47. Biodiesel washing may use the spray mist method or a "bubble washing" technique (shown here) in which water is added to the biodiesel and tiny air bubbles are blown through the water/biodiesel mixture.

taminants, it is drained off and temporarily stored before final treatment and drainage into the environment. This contaminated wash water is known to be toxic and must be treated prior to release.

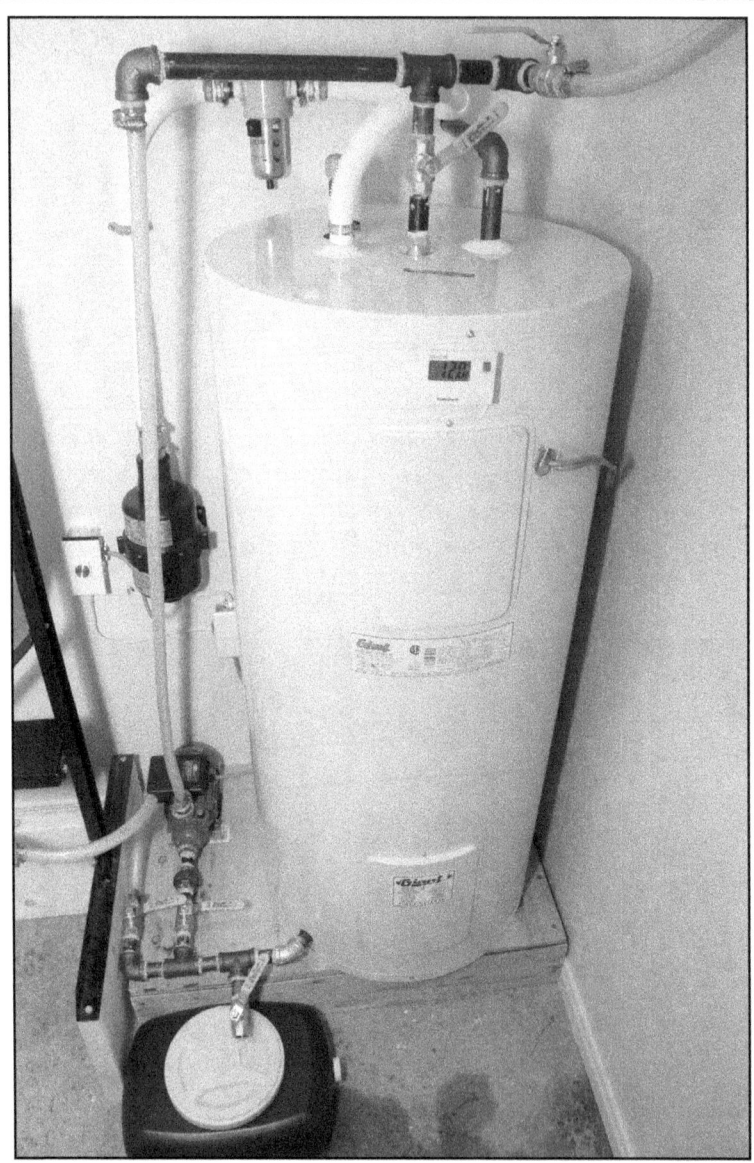

Figure 9-48. After the final wash water has been removed from the washing tank, the biodiesel is pumped into the dryer tank shown above.

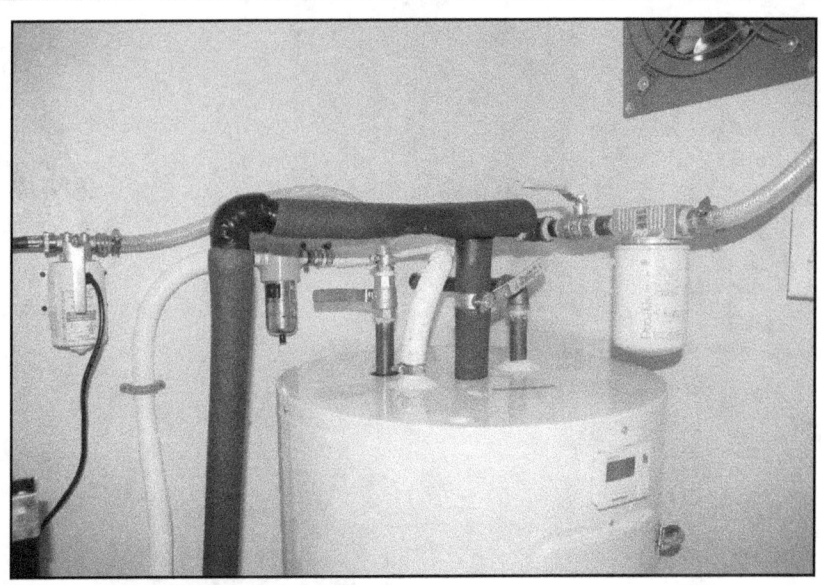

Figure 9-49. The fuel dryer is fitted with a heated air blower and filter arrangement which blows hot air through the biodiesel before venting outside to the atmosphere. This arrangement removes excess water from the fuel, ensures compliance with the ASTM limits for water and sediment, improves the biodiesel oxidation stability, and reduces the chances of microbial growth during storage.

Biodiesel Drying and Final Filtration

After the final wash water has been removed from the washing tank, the biodiesel is pumped into the dryer tank shown in Figure 9-48. The dryer uses an electric water heater arrangement and is equipped with a circulating pump. It is also fitted with a heated air blower and filter arrangement which blows hot air through the biodiesel (Figure 9-49) before venting outside to the atmosphere. This arrangement removes excess water from the fuel, aids in ensuring compliance with the ASTM limits for water and sediment, improves the biodiesel oxidation stability, and reduces the chances of microbial growth during storage.

Once the biodiesel has been heat-treated, it is pumped through a 20-micron fuel filter and is ready for quality testing and storage (Figure 9-50).

Figure 9-50. Once the biodiesel has been heat-treated, it is pumped through a 20-micron fuel filter and is ready for quality testing and storage.

Fuel Dispensing and Storage

With the biodiesel processor operating at full capacity, the system can process approximately 45 to 50 gallons (≈170 to 190 liters) of biodiesel per day, which is far in excess of our total requirements. However, processing, storing, and handling that much fuel does take a fair amount of space, labor, and care.

Figure 9-51. With the biodiesel processor operating at full capacity, the system can process approximately 45 to 50 gallons (≈170 to 190 liters) of biodiesel per day, which is far in excess of our total requirements. A fuel dispensing system can be as simple or complex as your needs dictate. Simple five-gallon fuel totes are inexpensive and work well but increase the amount of handling labor as well as the frequency of spills. An electric fuel dispensing system such as this model keeps the dispensing method in familiar territory.

The Fuel Dispensing Unit

A fuel dispensing system can be as simple or as complex as your needs dictate. Simple five-gallon (20-liter) fuel totes are inexpensive and work well but increase the amount of handling labor as well as the frequency of spills. An electric fuel dispensing system such as the model shown in Figures 9-51 and 9-52 keeps the dispensing method in familiar territory. This dispensing unit is available at most farm and auto supply stores and comprises an explosion-proof fuel pump that is fitted to a standard 55-gallon fuel drum. The pump may be driven by either 120-volt household power or a 12-volt supply connection for in-vehicle use. A special water-absorbing filter known as an *agglomerator* is connected to the pump discharge prior to feeding the fuel nozzle.

Incidentally, the 120-volt version can fill the Zero-Carbon Car fuel tank in less than a minute, making fill-ups a snap!

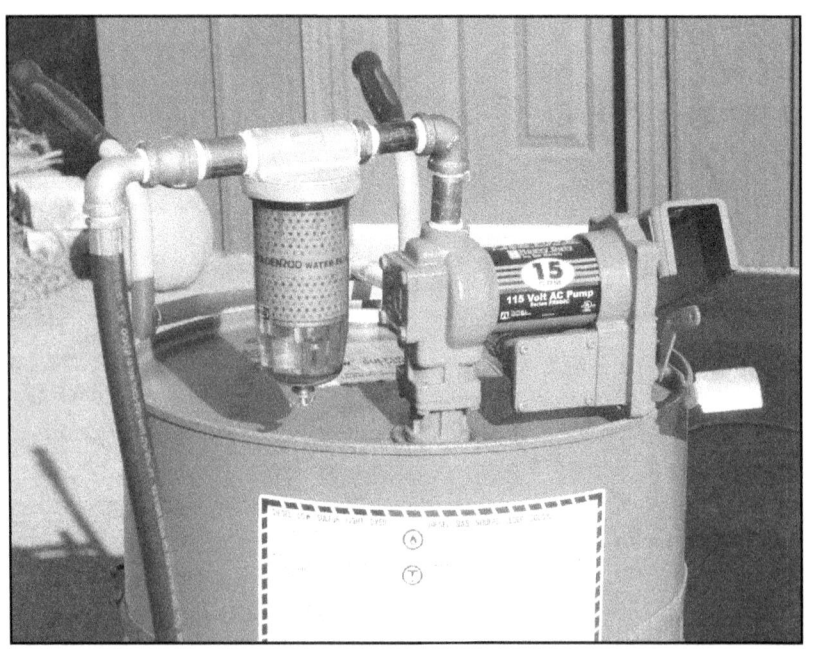

Figure 9-52. This dispensing unit is available at most farm and auto supply stores and comprises an explosion-proof fuel pump that is fitted to a standard 55-gallon (208-liter) fuel drum. The 120-volt version pump shown here can fill the Zero-Carbon Car fuel tank in less than a minute, making fill-ups a snap!

Figure 9-53 shows a detail view of a typical fuel dispensing unit. A 55-gallon fuel drum must be purchased or "rented" by paying a drum deposit charge. It is strongly recommended that only a new or recently used diesel fuel drum be adopted for your fuel dispensing system. The drum can be mounted on a drum dolly or a drum cart. A fully loaded fuel drum is very heavy and is difficult to move using standard drum carts; ideally the drum can be placed in a convenient central location and not moved.

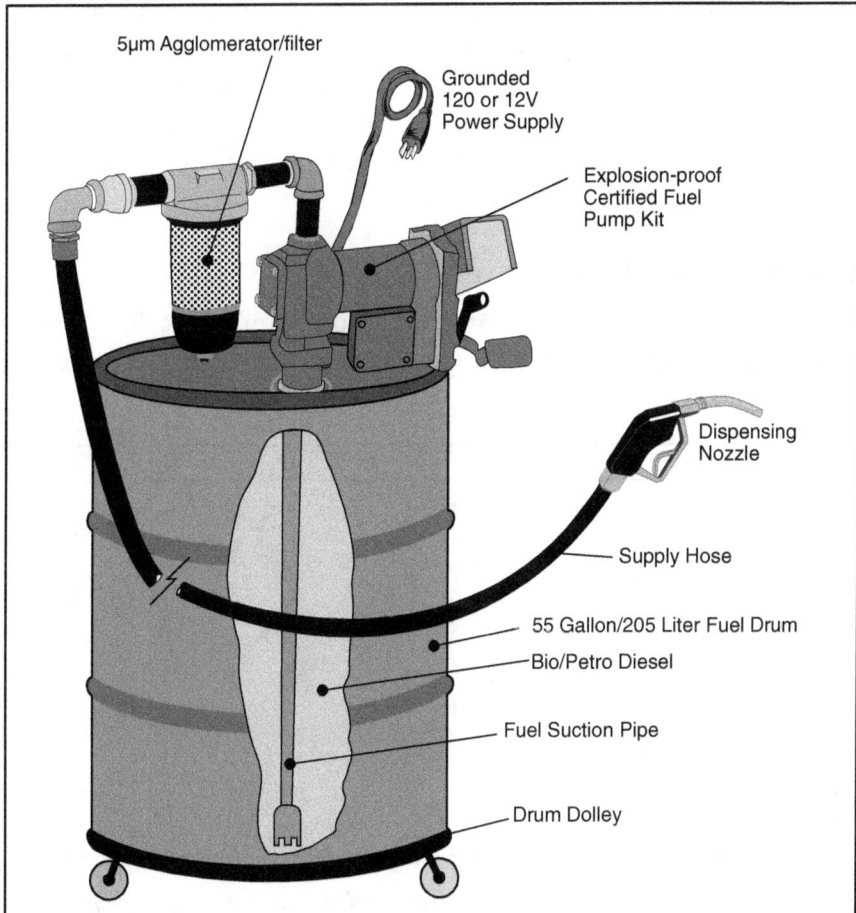

Figure 9-53. This image shows a detail view of a typical fuel dispensing unit. A 55-gallon fuel drum must be purchased or "rented" by paying a drum deposit charge. It is strongly recommended that only a new or recently used diesel fuel drum be adopted for your fuel dispensing system.

A commercial fuel dispensing pump is mounted to the drum using the threaded bung fitting. A suction pipe is lowered into the tank to draw fuel into the pump intake as shown.

The agglomerator filter is fitted on the discharge side of the pump as shown. Standard black pipe fittings of the same type used in the biodiesel facility are required. The agglomerator filter is a disposable filter that is designed to agglomerate (group together or coalesce) water droplets suspended in the fuel and cause them to fall to the bottom of the sight glass bowl. A small drain valve is located at the bottom of the glass bowl which allows for drainage of any accumulated water. (Do not believe for one second that having an agglomerator will lessen the need for diligent fuel handling and storage. These filters will only remove relatively large water droplets and are therefore intended as precautionary devices only.)

Cold Weather Issues

Biodiesel is subject to a gelling or freezing condition at its cold-temperature limits as discussed earlier. If you live in an area where cold weather

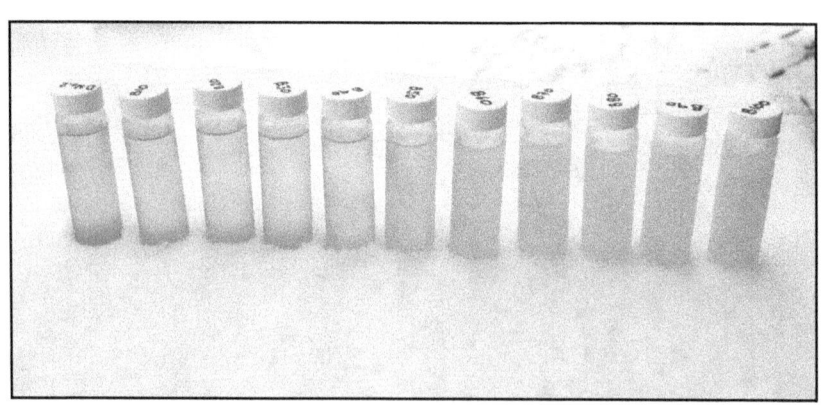

Figure 9-54. Biodiesel is subject to a gelling or freezing condition at its cold-temperature limits as discussed earlier. If you live in an area where cold weather is the norm, it will be necessary to adjust the concentration of biodiesel stored in the fuel dispensing system as a function of temperature. One very simple means of doing this is to create a "biodiesel blend thermometer" as shown here.

is the norm, it will be necessary to adjust the concentration of biodiesel stored in the fuel dispensing system as a function of temperature. One very simple means of doing this is to create a "biodiesel blend thermometer" as shown in Figure 9-54. A series of 11 glass vials is filled with mixtures of petro- and biodiesel. The vial at the left contains No. 2-D (straight diesel fuel), the next contains B10 (10% biodiesel/90% petrodiesel), ending with B100 (100% biodiesel) on the right. Each of the lids is marked with the appropriate concentration of fuel mixture.

The vials can be placed in a tin can and left in the same general location, out of direct sunlight, where you would normally store your diesel-powered vehicle. In the example shown in Figure 9-55, the No. 2-D sample (bottom vial) is free flowing and gel free, while the B100 (top vial) is frozen solid at 5° F (-15° C). At this temperature and using refined, cold-pressed, canola-based biodiesel, a concentration of B30 to B50 could easily be used, with greater concentrations possible if the vehicle were stored in a garage overnight.

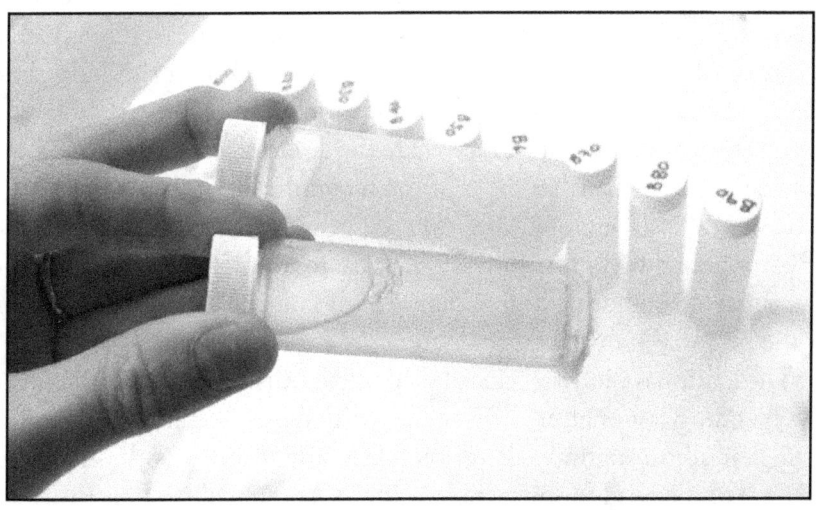

Figure 9-55. In this example, the No. 2-D sample (bottom vial) is free flowing and gel free, while the B100 (top vial) is frozen solid at 5° F (-15° C). At this temperature and using cold-pressed, canola-based biodiesel, a concentration of at least B30 could be used as evidenced by the lack of crystal formation or any sign of fuel thickening.

It is important to make a biodiesel blend thermometer since the ambient local temperature and feedstock oil composition of the biodiesel will greatly affect the concentration of biodiesel that can be mixed into the fuel tank. It is very difficult to accurately calculate the blend levels, and for this reason I keep a spare drum of No. 2-D fuel for splash blending on site.

An additional word of caution: note where your vehicles will be stored during the day. For example, the biodiesel blend thermometer will give an accurate reading assuming your vehicle is stored in a garage at night. However, it may be exposed to colder daytime temperatures if you are parked in a shaded outdoor parking lot all day long. It is always best to err on the side of caution and be a bit conservative when selecting the appropriate biodiesel blend level. There is nothing worse than having to call for a tow truck to pick up a stalled, fuel-starved vehicle because you pushed the blending limits.

Blending Biodiesel with Petrodiesel

Although blending biodiesel with petroleum diesel will increase fossil fuel consumption and greenhouse gas emissions, it must be remembered that PHEVs will typically use 85% less liquid fuel than similarly sized vehicles that operate on internal combustion engines only. If advanced Hypercar technologies are employed, the level of consumption could easily be reduced a further 50% of the remaining 15%.

It should also be noted that only areas with very cold weather need to worry about blending. Therefore total fossil fuel blending for cold flow issues will amount to less than 5% of current diesel consumption for the personal transportation sector.

The National Biodiesel Board (NBB) commissioned a report on cold flow blending issues after the State of Minnesota established a requirement that all on-highway diesel fuels contain at least 2% biodiesel. In response to the need for proper blending and other cold-temperature-related issues, the NBB established a Biodiesel Cold Flow Consortium to study the blending properties of biodiesel and report their results. The study evaluated both "splash" and "proportional" blending techniques, of which only splash blending is used by the micro- and small-scale producer.

The results of the Consortium testing showed that biodiesel must be kept at least 10° F (≈ 6° C) above its cloud point temperature to ensure successful, homogenous blending with petrodiesel. For those who would like to learn more about the cold flow blending study, a copy of the report can be downloaded from the National Biodiesel Board website.[28]

Summary

The use of waste or recycled oil feedstocks is a move in the right direction towards creating zero-carbon biodiesel. Unfortunately, feedstock oil makes up only 70.6% of the biodiesel production process, and the other inputs of chemical feedstocks and energy shown in Figure 9-56 must also be converted to carbon-neutral properties.

Electrical energy in the form of Bullfrog Power takes care of electricity, and carbon credits can be used to offset natural gas inputs at the commercial plant level.

The issue of alcohol is quite interesting, as approximately 20% of biodiesel is produced using methanol; it could also be called liquid methane, as it is principally derived from natural gas.

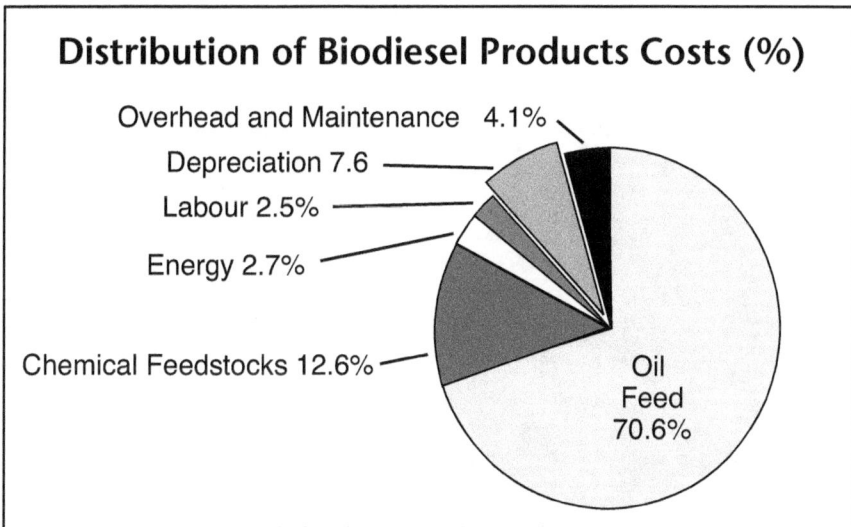

Figure 9-56. Although biodiesel can be made into a completely zero-carbon fuel, input energy and chemical feedstocks must be considered. (Source: Jon Van Gerpen et al, Building a Successful Biodiesel Business)[29]

The transesterification process does not care which alcohol is used, although methanol has been chosen for its simplicity to work with and low cost. As with anything derived from fossil fuels, carbon valuation will change this picture and alternative alcohols derived from renewable sources will become more interesting. An obvious candidate is cellulosic ethanol, described earlier in this chapter. Although I am still in the early stages of researching this input feedstock, I have discovered that there are no major impediments to the production of biodiesel using cellulosic ethanol. Stay tuned to the Zero-Carbon Car website for updates on this development work as I begin to close this last gap in liquid fuels research.

Because BEVs and PHEVs use 85% less liquid fuel than existing internal-combustion-powered models, fuel consumption used in personal transportation vehicles has the ability to drop dramatically. It will only take the right carbon policies to make it happen.

Looking into the future, it is not possible to say for certain what fuel, biological or fossil, will be used to power advanced high-efficiency vehicles. The likelihood is that there will be a mix of different choices, but sustainable biofuels will definitely be part of the supply mix.

Chapter 10
The Unveiling

After months of planning, design, and wiring along with a few bruised knuckles, the Zero-Carbon Car was finally ready to hit the road. It seemed fitting to bring the car to the Canada Science and Technology Museum on a warm summer evening for its unveiling, since this is also the site where the Electric Vehicle Council of Ottawa (EVCO) holds its monthly meetings.

Figure 10-1. After months of planning, design, and wiring along with a few bruised knuckles, the Zero-Carbon Car was finally ready to hit the road. The vehicle was brought over from the garage early so that it could remain under wraps, creating an air of excitement and mystery.

It was a beautiful evening, and because of the number of interesting projects that were being presented it was decided that the meeting would be held outdoors in the parking lot—quite a fitting decision given that there was the Zero-Carbon Car to present as well as a home-built battery-powered electric lawn mower and several electrically powered bicycles.

Rick Lane and his wife Micheline brought the vehicle over from the garage early so that it could remain under wraps, creating an air of excitement and mystery. EVCO members had been aware of the work that was going on, and a few had dropped by during the passing months to see the work in progress, but the finished product was still a surprise.

Once the group of about 25 people was assembled, the wrapping was removed and the vehicle presented for everyone to see. As expected, there was a lot to discuss about the car, transportation policy, biofuels,

Figure 10-2. Once the group of about 25 people was assembled, the wrapping was removed and the vehicle presented for everyone to see. As expected, there was a lot to discuss about the car, transportation policy, biofuels, and batteries.

Figure 10-3. Even though Rick Lane was involved in the design and development of the Zero-Carbon Car, he, like most people, wasn't too sure what to do with the charging plug!

and batteries. The questions went on and on, but the best part for most people was hopping in behind the wheel and taking off for a run around the building. Apart from the slight whirring from under the hood, the car was completely silent and ran flawlessly; it certainly captured the imagination of those who attended.

For me the best part of the evening was when someone asked, "So why can't I buy one of these cars now? It just seems so sensible."

And with a comment like that, I wonder how long it will take the rest of society and especially our elected officials to realize that the double-edged sword of transportation policy and climate change are destroying the very planet we depend on for life. We do not have the right to devastate or attempt to conquer this ecosystem. We do not own the earth; we are merely borrowing it, sharing the natural splendor with those who are not yet born. The Zero-Carbon Car is not a solution, simply a stopgap until the planet is devoid of personal automobiles.

Only time will tell if my vision for a more sustainable transportation plan will prevail.

Figure 10-4. The Zero-Carbon Car design team poses for a well-deserved shot with the car. The question now: what's next?

Figure 10-5. The questions went on and on, but everyone liked the car and wondered if I was planning on putting it into production.

Figures 10-6 and 10.7. The best part of the evening for most people was hopping in behind the wheel and taking off for a run around the parking lot. Apart from the slight whirring from under the hood, the car ran flawlessly and certainly captured the imagination of those who attended. For some, it was the first time they had ever been in an electrically powered vehicle, and they were amazed at the silence.

Figure 10-8. Now that the vehicle has been unveiled, the media are beginning to clamber through it, trying to understand not only the uniqueness of the vehicle but the bigger story behind transportation policy and the linkages with climate change.

Conclusion

What started out as a relatively simple project attempting to determine what vehicle technologies people would be using in the coming decades instead turned out into a giant web of dead ends, wrong turns, and political struggle.

Based on the theme of energy efficiency, this book should have been fairly simple to write. All I had to do was discuss which technologies would vastly improve fuel economy. But the further I went the deeper I found myself, like peeling the layers of an onion, inside a world that was trying to juggle the socioeconomic and political issues related to the matter of transportation policy and technology.

With a world population that is zooming upwards and completely out of proportion with the ability of the earth's ecosystems to support it, we are faced with a serious dilemma. As the level of personal entitlement and related energy consumption increases, there is neither the natural nor the financial capital to support billions of personal vehicles.

The world has to stop the madness of endless transportation privilege and take a lead from Europe, where high fuel taxes, better access to mass transit, "congestion" taxes, and carbon valuation are being used in an effort to reduce road transportation density. At least that is the hope.

With blockades by truckers protesting high fuel taxes and automobile advocacy groups in the UK demanding more and better roads (and having the majority of voters on their side), it is going to be a very difficult battle for politicians to enact any further "anti-driving" policy. Even the tiny steps being taken in England would be sufficient to impeach a politician in the United States.

The catalyst that will stop endless personal transportation growth may have to come from outside the neat walls of government. It may manifest itself as a massive terrorist attack on Saudi oil infrastructure or the bombing of the Strait of Hormuz by an angry anti-American regime in Iran, which could choke off 15 million barrels of oil per day, almost 20% of world supply.

Then again, it may come from the powerful and relentless pressures of Mother Nature. The flooding of New Orleans was just a coincidence, according to some. The complete failure of Australian wheat crops because of eight straight years without enough rainfall is just a "cycle," according to others. Australia is the world's second largest wheat producer and exporter and its 2007 crop will be 58% lower than last year's, send-

Figure C-1. For the foreseeable future we can continue to drive our cars, SUVs, and trucks, but as the paradigm of reality convergence is authenticated, society will slowly transition to evermore expensive, complex, and environmentally friendly automobiles such as the Zero-Carbon Car or a fleet of other as yet undreamed-of models.

Figure C-2. We have many choices to make as a society, and when we make the wrong ones it is often future generations who will suffer. Perhaps it is time to reverse that trend and seriously consider our actions and their impact on this planet—including an honest evaluation of our personal transportation needs.

ing wheat prices to record levels and global stocks to their lowest levels in almost 20 years.

How many of these coincidences will it take before people realize the severity of the situation? Will it be too late to take action once the worst effects of climate change hit and society can no longer afford to deal with the converging issues attacking it? The cost of climate-change-related incidents is not currently factored into the accounting books of governments, but once the linkage is made, personal transportation will be one of the first offenders to go.

People drive because they like to and because, despite its short comings, driving is convenient. Period. No one needs a big SUV or minivan to take the kids to hockey or dance lessons. Society has lived without them for millennia. The truth is we want our SUVs and minivans; they are our entitlements.

For the foreseeable future we can continue to have them, but as the paradigm of reality convergence is authenticated, society will slowly transition to evermore expensive, complex and environmentally friendly automobiles such as the Zero-Carbon Car or a fleet of other as yet undreamed-of models. Capital and operating prices will increase, as will the costs of infrastructure, and this will alienate an increasing number of people. Driving will become a concession to the rich or those so financially irresponsible they will mortgage their future for it.

I expect that society will not be able to play this game for much more than a few decades at most before a transition to a vastly different mass transit system begins to take hold, perhaps one based on the ideas I have presented in this book. People will lament and howl in protest as the transition occurs, but it will occur.

We have many choices to make as a society and when we make the wrong ones, it is often future generations who will suffer. Perhaps it is time to reverse that trend and seriously consider our actions and their impact on this planet—including an honest evaluation of our personal transportation needs.

ZCC Glossary
(Including Abbreviations)

Alternating Current (AC)
An electric current that reverses direction (polarity) on a periodic basis. In North America most household current is AC at 60 cycles per second.

Ampere (A)
The unit of measurement of electric current. It is proportional to the quantity of electrons flowing through a conductor past a given point in one second. It is analogous to cubic feet of water flowing per second.

Ampere-hour (Ah)
One ampere of current flowing for one hour. The basic unit of battery capacity.

Battery
An electrochemical cell that is capable of storing an electric charge through a chemical conversion process. Primary batteries may be used once and must be discarded when the cell is depleted. Secondary or rechargeable batteries may be regenerated or recharged multiple times.

Battery Electric Vehicle (BEV)
An electric vehicle that derives its electrical power from a battery bank exclusively.

Biodiesel
Biodiesel is a type of biofuel made by combining animal fat or vegetable oil (such as soybean oil) with alcohol and can be directly substituted for diesel fuel or heating oil.

Carbon Dioxide (CO_2)
A colorless, odorless, gas produced by burning fossil fuels, sometimes referred to as a greenhouse gas because it contributes to earth warming.

Continuously Variable Transmission (CVT)
The continuously variable transmission is a transmission in which the ratio of the rotational speeds of the input and output shafts of a vehicle

can be varied continuously within a given range, providing an almost infinite number of possible ratios.

Direct Current (DC)
An electrical current which flows only in one direction in a circuit. Batteries produce direct current.

Electric Motor
An electric motor converts electrical energy into mechanical energy and more correctly into mechanical torque.

Electric Vehicle (EV)
A vehicle that is propelled using electrical power exclusively. Electric vehicles are typically battery-powered, although Fuel Cell Vehicles are in effect, electrically-powered as well.

Energy Density
Is the amount of electrical energy stored per cubic metre of battery volume.

Fuel Cell Vehicle (FCV)
An electric vehicle that derives its electrical energy from a fuel cell and on-board fuel source. Most fuel cells use hydrogen as their fuel supply.

Greenhouse Gas
A gas in Earth's atmosphere that traps heat and can contribute to global warming. Carbon dioxide and methane are the two most important.

Hybrid Electric Vehicle (HEV)
A vehicle combining a battery-powered electric motor with a traditional internal combustion engine. The vehicle can run on either the battery or the engine or both simultaneously, depending on the design criteria.

Induction Motor
An AC motor in which the rotating armature has no electrical connections to it (ie. no brushes or commutator gear).

Inverter
An inverter is an electronic device that produces alternating current (AC) from direct current (DC).

Kilo-watt hour (kWh)
A unit of energy of work equal to 1,000 watt-hours. The basic measure of electric energy generation or use. A 100-watt light bulb burning for 10 hours uses one kilowatt-hour.

Lead-Acid Battery
The lead-acid battery is the most common rechargeable battery owing to its use as a source of energy for starting internal combustion engines. The battery is named for the main components, namely; lead and acid.

Lithium- ion (L-Ion)
One of the newer rechargeable battery technologies, Li Ion batteries can deliver 40% more capacity than comparably sized NiCad batteries and are one of the lightest rechargeable batteries available.

Nickel Cadmium Battery (NiCad)
A rechargeable battery with a nickel cathode and a cadmium anode. NiCad batteries have approximately twice the specific energy of lead-acid batteries.

Nickel Metal Hydride (NiMH)
A nickel-metal hydride battery is a type of rechargeable battery similar to a nickel-cadmium battery but has a hydrogen-absorbing alloy for the anode instead of highly-toxic cadmium.

Ohm (Ω)
The unit by which electrical resistance is measured. One ohm is equal to the current of one ampere which will flow when a voltage of one volt is applied.

Plug-In Hybrid Vehicle (PHEV)

A vehicle that is similar to a hybrid electric vehicle, but having a greatly enlarged battery bank and AC mains charging system. PHEVs are able to travel "reasonable" distances on battery power alone, but can rely on the internal combustion engine for extended trip distances.

Regenerative Braking

A system that generates electricity to recharge batteries while a vehicle is coasting or braking.

Specific Energy

The amount of electrical energy stored for every kilogram of battery mass.

Specific Power

Is the amount of power obtained per kilogram of battery.

Volt (V or E)

The basic unit of electrical potential. One volt is the force required to send one ampere of electrical current through a resistance of one ohm.

Watt (W)

A unit of electrical power, equal to the power developed in a circuit by a current of one ampere flowing through a potential difference of one volt.

Watt-hour (Wh)

The basic unit of measure for consumption of electric energy. One watt-hour is one watt of electricity used for a period of one hour.

Zero-Emission Vehicle (ZEV)

A zero emission vehicle, or ZEV, is one which has no tailpipe exhaust or carbon dioxide emissions.

Appendix A - Resource Guide

Councils and Societies:

American Council for an Energy Efficient Economy
Website: www.aceee.org
Phone: 202-429-8873
Publishes guides comparing the energy efficiency of appliances.

California Air Resources Board
Website: www.arb.ca.gov/
Promotes and protects public health, welfare and ecological resources through the effective and efficient reduction of air pollutants.

California Energy Commission
Website: www.energy.ca.gov
Phone: 916-654-4058
The CEC is the strongest supporter of grid inter-connected renewable energy systems in North America. Their website explores what is happening in California in this regard.

National Renewable Energies Laboratory (NREL)
Website: www.nrel.gov
Phone: 303-275-3000
The NREL is the national renewable energy research laboratory in the United States.

Natural Resources Canada
Website: www.nrcan.gc.ca
Phone: N/A
The Government of Canada hosts this website which includes the office of energy efficiency. Many resources are presented in this fact filled site.

Rocky Mountain Institute
Website: www.rmi.org
Phone: 970-927-3851
The Rocky Mountain Institute is a think tank regarding all energy efficiency issues. Their website contains a great deal of source information for books and applied research.

Underwriters Laboratories Inc.
Website: www.ul.com
Phone: 847-272-8800
Develops standards for the United States marketplace. Tests and administers safety certification work in North America.

Books
Boschert, Sherry, *Plug-in Hybrids*, New Society Publishers, 2006
Doucette, Clive, *Urban Meltdown*, New Society Publishers, 2007
Heintzman, Andrew and Solomon, Evan, et al, *Fueling the Future*, Anansi, 2003
McMahon, Darryl, *The Emperor's New Hydrogen Economy*, iuniverse, 2006
Monbiot, George, *Heat*, Doubleday Canada, 2006
Rifkin, Jeremy, *The Hydrogen Economy*, Tarcher Putnam, 2002
Romm, Joseph J., *The Hype About Hydrogen*, Island Press, 2005

Trade Publications and Magazines and News Services:
Home Power Magazine
Website: www.homepower.com
Phone: 800-707-6585
This magazine bills itself as "The hands-on journal of home-made power". Based in Oregon, it provides extensive details related to producing alternate energy.

Electrifying Times
www.electrifyingtimes.com

EV World online magazine
www.evworld.com

Green Car Congress
www.greencarcongress.com

Megawatt Motorworks
www.megawattmotorworks.com

Electric Vehicle Related Websites

Electric Auto Association
www.eaaev.org/

Plug-in America
www.pluginamerica.com

California Cars Initiative, promoting 100 MPG cars
www.calcars.org

Electric Vehicle Society of Canada
www.evsociety.ca/

Advanced Hybrid Vehicle Development Consortium
www.hybridconsortium.org

Plug In Partners, demonstrating demand for Plug-In Hybrids
www.pluginpartners.org

Sierra Club Angeles Chapter
www.Angeles.SierraClub.org

National Biodiesel Board
www.biodiesel.org

AC Propulsion, innovator of electric cars & drive systems
www.acpropulsion.com

Myers Motors, makers of the 3-wheel "NmG"
www.myersmotors.com

Tesla Motors, maker of high performance roadster
www.teslamotors.com

Hymotion, aftermarket plug-in hybrid conversions
www.hymotion.com/index.htm

Vehicle Conversion Companies

Advanced DC Motors
Phone: (315) 434-9303
Website: www.adcmotors.com

Electro Automotive
Phone: 831-429-1989
Website: www.electroauto.com

KTA Services
Phone: (909) 949-7914
Website: www.kta-ev.com

REV Consultants
Email: richardl@revconsultants.com
Website: www.revconsultants.com

Battery Manufacturers

Dyno Battery Inc.
Website: www.dynobattery.com
Phone: 206-283-7450

IBE Battery
Website: www.ibe-inc.com
Phone: 818-767-7067

Rolls Battery Engineering (USA)
Surrette Battery Company (Canada)
Website: www.surrette.com
Phone: 800-681-9914

Trojan Battery Company
Website: www.trojanbattery.com
Phone: 800-423-6569

U.S. Battery Manufacturing Company
Website: www.usbattery.com
Phone: 800-695-0945

Hydrogen Recombining Caps

Hydrocap Catalyst Battery Caps
Website: N/A
Phone: 305-696-2504

Generator Systems

Fischer Panda Generators
Website: www.fischerpanda.com
Phone: 800-508-6494

Generac Power Systems Inc.
Website: www.generac.com
Phone: N/A (sold through Home Depot)

Hardy Diesel & Equipment Inc.
Website: www.hardydiesel.com
Phone: 800-341-7027

Kohler Power Systems
Website: www.kohlerpowersystems.com
Phone: 800-544-2444

Energy Meters

Bogart Engineering
Website: www.borartengineering.com
Phone: 831-338-0616

Brand Electronics
Website: www.brandelectronics.com
Phone: 207-549-3401

Xantrex Technology Inc.
Website: www.xantrex.com
Phone: 360-435-2220

Miscellaneous

Automation Direct
Website: www.automationdirect.com

Bussmann
Website: www.bussmann.com
Phone: 314-527-3877
Fuses and electrical safety components.

Digi-Key (Canada and USA)
Website: www.digikey.com
Phone: 800-DIGI-KEY
Supplier of many electrical wiring components.

Electro Sonic Inc. (Canada)
Website: www.e-sonic.com
Phone: 800-56-SONIC
Supplier of many electrical wiring components.

Real Goods
Website: www.realgoods.com
Phone: 800-919-2400
Suppliers specializing in renewable energy systems.

Siemens Energy and Automation Inc.
Website: www.siemens.com
Phone: 404-751-2000
Fused disconnect switches and circuit breakers.

Xantrex Technology Inc.
Website: www.xantrex.com
Phone: 360-435-2220
DC circuit breakers, battery cables, power centers, metering shunts, fuses.

PLC Training and Simulator
www.plcs.net/contents.shtml

Appendix B - Engine Management Software Listing
Ladder Logic Software Source File Listing for Zero-Carbon Car - Release 1.0

7/23/2007   06   plc_car

Path: c:\sequence\electric_hybrid\plc_car\plc_car.prj
Save Date: 07/13/07 08:01:05
Creation Date: 02/10/07 11:47:36
PLC Type: 06
Class ID: DirectLogic 06 Series

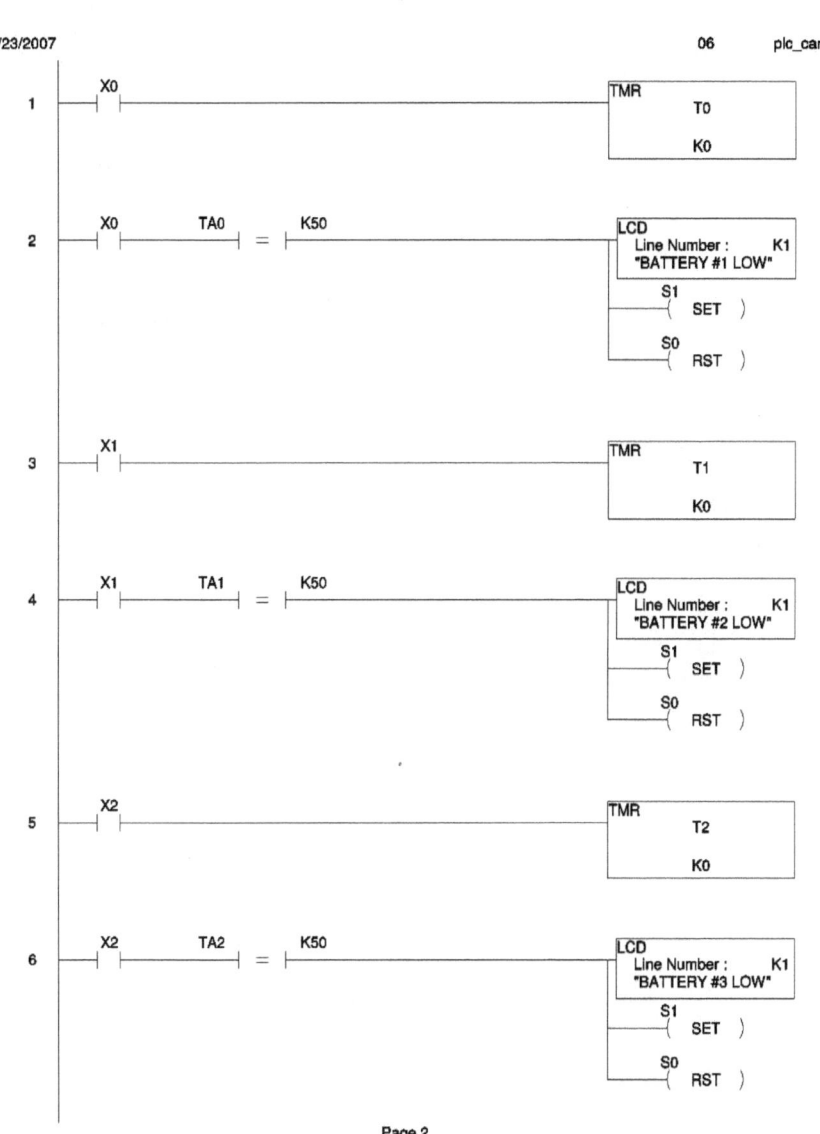

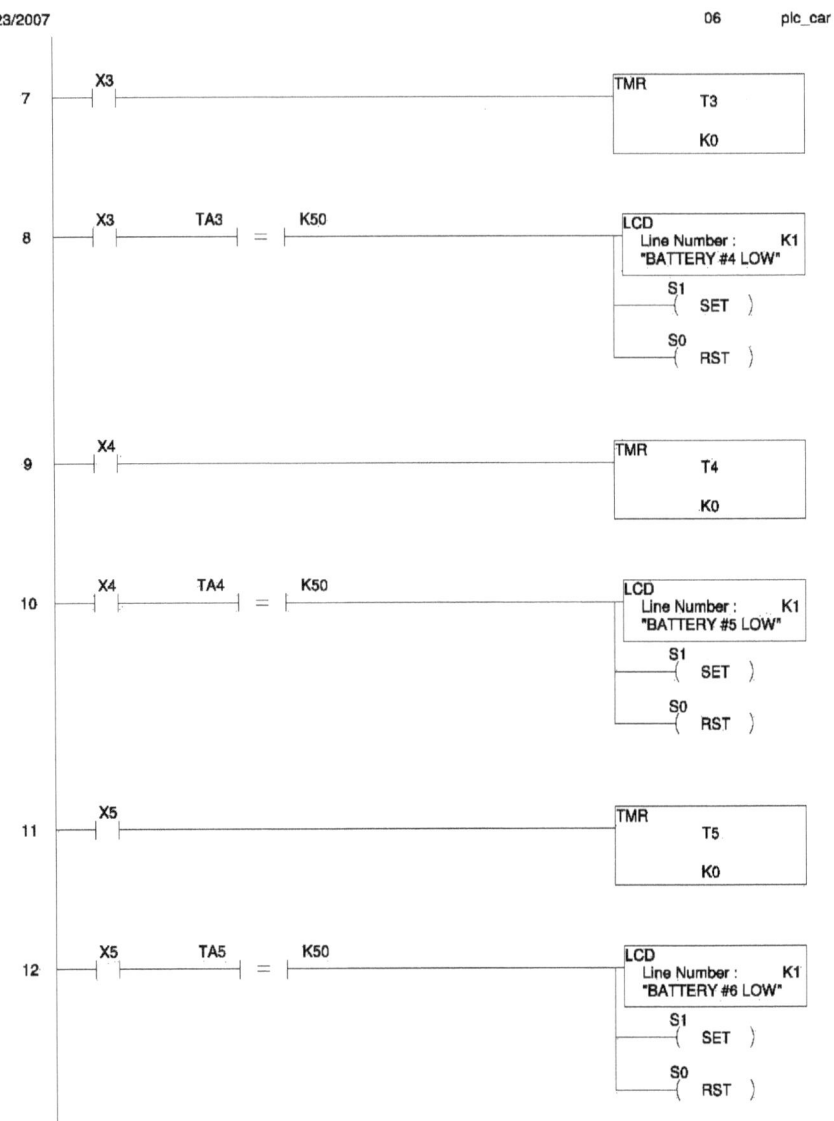

Plug-In Hybrid Vehicle Technology

511

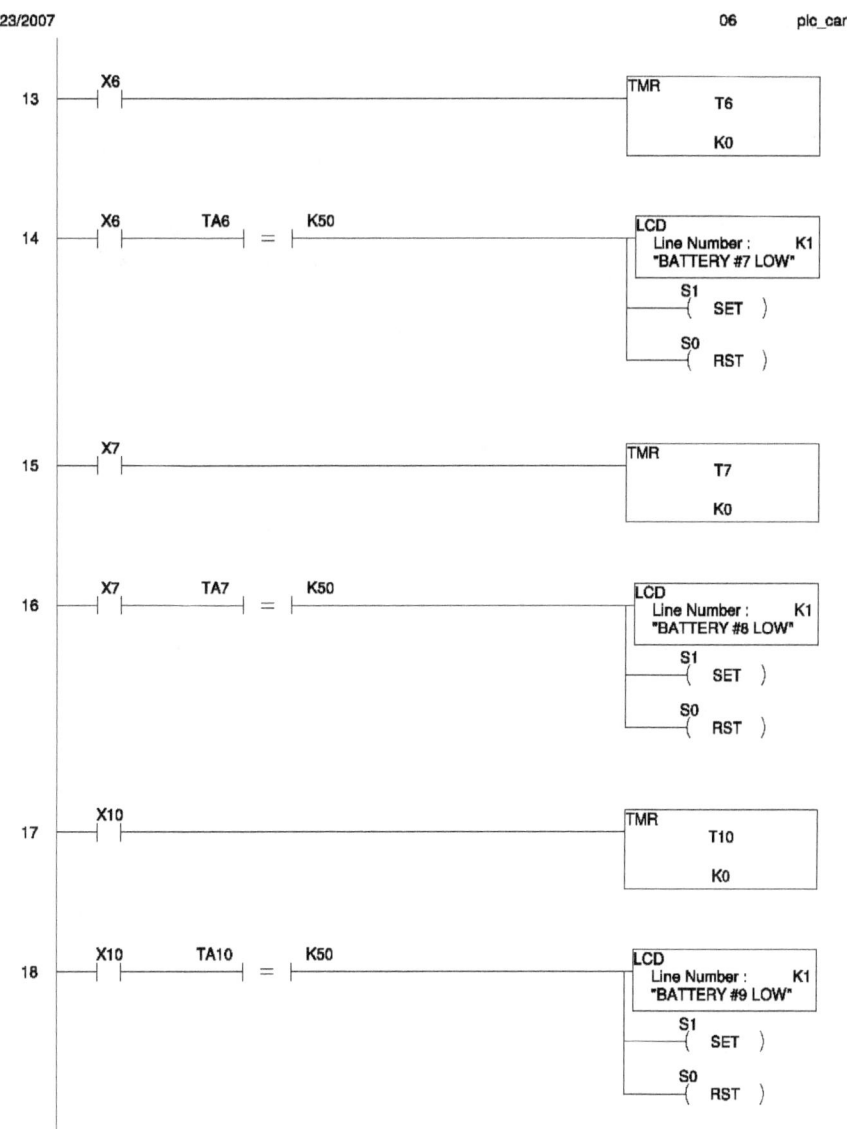

Page 4

Rung								
19	X11					TMR	T11	K0
20	X11	TA11	K50			LCD Line Number : K1 "BATTERY #10 LOW"	S1 SET	S0 RST
21	X11	TA11	K9000			TMR	T12	K0
22	X11	TA12	K9000			TMR	T13	K0
23	X11	TA13	K9000			TMR	T14	K0
24	X11	TA14	K9000			LCD Line Number : K2 "BATTERY FAULT"	Y0 RST	Y17 RST
								A
25	X0	X1	X2	X3	X4	X5	X6	

Plug-In Hybrid Vehicle Technology

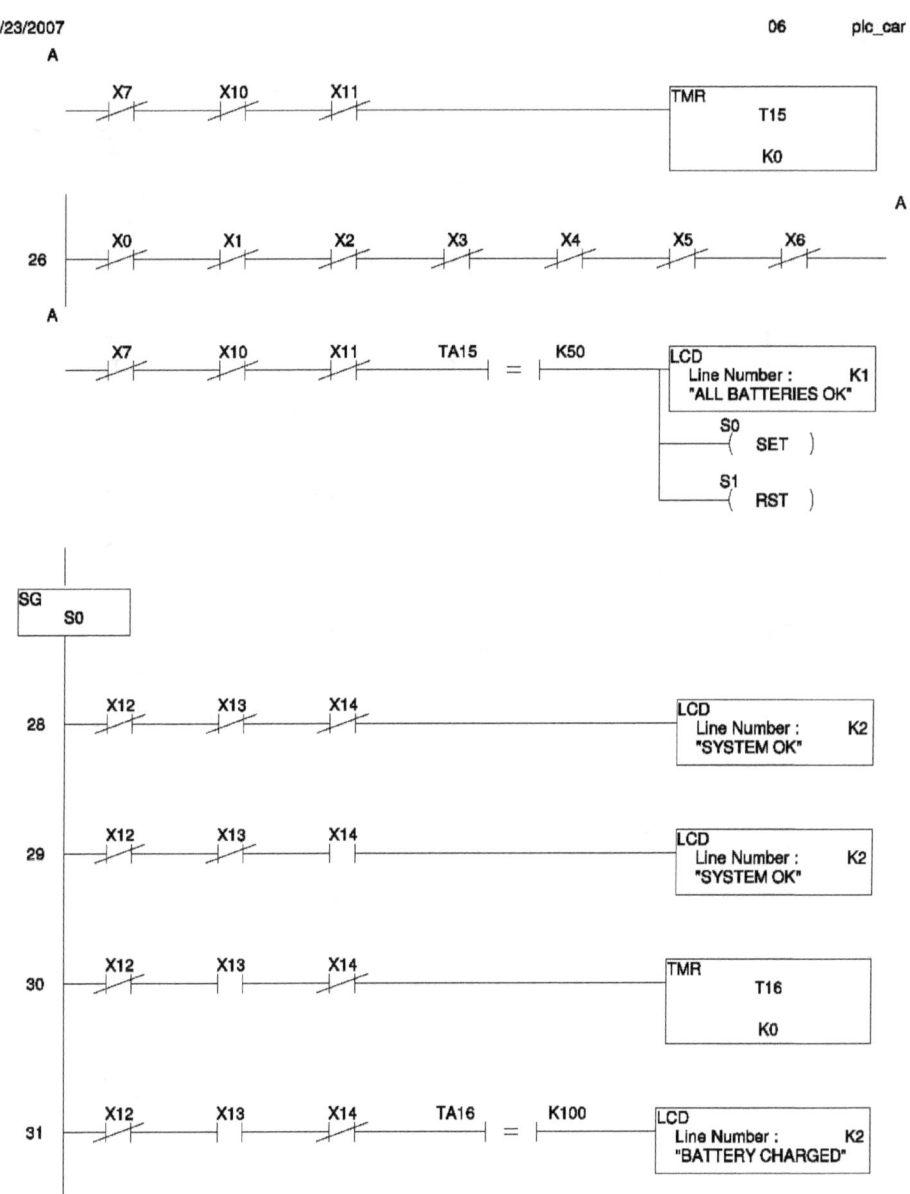

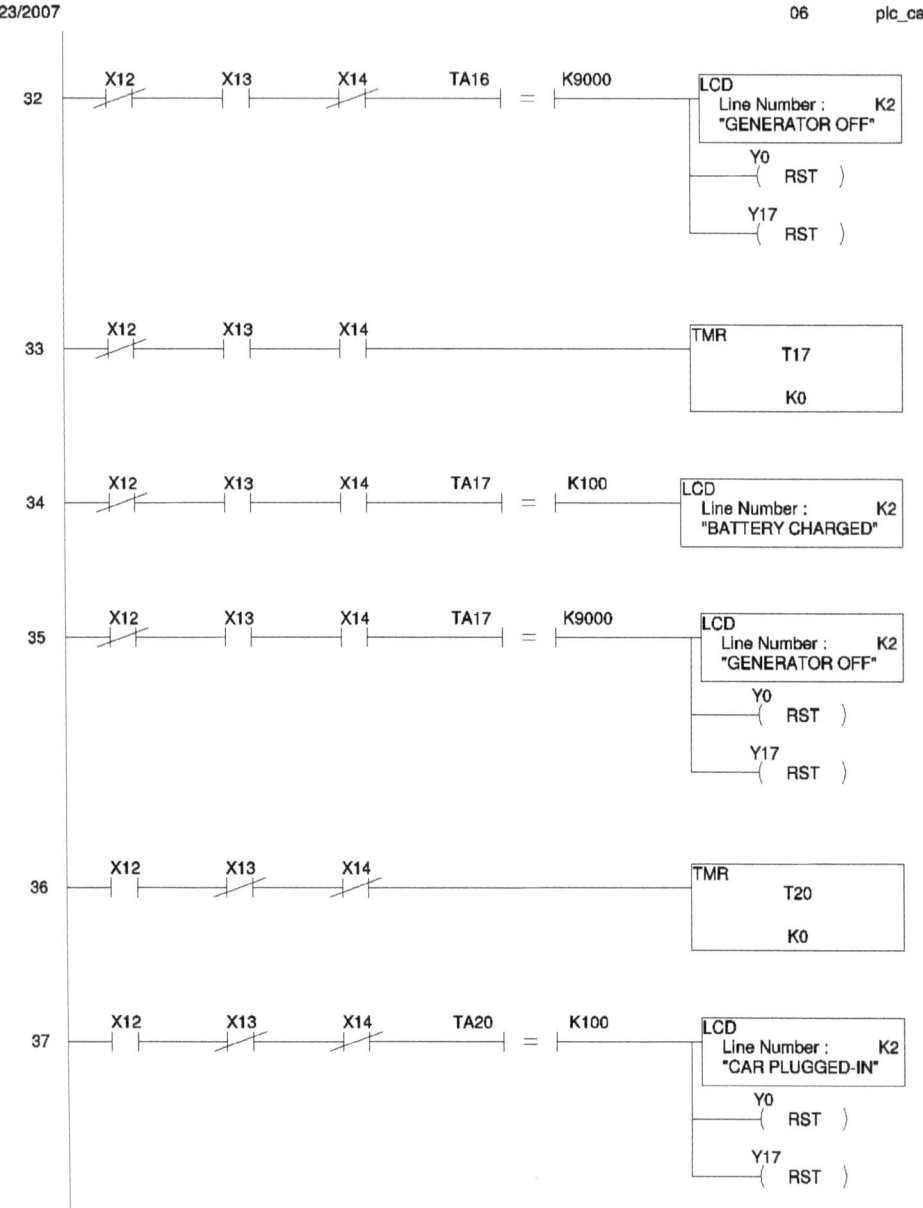

Plug-In Hybrid Vehicle Technology

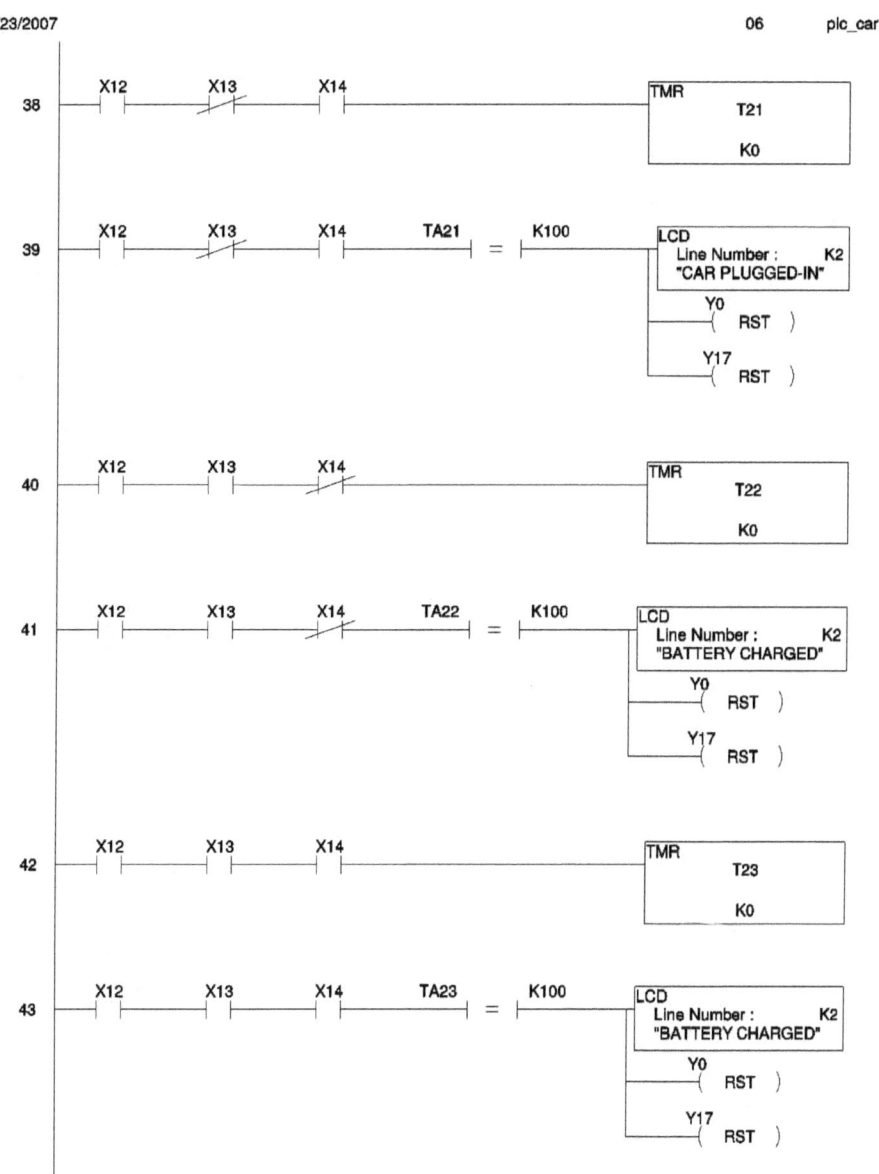

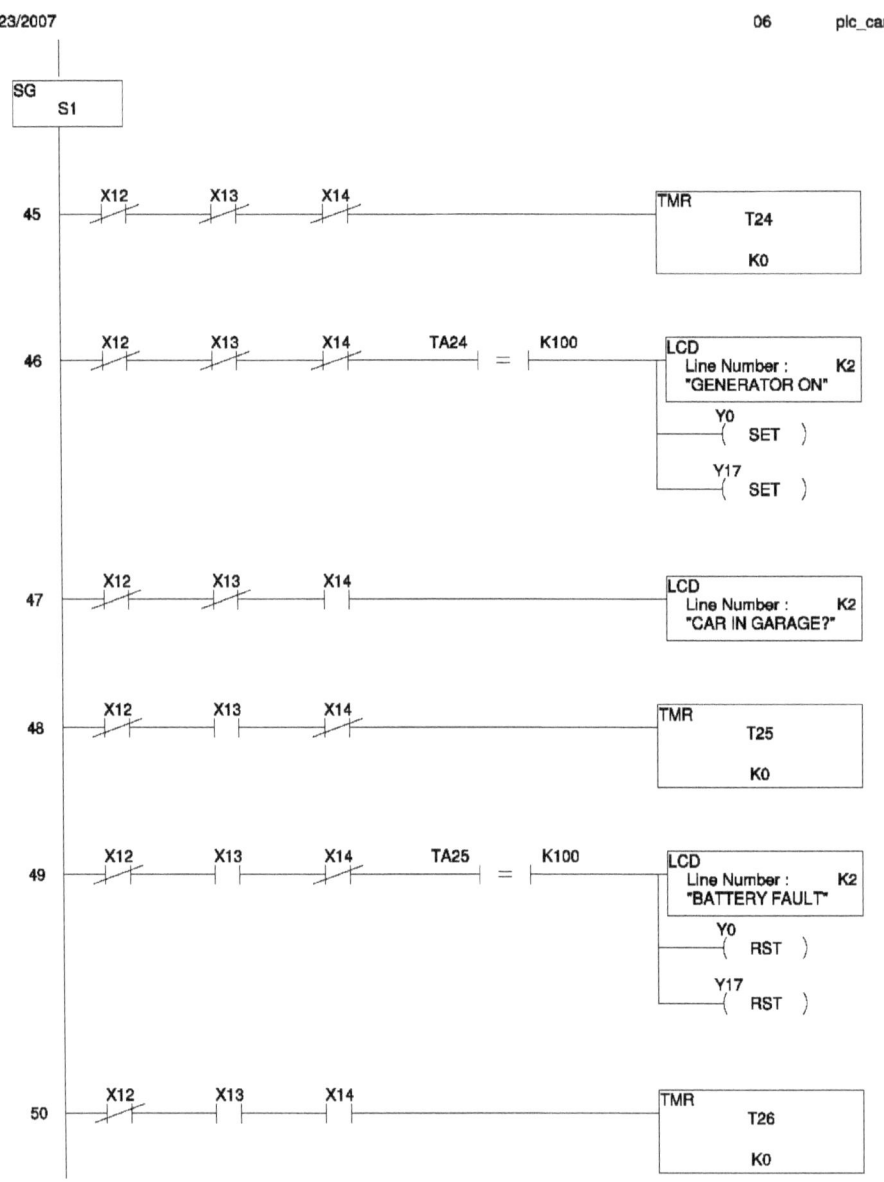

Plug-In Hybrid Vehicle Technology

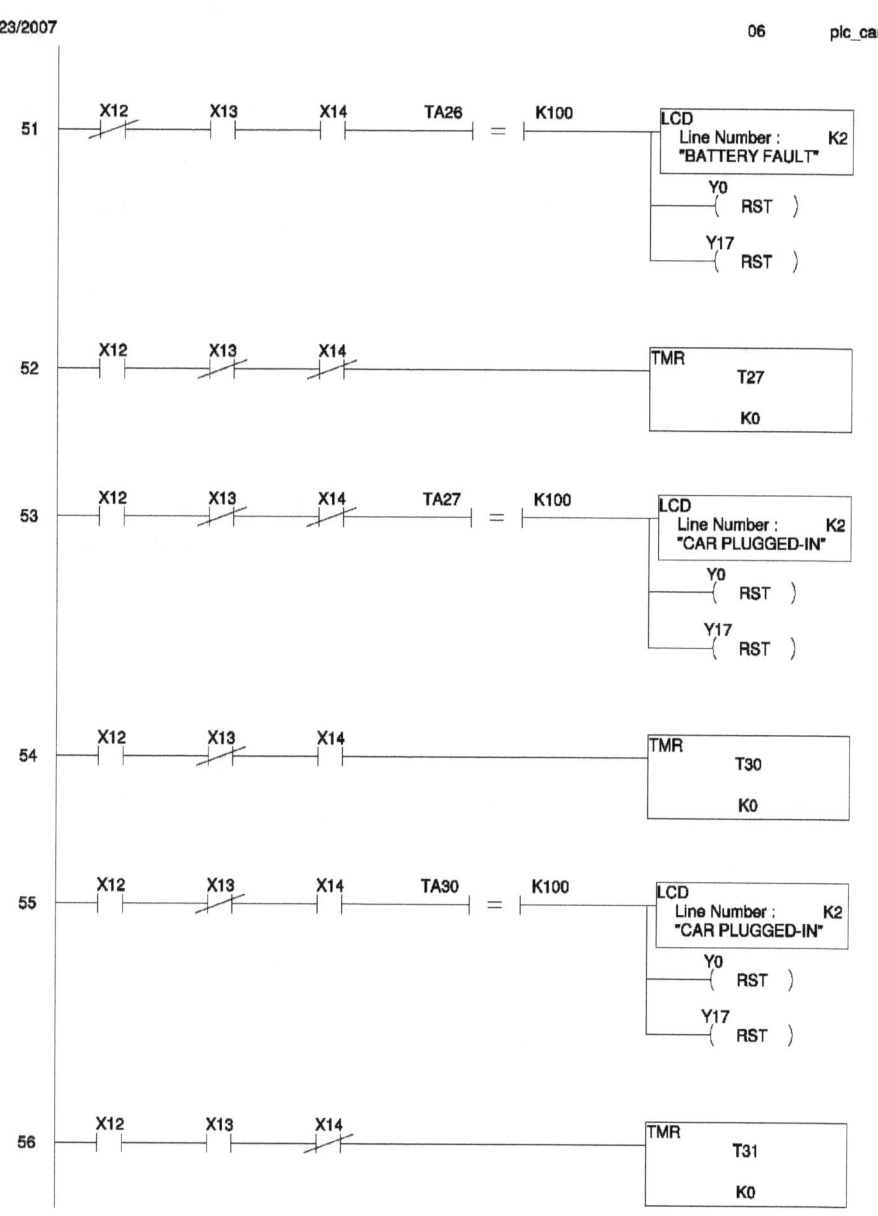

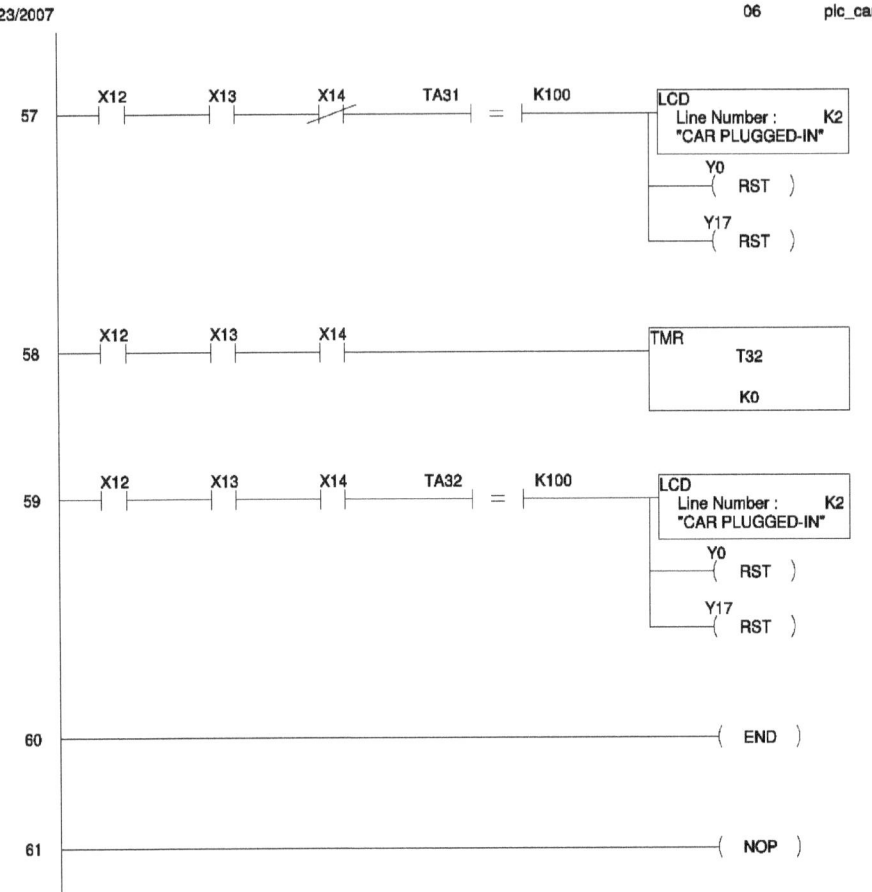

Appendix C - Zero-Carbon Car Specifications

Chassis: 2000 Mazda Miata Plug-in Hybrid Electric Vehicle
Weight: 3110 pounds (1413 kilograms)
Transmission: 5 speed manual (only 3rd and 4th gears plus reverse used)
Motor: Advanced DC Motors 9 inch model FB1-4001A
Horsepower: Rated 28 Hp, peak 85 hp (21 kW rated / 63.5 kW peak)
Torque: 90 foot-pounds
Battery Bank: Optima D750S Spiral Cells – Sealed Lead-Acid – 120 Volt
Controller: Curtis PMC 120V 500A #1231C-8601
DC to DC converter: Vicor VI-200 Series 120-13.8V 150Watt
Range:
  Electricity: 12.4 miles (20 kilometers) to 80% depth of discharge
  Consumption: 387 Wh per mile (240 Wh per kilomter)
  Biodiesel: Approximately 990 miles (1600 kilometers) per tank
  Consumption: 0.058 gallons (0.22 litres) per kWh

Speed mph/kph	Voltage (Volts)	Current (Amps)	3rd Gear Power horsepower/kW
37/60	110.9	112	16.6/12.4
43/70	111.5	128	19/14.2
50/80	111.7	170	25.5/19
55/90	111.6	204	30.5/22.8
			4th Gear
37/60	109.0	110	16/12
43/70	108.5	115	17/12.5
50/80	108.1	157	23/17
55/90	108.1	174	25/18.8

Table C-1. This table relates the speed of the Zero-Carbon Car to the battery bank voltage while under a given load shown in the third column in amps. The last column shows the actual power consumed by the electric motor in both horsepower and kilowatts.

Appendix D - LED Battery Voltage Indicator Schematic and Circuit Description

Circuit Design and Overview, © *Richard Hatherill, Aurial System*

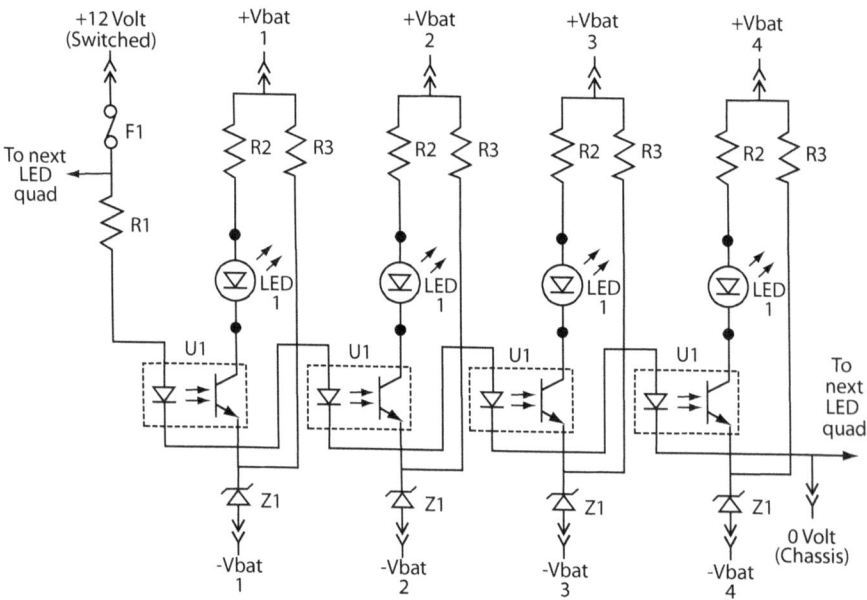

©2004 Richard Hatherill AurialSystems.com

Circuit Description

Overview

The circuit is designed to provide a visual indication of the state of a bank of batteries such as those used in a electric vehicle.

A set of LEDs (light emitting diodes), each corresponding to a battery in the bank, vary in light intensity depending on the voltage of each battery.

The individual LED intensities are reduced as a load is put on the batteries and the battery voltages drop. The LED intensities will fall and rise together if the batteries are matched, but the LED intensity of a weak battery will fall more than the others, and it will be immediately obvious that there is a problem with the battery.

Provision is made to disable the display when the batteries are on charge to prevent overdriving of the LEDs.

The LED display can be mounted so as to provide an instantaneous view of the state of the battery bank.

Operation

The design is based on groups of four LEDs or quads. A twelve battery bank would require three quads, a sixteen battery bank four quads.

Each indicator circuit is connected via sense leads to the positive and negative connection of its respective battery. It is best to use individual connections close to the terminals to avoid the effect of voltage drops in the heavy current wiring. It is also advisable to insert a 1 A pico-fuse into each sense lead where it connects to the battery, to guard against accidental short-circuits.

Each indicator circuit consists of a resistor, LED, opto-coupled transistor, and zener diode connected in series, together with a parallel zener diode bias current resistor.

The LED drive current for the opto-coupler is normally off until the display is to be activated. The display is turned on by putting enough drive current (about 80 mA) into the opto-coupler LED to insure that the output transistor is driven into saturation. This drive current is shared by four opto-couplers such that their combined LED forward voltage drop adds up to about 6 volts, a resistor connected to the 13.5 volt vehicle auxiliary battery via the 'ignition' switch sets the required current.

The combined voltage drop of the display LED and zener diode is set to around the discharge cut-off voltage of the battery (nominally 10 volts for a twelve volt battery).

The display LED current varies from around 20 mA when the battery is at 13 volts to close to zero at 10 volts.

The zener bias current resistor ensures that the zener voltage remains reasonably constant over the full LED current range.

Comments

I am using the 12 volt version of this circuit in my Honda CRX electric conversion, and have found it a far easier way of monitoring the battery pack state while driving than using the digital voltmeter.

Please let me know if you use this design, any feedback would be welcome. You may also contact me if you require assistance.

This design is not patented, but if anybody intends the commercialize the design I would appreciate them contacting me to discuss a reasonable royalty arrangement.

Richard Hatherill
Aurial Systems

Parts List

Note: The design is based on sections of four-LED (quads). Twelve batteries would require 3 quads, sixteen 4 quads.

6 Volt Batteries

Component Number	Description	Part Number	Quantity
F1	Fuse	1 A	1 Per Unit
R1	Resistor	100 ohm, 1 W	1 Per Quad
R2	Resistor	33 ohm, 1/4 W	4 Per Quad
R3	Resistor	200 ohm, 1/4 W	4 Per Quad
LED 1	High Intensity Green LED	As preferred	4 Per Quad
U1	Opto-coupler *Note: CNY17-3 would work but -4 is better*	Fairchild CNY17-4	4 Per Quad
Z1	Zener Diode 3.3 V, ½ W	Fairchild 1N5226B	4 Per Quad

12 Volt Batteries

Component Number	Description	Part Number	Quantity
F1	Fuse	1 A	1 Per Unit
R1	Resistor	100 ohm, 1 W	1 Per Quad
R2	Resistor	120 ohm, 1/4 W	4 Per Quad
R3	Resistor	680 ohm, 1/4 W	4 Per Quad
LED 1	High Intensity Green LED	As preferred	4 Per Quad
U1	Opto-coupler *Note: CNY17-3 would work but -4 is better*	Fairchild CNY17-4	4 Per Quad
Z1	Zener Diode 8.2 V, ½ W	Fairchild 1N5237B	4 Per Quad

Appendix E
Battery Voltage Monitor
PCB Layout & Part Locator
(Typical for all channels)

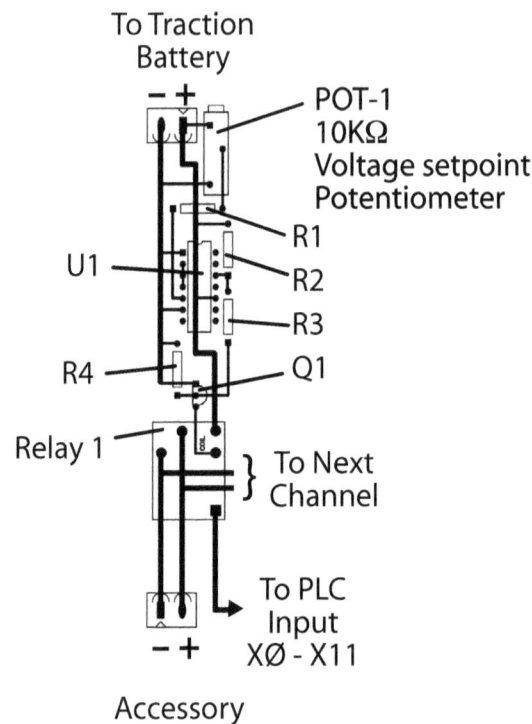

R1-R4 : 10KΩ 1/4W
U1   : LM322N
Relay 1: AZ8-1CH-12DE
Q1   : MPSA13
POT   : Trimpot 3006P

Notes

Introduction
1. International Organization of Motor Vehicle Manufacturers, The World's Automotive Industry: Some Key Figures (November 2006), http://www.oica.net/htdocs/statistics/OICA_depliant-final.pdf
2. Goethe-Institut, Germany—Motor Mecca (March 2007), http://www.goethe.de/ges/wrt/dos/aut/zuk/en2161667.htm
3. U.S. Department of Transportation, Bureau of Transportation Statistics, Transportation Statistics Annual Report (November 2005), http://www.bts.gov/publications/transportation_statistics_annual_report/2005/
4. Ibid.
5. Ibid.
6. International Organization of Motor Vehicle Manufacturers, The World's Automotive Industry: Some Key Figures (November 2006), http://www.oica.net/htdocs/statistics/OICA_depliant-final.pdf
7. Pricewaterhousecoopers Global Automotive Shareholder Value Awards, page 5. http://64.233.167.104/search?q=cache:Th7zqOhXAggJ:www.pwc.com/Extweb/pwcpublications.nsf/docid/5D6579CABB10D5A98525726100717823/%24FILE/2006GSHVAfinal.pdf+pricewaterhousecoopers+automotive+manufacturing+forecast&hl=en&ct=clnk&cd=1&gl=ca&client=firefox-a
8. Susan A. Shaheen and Elliot Martin, Assessing Early Market Potential for Carsharing in China: A Case Study of Beijing (University of California, Davis, 2006), http://www.carsharing.net/library/UCD-ITS-RR-06-21.pdf
9. U.S. Department of Transportation, Bureau of Transportation Statistics, Transportation Statistics Annual Report (November 2005), http://www.bts.gov/publications/transportation_statistics_annual_report/2005/
10. Ibid.
11. The Economist (May 5, 2007): 75.
12. U.S. Department of Transportation, Bureau of Transportation Statistics, Transportation Statistics Annual Report (November 2005), http://www.bts.gov/publications/transportation_statistics_annual_report/2005/

Chapter 1
1. International Organization of Motor Vehicle Manufacturers, The World's Automotive Industry:Some Key Figures (November 2006), http://www.oica.net/htdocs/statistics/OICA_depliant-final.pdf
2. Midori Matsuoka, "Asia Leads Global Car Ownership Aspirations," ACNielsen. Trends & Insights, www2.acnielsen.com/pubs/2005_q1_ap_car.shtml
3. Ibid.
4. Durhl Caussey, "2007 Hummer H2 SUV," The Epoch Times, February 11, 2007, http://en.epochtimes.com/news/7-2-11/51592.html
5. "Automobile Dealers—Luxury, Hoover's, 2007, http://www.hoovers.com/automobile-dealers---luxury/--ID__32--/free-ind-fr-profile-basic.xhtml
6. U.S. Department of Transportation, Bureau of Transportation Statistics, Highlights of the 2001 National Household Travel Survey, "Daily Passenger Travel," http://www.bts.gov/publications/highlights_of_the_2001_national_household_travel_survey/html/section_02.html
7. Ibid.

8. Ibid.
9. U.S. Department of Transportation, Bureau of Transportation Statistics, Highlights of the 2001 National Household Travel Survey, "Long-Distance Travel," http://www.bts.gov/publications/highlights_of_the_2001_national_household_travel_survey/html/section_03.html
10. Ibid.
11. Ibid.
12. U.S. Department of Transportation, Bureau of Transportation Statistics, Transportation Statistics Annual Report (November 2005), http://www.bts.gov/publications/transportation_statistics_annual_report/2005/
13. Energy Information Administration, Energy Information Sheets Index, Petroleum Products Consumption, (October 2006), http://www.eia.doe.gov/neic/infosheets/petroleumproductsconsumption.html
14. U.S. Department of Transportation, Bureau of Transportation Statistics, Transportation Statistics Annual Report (November 2005), http://www.bts.gov/publications/transportation_statistics_annual_report/2005/html/chapter_02/cost_of_owning_and_operating_an_automobile.html
15. Population Resource Centre, A Population Perspective of the United States, (2004), http://www.prcdc.org/summaries/uspopperspec/uspopperspec.html
16. Heather L. MacLean & Lester B. Lave, "A Life-Cycle Model of an Automobile," Environmental Policy Analysis v 3, no.7 (1998): 22A-330A. Summary available at: http://www.ilea.org/lcas/macleanlave1998.html
17. Ibid.
18. Greenpeace Report, The Environmental Impact of The Car, 1992.
19. U.S. Department of Transportation, Bureau of Transportation Statistics, Transportation Statistics Annual Report (November 2005), http://www.bts.gov/publications/transportation_statistics_annual_report/2005/html/chapter_02/vehicle_loadings_on_the_interstate_highway_system.html

Chapter 2

1. Ross Moore, "Consumers and Business Must Brace Themselves for Higher Parking Rates," Collier International (July 25, 2007), http://www.colliers.com/Markets/USA/News/2007ParkingRelease
2. Current U.S. Oil consumption is approximately 20.8 million barrels of oil per day or 7.6 billion barrels per year. Proven U.S. oil reserves are 21.4 billion barrels, yielding a "depletion to zero supply" time of 2.81 years assuming no imports of oil were used to support domestic supplies, http://www.eia.doe.gov/emeu/cabs/Usa/Oil.html
3. Ibid.
4. Energy Information Administration, Country Analysis Briefs, "Canada" (April, 2007), http://www.eia.doe.gov/emeu/cabs/Canada/Oil.html
5. Energy Information Administration, Country Analysis Briefs, "Canada" (April, 2007), http://www.eia.doe.gov/emeu/cabs/Canada/NaturalGas.html
6. Rob Woronuk, Basin Deliverability in the WCSB, Rob Woronuk, Senior Analyst, Canadian Gas Potential Committee, http://www.canadiangaspotential.com/papers_presentations/futurepotentialWCSB-woronuk.doc
7. Ibid.
8. Energy Information Administration, Country Analysis Briefs, "Mexico" (January, 2007), http://www.eia.doe.gov/emeu/cabs/Mexico/Oil.html

9. Matthew R. Simmons, Twilight in the Desert: The Coming Saudi Oil Shock and the World Economy (New Jersey: John Wiley & Sons, 2005), 290.
10. Energy Information Administration, Country Analysis Briefs, "Mexico" (January, 2007), http://www.eia.doe.gov/emeu/cabs/Mexico/Oil.html
11. "Worldwide Look at Reserves and Production," Oil and Gas Journal, December 18, 2006, 24-25. North American total proven reserves including oil sands, as of January 1, 2007, are 213 billion barrels. North American aggregate oil consumption is 9.2 billion barrels per year, yielding a "deplete to zero" time of 23.1 years. If only conventional oil sources are considered, proven reserves are 61 billion barrels, leading to full depletion in 6.6 years. These calculations assume no change in oil consumption or reserve capacity.
12. Energy Information Administration, Country Analysis Briefs, "Saudia Arabia," (February, 2007), http://www.eia.doe.gov/emeu/cabs/Saudi_Arabia/Oil.html
13. Ibid.
14. Matthew R. Simmons, Twilight in the Desert: The Coming Saudi Oil Shock and the World Economy (New Jersey: John Wiley & Sons, 2005), 283.
15. Patrick Clawson and Simon Henderson, "Reducing Vulnerability to Middle East Energy Shocks: A Key Element in Strengthening U.S. Energy Security," The Washington Institute for Near East Policy, (November, 2005) http://www.washingtoninstitute.org/templateC04.php?CID=227.
16. Energy Information Administration, Country Analysis Briefs, "China," (August, 2006), http://www.eia.doe.gov/emeu/cabs/China/Oil.html
17. Ibid.
18. A quadrillion is a stunningly large number that is meaningless for most people to comprehend: a one followed by 24 zeros. A Btu or "British thermal unit" is the amount of heat energy necessary to raise the temperature of one pound of water one degree Fahrenheit. It was originally used to estimate the amount of heat energy in one match. Equalizing energy magnitude to a common root allows cross comparison in relative total energy consumption without having to worry about differing units normally used for these measurements. Source: U.S. Energy Information Administration, International Energy Outlook 2007, http://www.eia.doe.gov/oiaf/ieo/world.html
19. Energy Information Administration, International Energy Outlook 2007, http://www.eia.doe.gov/oiaf/ieo/oil.html
20. Ibid.
21. Ibid.
22. Ibid.
23. Patrick Brethour, "Only conservation efforts will keep a lid on energy costs: IEA," The Globe and Mail, 3 December 2005, B6.
24. Energy Information Administration, Country Analysis Briefs, "United States," (November 2005), http://www.eia.doe.gov/emeu/cabs/Usa/Oil.html
25. Ibid.
26. Tom Bergin, "Shell Cuts Oil Reserves Again," Energy Bulletin, February 2, 2005, http://www.energybulletin.net/4220.html.
27. Global Upstream Performance Review and Andrew Barr / National Post

Chapter 3
1. Ontario Medical Association, The Illness Costs of Air Pollution, (June 2005), http://www.oma.org/phealth/smogmain.htm,.

2. Micron is the short form for micrometer, a unit of measure that is one millionth of a meter, or one thousandth of a millimeter. A human hair is typically 100 microns in diameter.
3. Susan Watts, A Coal-Dependent Future," BBC News (March 9, 2005), http://news.bbc.co.uk/2/hi/programmes/newsnight/4330469.stm
   Although electrical energy demand is increasing and many countries are attempting to develop their renewable and clean-power generation technologies, high demand and energy inefficiency are creating a huge market for traditional power-generating companies and their fossil fuel-based products.
4. "A Plan to Keep Carbon in Check," Scientific American, September 2006, 50.
5. Intergovernmental Panel on Climate Change, Climate Change 2001: Synthesis Report, http://www.grida.no/climate/ipcc_tar/vol4/english/index.htm. See also updated report, Climate Change 2007: The Physical Science Basis at http://www.ipcc.ch
6. "Changing Science: Climatology," The Economist, December 10, 2005, 91.
7. Intergovernmental Panel on Climate Change, Climate Change 2007: The Physical Science Basis, http://www.ipcc.ch.
8. Most major reports, including the IPCC report described earlier, suggest holding carbon dioxide emission levels at or below those of the year 2000. Author George Monbiot proposes in his book Heat a simpler concept: reduce current carbon emissions by 90% to achieve the same goal.
9. The Kyoto Protocol was adopted at the Third Session of the Conference of the Parties (COP) to the UN Framework Convention on Climate Change (UNFCCC) in 1997 in Kyoto, Japan. It contains legally binding commitments in addition to those included in the UNFCCC. Country signatories to the Protocol agreed to reduce their anthropogenic emissions of greenhouse gases (CO_2, CH_4, N_2O, HFCs, PFCs, and SF_6) by at least 5 % below 1990 levels in the commitment period 2008 to 2012.
10. Professors Robert Socolow and Stephen Pacala of the Carbon Mitigation Initiative at Princeton University have proposed a "wedge theory" wherein society would have to reduce emissions by seven wedges to stabilize emissions at current levels, which are already twice the pre-industrialized values. One wedge equates to the equivalent of one billion tons of averted emissions 50 years from now. Their report is outlined in "A Plan to Keep Carbon in Check," Scientific American, September 2006, 50.
11. Each liter of gasoline, weighing approximately 756 grams, yields 2.43 kilograms of carbon dioxide emissions. When burned, the carbon and hydrogen of gasoline disassociate, with the carbon bonding to atmospheric oxygen. A carbon atom has an atomic weight of 12 and each oxygen atom has a weight of 16, resulting in a combined atomic weight of 44 for the carbon dioxide molecule. The carbon content of gasoline is 87% of the total weight (87% of 0.756 kg/liter = 0.658 kg carbon/liter). The ratio of CO_2 to carbon by mass is 44/12 or 3.7. Therefore 3.7 ratio carbon to CO_2 x 0.658kg of carbon yields 2.43 kg of carbon dioxide per liter of gasoline burned. Simple!
12. Assuming a carbon dioxide price of US$50 per short ton, this works out to $0.025 per pound of carbon dioxide emissions. Each gallon of gasoline burned produces 20 pounds of CO_2. $0.025 per pound of CO_2 x 20 pounds of CO_2 emitted per gallon of gasoline yields US$0.50 per gallon carbon tax.

13. U.S. Department of Transportation, Federal Highway Administration, *Highway Statistics 2003* (Washington, DC: 2004), table HM-64

Chapter 4
1. "Far Away Yet Strangely Personal," *The Economist* (August 25, 2007), http://www.economist.com/business/displaystory.cfm?story_id=9687655
2. 1,361 kilograms of carbon emissions for the trip equates to 1.361 metric tonnes. Multiplied by $500 per tonne = $680.00.
3. "An American Epidemic," *The Economist*, (February 17, 2007), http://www.economist.com/world/na/displaystory.cfm?story_id=8715403
4. Ibid.
5. Americans choose public transit for 1.5% of all trips, while personal automobiles account for 86.5%. Therefore 86.5% / 1.5% = 58 times preference for driving over mass transit.
6. David Jamieson, Transport Minister, Hansard column 786W, 8 July 2004. http://www.parliament.the-stationery-office.co.uk/pa/cm200304/cmhansrd/vo040708/text/40708w05.htm
7. Duncan Austin et al., *Changing Drivers: The Impact of Climate Change on Competitiveness and Value Creation in the Automotive Industry*, 2003, http://pdf.wri.org/changing_drivers_full_report.pdf.
8. John B. Heywood, "Fuelling Our Transportation Future," *Scientific American*, September 2006, 60.
9. Toyota, "Sequoia 2007," http://www.toyota.ca/cgi-bin/WebObjects/WWW.woa/12/wo/Home.Vehicles.Sequoia-MCw6JZ4YkYjXctkBc_QzBmM/3.7?fmg%2fsequoia%2fintro%2ehtml
10. US Department of Transportation, *Transportation Statistics Annual Report*, November 2005 Page 26.
11. Total passenger miles of travel (pmt) excludes travel by walking, bicycling, boat, and heavy truck. Independent rounding and source data errors cause percentages not to add to 100.
12. US Department of Transportation, Bureau of Transportation Statistics, *2001 National Household Travel Survey Data* (February 2004),
13. Ibid.
14. Daniel Pulliam, "IRS Raises Mileage Rate to 48.5 Cents a Mile," *Government Executive.com* (September 9, 2005), http://www.govexec.com/story_page.cfm?articleid=32204&dcn=todaysnews
15. Pace, "How Much Does it Cost to Drive? (2006)," http://www.pacebus.com/sub/vanpool/cost_of_driving.asp
16. George Monbiot, *Heat: How to Stop the Planet from Burning* (Doubleday Canada, 2006), 185.
17. Maersk Line, "Emma Maersk," http://www.maerskline.com/link/?page=brochure&path=/about_us/facts_and_figures/emma_facts
18. The Maersk Line website states that the *Emma Maersk* can travel 66 kilometers using 1 kWh of energy per ton of cargo. A jumbo jet can travel only 500 meters using the same amount of energy and carrying the same load.
19. Monbiot quotes George Marshall of the Climate Outreach Information Network as saying that each passenger on a fully loaded *QEII* is responsible for 9.1 tonnes of carbon emissions. At $500 per tonne of carbon, the result is $4,550 per passenger.

20. The Maersk Line, "Emma Maersk," http://www.maerskline.com/link/?page=brochure&path=/about_us/facts_and_figures/emma_facts
21. Australian Uranium Association, Nuclear-powered Ships, (March 2007), http://www.uic.com.au/nip32.htm
22. Ibid.
23. Ibid.
24. US Department of Transport, , Bureau of Transportation Statistics, *Transportation Statistics Annual Report* (November 2005), Report Summary: http://www.bts.gov/publications/freight_in_america/html/executive_summary.html
25. Ibid.
26. Ibid.
27. U.S. Department of Transport, , Bureau of Transportation Statistics, *Transportation Statistics Annual Report* (November 2005), 40-41, and Figures 1-19 and 1-20, http://www.bts.gov/publications/transportation_statistics_annual_report/2005/html/appendix_b/html/table_01_19.html
28. Ibid.
29. Ibid.
30. FedEx, Newsroom, http://news.van.fedex.com/node/506
31. http://www.maerskline.com/link/?page=brochure&path=/about_us/facts_and_figures/emma_facts
32. http://www.maerskline.com/link/?page=brochure&path=/our_services/containers
33. Channels, "Communications Help Port of Rotterdam Container Operator Speed Cargo Around the Globe," Channels Closeup, www.macom-wireless.com/Channels/volume2iss3/ecr6067_Port%20of%20Rotterdam.pdf

Chapter 5

1. US Department of Energy, "Advanced Technologies & Energy Efficiency," http://www.fueleconomy.gov/feg/atv.shtml
2. BTG Biomass Technology Group, "Flash Pyrolysis," (August5, 2004), http://www.btgworld.com/technologies/pyrolysis.html
3. Natural Resources Canada, "Cathodic Protection," (December 1, 2004), http://www2.nrcan.gc.ca/picon/topics/cathodicprotection_e.asp
4. US Department of Energy, "Advanced Technologies & Energy Efficiency," http://www.fueleconomy.gov/feg/atv.shtml
5. Glenn Ebert, The Physics Factbook, "Energy Density of Gasoline," (2003), http://hypertextbook.com/facts/2003/ArthurGolnik.shtml. Energy density of liquid fuels is more commonly annotated in megajoules to express heat energy. Hereinafter, all fuel energy contents will be expressed in kilowatt-hours (kWh) to allow for ease of comparing with electrical energy.
6. In this example, the passenger represents 10% of the total vehicular mass. Assuming that the Smart™ car has total vehicle efficiency of 8.4%, the energy attributed to moving the passenger is 1/10th of 8.4% = 0.84%
7. Rocky Mountain Institute, "What is a Hypercar® Vehicle?" (2007), http://www.rmi.org/sitepages/pid191.php
8. Monash University Accident Research Centre, "Vehicle Mass, Size and Safety," (November 15, 2005), http://www.monash.edu.au/muarc/reports/atsb133.html
9. Fiberforge, (2007), www.fiberforge.com
10. http://www.fiberforge.com/DOWNLOADS/FiberforgeBrochure.pdf

11. National Renewable Energy Laboratory, "Life Cycle Assessment of Hydrogen Production via Natural Gas Steam Reforming," (February 2001), http://www.nrel.gov/docs/fy01osti/27637.pdf
12. Ibid.

Chapter 6

1. United States Council for Automotive Research, "FreedomCAR and Fuel Partnership," (2006), http://www.uscar.org/guest/view_partnership.php?partnership_id=1
2. Electric Power Research Institute, Comparing the Benefits and Impacts of Hybrid Electric Vehicle Options for Compact Sedan and Sport Utility Vehicles, Report No. 1006892, (July 2002), http://www.evworld.com/library/EPRI_sedan_options.pdf
3. Wikipedia, "Citicar," (June 5, 2007), http://en.wikipedia.org/wiki/Citicar
4. An electric vehicle requiring 250 watt-hours of energy per mile would require 1kWh for every 4 miles. Assuming a delivered price of 10 cents per kWh and an "outlet-to-wheels" conversion efficiency of 30%, our example car would require 1.3 kWh of gross energy, costing 13 cents, to drive 4 miles, or approximately 3.25 cents per mile. A gasoline-powered vehicle that gets 25 miles per gallon would cost 12 cents per mile to operate, assuming a fuel cost of $3.00 per gallon.
5. Colin Angus Vincent and Bruno Scrosati, Modern Batteries: An Introduction to Electrochemical Power Sources (London: Arnold, 1997).
6. Assuming a BEV or modified Toyota Prius operates on 200 Wh per mile and that a 215-watt solar-electric panel has been added to the vehicle, the increase in battery charge in real-world conditions will be far less than indicated. A PV panel will only achieve peak output under laboratory conditions, and as a result of oblique sun angle and the charge efficiency and other factors, I would expect a drop in claimed mileage. Obviously this range extension would be reduced to zero on cloudy days or if the vehicle were parked in the shade. 215 Wpeak x 0.8 derating factor x 4 sun-hours per day x 30% derating for non-perpendicular sun angle x 28% system charging losses = 347 watts to drive the vehicle. Assuming 200 Wh per mile, this equates to 1.7 miles given these optimum sample values.
    1. 215 Wpeak is measured at 1,000 watts of solar energy under laboratory conditions. This will almost never occur in the real world. A derating factor of 0.8 is very conservative given that the panel's electrical power rating will decline further as a result of heat buildup on the vehicle roof.
    2. 4 sun-hours per day is the mean number of sun hours per day all year in North America. This value can vary dramatically but will typically be much lower than this, especially if the vehicle is parked in shady areas which typically exist in urban centers.
    3. Maximum PV power output is provided only when the panel is perpendicular to the sun's rays. This will almost never occur on a moving car, especially after solar noon when the sun drops in the sky.
    4. Charging system losses relate to the energy lost in converting electrical power to chemical storage in the vehicle's battery bank.
7. Electric Power Research Institute, Comparing the Benefits and Impacts of Hybrid Electric Vehicle Options for Compact Sedan and Sport Utility Vehicles, Report No. 1006892, (July 2002), http://www.evworld.com/library/EPRI_sedan_options.pdf

8. Ibid.
9. A PHEV that was recharged with renewable electricity from sources such as wind or solar-electric power and fueled with renewable fuels from appropriate sources would consume no gasoline and would therefore have infinite gasoline fuel economy. See Chapters 8 and 9 for further reading.
10. Audi, "Audi Q7 Hybrid," http://www.audi.co.uk/audi/uk/en2/experience/Studies/q7_hybrid_concept/hybrid_history.html
11. CBC News, "Nortel—The Wild Ride of Canada's Most-Watched Stock," (December 1, 2006), http://www.cbc.ca/news/background/nortel/stock.html
12. CBC News, "NBN Insight: Ballard Power Systems (November 10, 1000), http://www.cbc.ca/money/story/1999/04/27/ballard990427.html
13. Ibid.
14. http://www6.autonet.ca/Environment/story.cfm?story=/Environment/2007/05/16/4185783-cp.html
15. Pamela L. Spath and Margaret K. Mann, "Life Cycle Assessment of Hydrogen Production via Natural Gas Steam Reforming," National Renewable Energy Laboratory (February 2001), http://www.nrel.gov/docs/fy01osti/27637.pdf
   On an energy balance basis, each unit of methane consumed in the production process yields only 0.66 units of gaseous hydrogen. Further energy is consumed in the compression or liquefaction process and the US National Renewable Energy Laboratory estimates a final net energy ratio of 59%.
16. Toyota, "Hybrid Synergy View Newsletter" (Fall 2005), http://www.toyota.com/html/hybridsynergyview/2005/fall/hybridorhydrogen.html
17. Joseph J. Romm, The Hype About Hydrogen: Fact and Fiction in the Race to Save the Climate (Washington, DC: Island Press, 2004), 76.
18. Ulf Bossel and Baldur Eliasson, Energy and the Hydrogen Economy (January 8, 2003), www.hyweb.de/News/Bossel-Eliasson_2003_Hydrogen-Economy.pdf
19. Joseph J. Romm, The Hype About Hydrogen: Fact and Fiction in the Race to Save the Climate (Washington, DC: Island Press, 2004), 94.
20. Toyota, "Hybrid Synergy View Newsletter" (Fall 2005), http://www.toyota.com/html/hybridsynergyview/2005/fall/hybridorhydrogen.html

Chapter 7.2
1. Saft Industrial Battery Group, "STM Module," http://www.saftbatteries.com/130-Catalogue/PDF/data_stm_en.pdf
2. Electrical Power Research Institute, *Advanced Batteries for Electric-Drive Vehicles* (May 2004), http://www.evworld.com/library/EPRI_adv_batteries.pdf

Chapter 7.5
1. AutomationDirect, "Products" (2007), http://web1.automationdirect.com/adc/Shopping/Catalog/PLC_Hardware/DirectLogic_06/PLC_Units

Chapter 8
1. Electric Power Research Institute, "Technology Primer: The Plug-in Hybrid Electric Vehicle" (2007), http://www.epri-reports.org/Otherdocs/PHEV-Primer.pdf.
2. Electric Power Research Institute, "Welcome to EPRI Electric Transportation" (2007), http://archive.epri.com/et//index.html
3. US Census Bureau, "Industry Statistics Sampler, NAICS 4471 Gasoline Stations," (2002), http://www.census.gov/econ/census02/data/industry/E4471.HTM

Chapter 9

1. Husky Energy, "Better for your car. Better for the environment" (2006), www.huskyenergy.ca/products/downloads/Ethanol.pdf
2. Canadian Renewable Fuels Association, "Ethanol: Reducing Greenhouse Emissions by the Equivalent of 200,000 Cars," http://www.greenfuels.org/
3. Husky Energy, "Better for your car. Better for the environment," (2006), www.huskyenergy.ca/products/downloads/Ethanol.pdf
4. Canadian Renewable Fuels Association, "Ethanol."
5. Murtagh & Associates, *The Online Distillery Network for Distilleries and Fuel-Ethanol Plants Worldwide* (2007), http://www.distill.com
6. Kurt K. Klein, *The Biofuels Frenzy: What's in it for Canadian Agriculture?* (March 28, 2007), http://www.aic.ca/whatsnew_docs/Klein%20Final%20%234.pdf.
7. National Biodiesel Board, "Specification for Biodiesel (B100)—ASTM D6751-07b" (March 2007), http://www.biodiesel.org/pdf_files/fuelfactsheets/BDSpec.PDF.
8. National Biodiesel Board, "National Biodiesel Board, DOE, USDA Officials Dispute Biofuels Study," (July 21, 2005), http://www.biodiesel.org/resources/pressreleases/gen/20050721_pimentel_response.pdf
9. Mercedes-Benz, "**Mercedes-Benz Debuts High-Tech Diesel Car For Canadian Market," (April 8, 2004),** http://www.mercedes-benz.ca/index.cfm?NewsID=144&id=3246.
10. National Biodiesel Board, "Fuel Fact Sheets" (2007), http://www.biodiesel.org/resources/fuelfactsheets.
11. Martin Mittelbach and Claudia Remschmidt, *Biodiesel: The Comprehensive Handbook* (Martin Mittlebach: Vienna, 2004), 1.
12. Martin Mittelbach and Claudia Remschmidt, *Biodiesel: The Comprehensive Handbook* (Martin Mittlebach: Vienna, 2004)..
13. Ibid.
14. Ibid.
15. Martin Mittlebach, in discussion with the author, Ottawa, 2005.
16. Note that the ASTM D6751 standard for biodiesel makes reference to the fact that there is little practical knowledge about using high concentrations of biodiesel. While this may be true in the commercial markets, there has been an "underground" economy using biodiesel at 100% concentrations for many years. For further information on this topic, consult the author's book *Biodiese: Basics and Beyond*, Aztext Press, 2006.
17. Martin Mittelbach and Claudia Remschmidt, *Biodiesel: The Comprehensive Handbook* (Martin Mittlebach: Vienna, 2004), 6.
18. Green Trust, "Biodiesel," (2000), http://www.green-trust.org/biodiesel1.htm
19. Union for the Promotion of Oil and Protein Plants, "Status Report Biodiesel: Production and Marketing in Germany 2005," http://www.biodiesel.org/resources/reportsdatabase/reports/gen/20050601_gen358.pdf
20. National Biodiesel Board, "National Biodiesel Board, DOE, USDA Officials Dispute Biofuels Study," (July 21, 2005), http://www.biodiesel.org/resources/pressreleases/gen/20050721_pimentel_response.pdf,
21. http://www.biodiesel.org/resources/reportsdatabase/reports/gen/19980501-gen-203.pdf
22. Union for the Promotion of Oil and Protein Plants, "Status Report Biodiesel: Production and Marketing in Germany 2005." http://www.biodiesel.org/resources/

reportsdatabase/reports/gen/20050601_gen358.pdf
23. In conversation with the author.
24. Every manufacturer has a fuel warranty policy concerning not only biodiesel but any off-specification fuel. Damage caused to a fuel system through the use of biodiesel will not be covered under warranty. Likewise, damage caused by sand, dirt, or other contaminants in petroleum diesel is not covered under warranty either.
25. Automotive Industry Data, (http://www.eagleaid.com/index.htm).
26. Diesel Technology Forum, "Demand for Diesels: The European Experience," (July 2001), http://www.dieselforum.org/fileadmin/templates/whitepapers/EuropeanExperience.PDF.
27. Ibid.
28. http://www.biodiesel.org/resources/reportsdatabase/reports/gen/20050728_Gen-354.pdf.
29. Van Gerpen, Jon et al, *Building a Successful Biodiesel Business*, January 2005, pp. 172, Figure 35.

Index

A

A.P. Moller-Maersk Group. *See* MMG
air travel 21, 22
Alternating Current (AC) 273
alternative-powered vehicles 106
Amédée Bolée 106
Amory Lovins 163
Audi 221, 532

B

Ballard Power Systems Inc. 232, 233, 532
Batteries
    Elasticity of voltage 277
    Operating Temperature 282
battery-swapping stations 211
battery cables 291, 297, 299, 508
battery charging 210, 218, 288, 294, 302, 357, 358, 360, 361
battery electric vehicles 199, 217, 218, 243, 307, 314. *See also* BEV
Battery Voltage Monitor 354, 355, 557. *See also* BVM
BEV 200, 201, 202, 203, 205, 207, 208, 209, 210, 211, 212, 213, 214, 218, 219, 221, 223, 245, 267, 273, 287, 305, 311, 391, 411, 412, 499, 531
Biodiesel Drying 479, 556
Biodiesel Washing 475, 556
Birol, Fatih 64
blending biodiesel 486
BMW Hydrogen 7 240, 241, 242
Bullfrog Power Inc. 407
BVM 354, 355, 357, 359, 361. *See also* battery voltage monitor

C

California Cars Initiative 223, 505
Carbon capture and sequestration 405. *See also* CCS
carbon dioxide 83, 84, 85, 87, 500
carbon dioxide taxation 99
carbon life cycle 95, 180, 220
carbon offset program 407
carbon price 98, 248
carbon tax 37, 38, 96, 98, 101, 112, 146, 162, 169, 172, 194, 236, 392, 412, 528
carbon taxation 100, 122, 161
CCS 402, 403, 404, 405, 406
Chelsea Sexton 222
Chevrolet Volt 223, 225, 227, 228
China 2, 3, 11, 58, 59, 60, 83, 88, 91, 100, 117, 154, 155, 407, 527
Chrysler 132, 186, 232
Citicar 531
climate change 4, 5, 28, 45, 46,

80, 86, 87, 89, 91, 92, 93, 103, 111, 115, 135, 139, 140, 143, 144, 163, 164, 166, 167, 230, 392, 429, 491, 494, 497, 559
Cold Flow Issues 442, 556
Common Rail Direct Injection 446, 454, 455, 556
Commuter Cars Corporation 215
commuting 2, 4, 30, 34, 38, 42, 48, 99, 100, 111, 112, 117, 120, 121, 132, 133, 150, 208, 558
compact fluorescent lamp 92, 93
continuously variable transmission. See CVT
current 267
CVT 193, 194, 499

D

DC 270
diabetes 117, 118, 119, 559
Displacement on Demand 195
domestic oil production 23, 24, 28, 48
driving range 107, 200, 205, 207, 209, 211, 241, 243
Dynamometer Testing 374, 555

E

E-flex 225, 226, 227, 238
ECT 158, 159, 160, 161
Electrical Circuit
    Operation of 275
    electrical grid 79, 316, 319, 392, 393, 395, 543

Electricity
    current 267
    voltage 267
electric motor 119, 133, 186, 189, 190, 191, 192, 195, 197, 204, 205, 207, 208, 219, 221, 225, 227, 253, 254, 264, 274, 309, 311, 312, 315, 316, 317, 500, 519
Electric Power Research Institute 305, 401, 531, 532. See also EPRI
Electric Vehicle Council of Ottawa 252, 489
Emma Maersk 146, 147, 154, 159, 161, 529, 530
energy consumption 46
Energy Demand of Transportation 23
EPRI 305, 401, 402, 531, 532
equalization charge 303
Ethanol 416, 417, 419, 421, 422,
ethanol 27, 29, 36, 91, 94, 95, 97, 110, 226, 317, 415, 416, 417, 418, 419, 420, 421, 422, 423, 424, 425, 428, 434, 436, 459, 488, 533, 549, 556
European Combined Terminals. See ECT
EV1 132, 133, 134, 135, 136, 137, 200, 213, 221, 222, 223

F

Fatih Birol 64
FedEx 150, 151, 152, 153, 154, 155, 156, 157, 160, 530

Fischer Panda 317, 318, 319, 322, 326, 507, 549
flying 112, 121, 140, 142, 146, 407
Ford 44, 47, 89, 107, 176, 186, 199, 204, 224, 232, 400, 417
Formula One 177, 178, 179
fossil fuels 79
Fred Green 201, 204, 212, 214, 252, 253, 255, 257
freight 24, 31, 34, 35, 36, 39, 91, 99, 122, 148, 149, 151, 152, 153, 154, 155, 156, 157, 158, 159, 160, 161, 162, 166, 167, 169, 187, 247, 530
Fuel Cell Vehicles 245, 500
fuel economy 27, 46, 48, 90, 91, 122, 123, 124, 125, 128, 130, 135, 171, 174, 179, 184, 188, 193, 194, 195, 196, 219, 220, 221, 242, 424, 427, 495, 532, 558
fuel efficiency standards 38
Fuel Injection 452, 453, 556

G

General Motors 102, 107, 122, 123, 124, 125, 132, 134, 137, 166, 180, 181, 182, 186, 195, 196, 200, 213, 221, 223, 225, 226, 227, 229, 238, 239, 245, 417, 549
generator 123, 124, 189, 191, 192, 193, 225, 226, 227, 255, 256, 259, 260, 261, 268, 274, 290, 315, 316, 317, 318, 319, 320, 321, 322, 323, 327, 328, 329, 330, 332, 333, 352, 360, 361, 362, 363, 368, 456
Geophysics 67
geopolitics 44
George Monbiot 89, 146, 528, 529
George W. Bush 27
global warming 4, 6, 83, 84
glycerin 430, 457, 458, 459, 460, 476
glycerol 430, 460. *See also* glycerine
Google 201, 220, 222, 223, 224, 226, 411
green car 125
Green, Fred 201, 204, 212, 214, 252, 253, 255, 257
greenhouse gas emissions 5, 28, 32, 53, 63, 80, 83, 85, 87, 89, 90, 91, 112, 116, 119, 124, 129, 141, 142, 172, 173, 179, 221, 222, 224, 228, 236, 237, 244, 246, 389, 416, 418, 419, 421, 441, 486, 558, 559

H

Hatherill, Richard 349, 354, 355, 520, 522
highway congestion 19
Honda 125, 130, 136, 173, 192, 200, 234, 412, 549
Hubbert 65, 66
Hummer 166
Hy-wire 180, 181, 182, 229
hybrid 123, 124, 125, 127, 130, 173, 187, 189, 192, 194, 195, 196, 219, 221, 412, 500, 502,

505, 519, 531, 532
hydrogen 7, 29, 30, 32, 91, 110, 229, 234, 235, 236, 238, 239, 240, 241, 242, 243, 245, 248, 288, 507, 528, 531, 532
hydrogen-powered vehicles 183, 229, 230
Hymotion 224, 225, 254, 411, 505
Hypercar 275, 486

I

infrastructure 7, 11, 16, 25, 28, 30, 31, 32, 33, 34, 35, 36, 38, 41, 43, 46, 51, 53, 56, 57, 64, 91, 96, 98, 100, 101, 102
Inputs and Outputs 352, 554
Intercity Transport 157
Intergovernmental Panel on Climate Change 86, 528
internal combustion engines 219, 421

J

James Kunstler 28

K

Kunstler, James 28
KVR Performance 374, 375
Kyoto Protocol 5, 51, 89, 528

L

ladder logic 334, 336, 341, 342
Lane, Rick 252, 253, 258, 290, 297, 313, 321, 322, 350, 372, 373, 490, 491
lead-acid battery 209, 212, 213

Lithium ion 209, 306
Local Transport 157
logic tables 261
long-distance trips 20, 22
Lovins, Amory 163

M

Marine Transportation 161
Matthew Simmons 55, 64, 68, 69, 72
Mercedes 125, 127, 129, 177, 426, 441, 442, 447, 452, 533, 549
Methanol 459
Mexico 51, 55, 56, 100, 145, 149, 150, 417, 420, 526, 527
Milburn Electric Car 199
MMG 153, 154, 155, 156, 157
Monbiot, George 89, 146, 528, 529
motor 25, 26, 80, 96, 117, 119, 133, 186, 189, 190, 191, 192, 193, 195, 197, 204, 205, 207, 208, 213, 216, 219, 221, 225, 227, 252, 253, 254, 255, 257, 264, 274, 288, 289, 290, 293, 309, 310, 311, 312, 313, 314, 315, 316, 316, 317, 318, 319, 323, 325, 338, 349, 361, 368, 376, 379, 380, 381, 382, 420, 428, 437, 456, 500, 519
motor controller 207, 290, 313, 314
Myers Motors 214, 505

N

national energy supply 28

natural gas 25, 54, 55, 56, 57, 59, 61, 63, 66, 67, 69, 79, 83, 110, 183, 184, 234, 235, 236, 243, 247, 389, 404, 420, 459, 487
Nickel-metal hydride 209
nitrogen oxide 6, 122, 441
NmG 214, 505
Nuclear power 390

O

oil reserves 48, 49, 54, 55, 56, 59, 62, 64, 71, 73, 526
oil sands 53, 527
OPEC 45, 50, 62, 63, 64, 65

P

parallel hybrid 189, 190
passenger miles of travel (pmt) 138, 529
Peak Oil 5, 64, 65, 66, 67, 68, 100
Persian Gulf 56, 57
personal vehicles 2, 18, 19, 20, 22, 23, 34, 42, 101
PHEV 219, 220, 221, 222, 223, 224, 225, 226, 227, 228, 246, 247, 251, 253, 254, 255, 257, 265, 267, 273, 305, 311, 402, 410, 411, 412, 447, 502, 532
PLC 260, 261, 262, 263, 328, 333, 334, 335, 336, 337, 338, 339, 340, 341, 342, 343, 344, 345, 346, 347, 349, 350, 351, 352, 353, 354, 355, 356, 357, 358, 359, 360, 361, 362, 363, 364, 365, 366, 367, 368, 369, 370, 508, 532, 549, 554. *See also* programmable logic controller
plug-in hybrid 7, 8, 204, 223, 225, 251, 253, 307, 317, 402, 411, 505, 558, 559
Plug-In Hybrid Electric Vehicle 219. *See also* PHEV
Plug In America 222, 223
population 7, 20, 22, 26, 35, 41, 42, 63, 79, 88
Prius 125, 129, 131, 186, 187, 188, 192, 210, 220, 222, 223, 224, 225, 226, 228, 246, 247, 254, 286, 287, 410, 411, 412, 531
Prius+ 223, 228
Production of vehicles 3, 7, 29, 54, 59, 100
programmable logic controller 260, 261, 333. *See also* PLC
public transit 2, 18, 36, 120, 529
public transportation 2

R

regenerative braking 124, 135, 173, 189, 193, 194, 195, 212, 213, 221, 316
Richard Hatherill 349, 354, 355, 520, 522
Rick Lane 252, 253, 258, 290, 297, 313, 321, 322, 350, 372, 373, 490, 491
Roadster 209, 210, 216, 217, 252
Rocky Mountain Institute 163, 503, 530

Rudolf Diesel 429, 445

S
Saudi Arabia 51, 57, 64, 71, 72
series hybrid 189, 191
Sexton, Chelsea 222
Sign It in Vinyl 383
Simmons, Matthew 55, 64, 68, 69, 72
smog 77, 79, 80, 81, 82, 83, 89, 93, 96, 115, 143, 186, 195, 219, 391, 403, 429, 434, 448
software development 340
SPR 71
Stanley Steamer 106, 107
Strategic Petroleum Reserve (SPR) 28, 71
sulfur 5, 6, 82, 83, 278, 281, 403, 428, 437, 439, 440, 441

T
Tango 136, 215
telecommuting 99, 111, 112
telepresence 112, 117, 140, 147
Tesla 209, 210, 216, 217, 252, 311, 505, 549
Tesla Roadster 209, 210, 216, 217, 252, 553
TGV 144
total number of vehicles 1
touch screen 261, 262, 344, 345, 347, 348, 359, 362, 365, 366, 368, 369, 370, 549
Toyota 125, 129, 131, 186, 187, 188, 192, 200, 210, 220, 222, 223, 224, 225, 226, 227, 246, 247, 254, 286, 287, 306, 410, 411, 412, 529, 531, 532
Train à Grande Vitesse. *See also* TGV
transportation energy consumption 26, 28
Truth Table 363, 364

U
U.S. oil imports 51
urban sprawl 41, 42
urban vehicular density 41

V
Vehicle-to-Grid 397
vehicle's "energy cycle" 29
vehicle mass 176, 177, 265
vehicular production 3
Volkswagen 15, 125, 127, 201, 209, 210, 212, 252, 253, 255, 257, 328, 444
voltage 267

W
Warranty Issues 444, 556
Western Canada Sedimentary Basin (WCSB) 53
Who Killed the Electric Car? 136, 222
World Health Organization 6
WVO 464, 465, 466, 467, 468, 469, 470, 471, 473, 474, 475, 556. *See also* waste vegetable oil

Z
ZENN 134, 137, 175, 212, 214
zero-carbon biodiesel 317, 415,

447, 487, 549
Zero-Carbon Electricity 389, 391, 393, 395, 397, 399, 401, 403, 405, 407, 409, 410, 411, 413, 555
Zero-Carbon Liquid Fuels 415, 417, 419, 421, 423, 425, 427, 429, 431, 433, 435, 437, 439, 441, 443, 445, 447, 449, 451, 453, 455, 457, 459, 461, 463, 465, 467, 469, 471, 473, 475, 477, 479, 481, 483, 485, 487, 556
Zero-Emission Vehicles 183

About the Author

Bill takes the wheel of The Zero-Carbon Car

William (Bill) Kemp is the V.P. Engineering for an energy sector corporation, where he leads the development of low-environmental-impact hydroelectric and agricultural biogas/biomass power systems. Bill is a leading expert in small- and mid-scale renewable energy technologies. He has been involved in the construction of hundreds of megawatts of renewable energy projects in developed and developing countries around the world. In addition, Bill is the chairman of electrical safety standards committees with the Canadian Standards Association and is a member of a working group which develops renewable energy policy for the province.

He is the award-winning author of four books distributed by New Society Publishers: *The Renewable Energy Handbook*; *$mart Power: An Urban Guide to Renewable Energy and Efficiency*; and *Biodiesel: Basics and Beyond*. His fourth book, *The Zero-Carbon Car*, was published in the fall of 2007. Bill is also a co-author of the David Suzuki Foundation report *Smart Generation: Powering Ontario with Renewable Energy*.

As the Executive Director of www.solutionsforsustainability.ca, Bill lectures frequently on such topics as sustainable living, resource management, renewable energy, and the societal value and development of enabling policies for a low-carbon and sustainable future. He has appeared on dozens of television, radio, magazine, news, and podcast programs and documentaries, working to engage North Americans in following an environmentally sound, sustainable pathway.

He and his wife Lorraine built their own home in Ontario, Canada using local, low-environmental-impact materials. They produce zero-waste-stream biodiesel for their transportation needs and lead a low-carbon lifestyle off the electrical grid, producing the power for their horse farm from renewable resources. Bill makes his day-to-day living a test ground for his writing, practising everything he preaches and acting as a role model for others.